R00218 74927

CHICAGO PUBLIC LIBRARY
HAROLD WASHINGTON LIBRARY CENTER
R0021874927

D1714682

DANIEL BERNOULLI JOHANN BERNOULLI
(1700–1782) (1667–1748)

*Engravings by Jacob Hain after paintings by Rudolph Huber
Courtesy Universitäts-Bibliothek, Basel*

HYDRODYNAMICS
BY
DANIEL BERNOULLI

&

HYDRAULICS
BY
JOHANN BERNOULLI

TRANSLATED FROM THE LATIN BY
THOMAS CARMODY and HELMUT KOBUS

PREFACE BY
HUNTER ROUSE

Under the Auspices of the
Iowa Institute of Hydraulic Research

DOVER PUBLICATIONS, INC.
NEW YORK

Copyright © 1968 by Dover Publications, Inc.
All rights reserved under Pan American and
International Copyright Conventions.

Published in Canada by General Publishing Company, Ltd.,
30 Lesmill Road, Don Mills, Toronto, Ontario.
Published in the United Kingdom by Constable and Company, Ltd.,
10 Orange Street, London W.C. 2.

This volume, first published by Dover Publications, Inc., in 1968, contains new English translations by Thomas Carmody (presently Associate Professor of Civil Engineering, University of Arizona, Tucson) and Helmut Kobus (presently Gruppenleiter der Gruppe Wasserbau, Versuchsanstalt für Wasserbau und Schiffbau, Berlin) of:

Hydrodynamica, by Daniel Bernoulli, as published by Johann Reinhold Dulsecker at Strassburg in 1738;

Hydraulica, by Johann Bernoulli, as published by Marc-Michel Bousquet et Cie. at Lausanne and Geneva in 1743.

This volume also contains a Preface to the English Translations by Hunter Rouse, Dean, College of Engineering, The University of Iowa, Iowa City.

Library of Congress Catalog Card Number: 68-11668

Manufactured in the United States of America

DOVER PUBLICATIONS, INC.
180 Varick Street
New York, N.Y. 10014

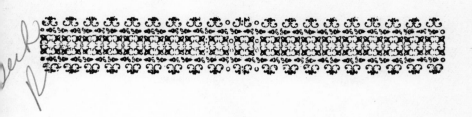

CONTENTS

	Page
PREFACE TO THE ENGLISH TRANSLATION BY HUNTER ROUSE	vii
HYDRODYNAMICS BY DANIEL BERNOULLI	1
HYDRAULICS BY JOHANN BERNOULLI	351
INDEX TO BOTH TREATISES	458

CELSISSIME ATQUE SERENISSIME
PRINCEPS,
DOMINE GRATIOSISSIME.

On ausus fuissem Sereniſſimo Nomini Tuo *Hydrodynamicam* hanc inſcribere, niſi illa Academiæ Scientiarum, ſub umbone Tuo Petropoli florentis, conſilio & ſubſidiis a me conſcripta fuiſſet. Novimus quantum Tibi, Sereniſſime Princeps, Magnanime Academiæ Protector, poſt Auguſtam illam orbis borealis Palladem, debeamus, idque cum toto orbe literato, qui præclara ſibi porro ab Academia, amœnis benevolentiæ Tuæ radiis colluſtrata, pollicetur, pia & immortali recolemus memoria. Florebit in æternitatis ſacrario apud Ruſſicam gentem Tuorum in illam meritorum

(Reproduction of first page of dedication to Daniel Bernoulli's Hydrodynamics. *See page xix for translation.)*

PREFACE
TO THE ENGLISH TRANSLATIONS

In the belief that students attaining the doctoral level should know something about the background of their profession, the writer began in 1960 to offer a graduate course at the University of Iowa on the history of hydraulics. Instead of attending lectures, every student was expected to read the Institute book on the subject [1], select a lesser-known investigator from the past in each of his three required doctoral languages (English, of course, included), and submit original monographs summarizing the respective lives and works. If he so preferred, a student could prepare instead one monograph on a single subject, such as pipe resistance, the roots of which would carry him into several source languages. Furthermore, a student whose language background was sufficiently broad to need no further exercise could concentrate on a single national literature, like Russian hydraulics.

The lack of Latin in the writer's background—which had proved particularly troublesome when he was seeking to digest early treatises on hydraulics—led him to suggest to two students with foreign-language upbringing (the one, Italian; the other, German) who were able as well to read Latin that they compare the works on fluid motion of the two Bernoullis, Johann and Daniel. The idea soon grew to the point of involving the complete translation of the two present books, in part as the regular course requirement, in part as salaried employment with the Iowa Institute of Hydraulic Research, and in no small part as a labor of love. To the American-born Thomas Carmody (who really began the undertaking) fell the task of actual translation; but to the German-born Helmut Kobus (who had studied far more Latin) fell that of checking meticulously every word and thought and of preparing the manuscript for the printer after his fellow translator had left for another university. Kobus, in turn, finally left the country, and it remained to Carmody (and the writer) to read proof.

Why the Bernoullis' works should have been singled out for translation seems at first thought rather obvious, if only because of the frequency with which the name Bernoulli is on a hydraulician's lips. But it is only Daniel to whom one is making reference, and the word is gradually spreading that the theorem bearing his name is nowhere to be found in his habitually cited *Hydrodynamica* [2]. Not until the last few years has mention of either the work *Hydraulica* [3] or its author Johann Bernoulli appeared in fluids literature with any frequency whatever, and this almost exclusively in the writings of C. Truesdell [4]. It is Truesdell's thesis that, whereas Daniel has received too much credit for the formulation of the Bernoulli theorem, Johann has received too little. Readers who have not studied Latin, and who may never have the chance of seeing the original works, can now judge the matter for themselves. They can also marvel at the many familiar concepts which Daniel did originate and for which he has received almost no credit at all.

To understand the rather curious relationship between Johann Bernoulli and his son Daniel, one must know something of the family itself [5]. Basel had become a university town in 1460, a center of early printing that attracted such Renaissance writers as Erasmus and Paracelsus, a refuge for Huguenots during the Reformation of 1530, and finally by the seventeenth century a very literate city of strong family ties. To this city, in 1622, came a Huguenot from Antwerp by the name of Bernoulli. He established himself as a merchant and raised sons who also became merchants. One of these fathered a dozen children, of whom four lived: a mathematician, an artist, another mathematician, and at last a merchant. The oldest was Jakob Bernoulli (1654–1705), who became professor of mathematics at the university and finally rector. The second mathematician was Johann Bernoulli (1667–1748), who was trained by his brother, worked with the French mathematician L'Hôpital at Paris, taught mathematics for ten years in Holland, and then succeeded his brother as professor at Basel. It was he who had as sons Nikolaus, Daniel, and Johann II.

Though Jakob was an extremely able mathematician in his own right, his great contribution to the present history was the education of his younger brother, who in turn taught his own sons. Unfortunately, friction developed and steadily increased between the brothers, and their early collaboration eventually changed to rivalry. Johann's bitterness was increased by L'Hôpital's publication in his own name of various discoveries communicated to him in Johann's many letters—not to mention an entire course of instruction which he had given him in Paris—but he nonetheless became upon Newton's death the foremost mathematician in the world. Because of his close association with Leibniz (it was Johann who first applied the new calculus—and, in fact, introduced the word "integral"), he sided eloquently with him against Newton in the fluxion–calculus controversy, to the chagrin of the Royal Society.

Daniel Bernoulli (1700–1782) was born at Groningen near the middle of his father Johann's ten-year Dutch professorship. He studied under his father at Basel after the latter returned to fill the chair left vacant by Jakob's death. In his twenties Daniel spent seven or eight years as professor of mathematics at St. Petersburg, a period darkened only by the early death there of his older brother Nikolaus, also a mathematician. Daniel won or shared, in the course of his life, ten prizes awarded by the Paris Académie des Sciences for the solution of designated problems. The first of these, received at the age of twenty-four, involved the design of a clepsydra for the exact measurement of time at sea. The third, for a paper on tides, was shared with Euler and Maclaurin. One which he divided with his father dealt with the inclination of the planetary orbits, and another, shared with Johann II, was on the best form of anchors. Still another had to do with the nature and cause of ocean currents. In each of these he displayed considerable mathematical ability, to be sure, but above all a keen physical perception and the ingenuity to produce a solution regardless of the method used [1].

Daniel's *Hydrodynamica* was begun in 1729, during his sojourn in Russia, and an uncompleted manuscript of it was left at St. Petersburg when he returned to Basel four years later. In the course of its revision and completion, he wrote for permission to dedicate it to the Empress of Russia, as an acknowledgment of his debt to that country. When the book was finally published in Germany in 1738, he requested that the Russian manuscript be destroyed, but it is still preserved in the files of the Soviet Academy of Science (as remarked in the Preface of a Russian translation of the German edition that recently appeared [6]).

A younger colleague of Daniel's, Leonhard Euler (1707–1783), had also studied mathematics under Johann and, largely as the result of Daniel's influence, was also invited to St. Petersburg by Catherine I. There he became professor of mathematics when Daniel returned to Basel. Euler was eventually to surpass the great Johann Bernoulli as a mathematician (fifty pages were finally required in his eulogy to list merely the titles of his writings), and hence it is hardly surprising that not only Daniel but also Johann was to defer to his judgment at even so early a date. Daniel wrote him near the end of 1734 that he had arranged with Dulsecker in Strasbourg to publish his new work, but not for three years could he report that it was nearly finished. Early in 1738 he sent several copies to St. Petersburg, but by the end of the year Euler wrote that they had not yet arrived. On March 7 of the following year Daniel again protested the total lack of news as to their fate [7].

On the very same day Daniel's father notified Euler that he was sending him the first part of his own manuscript *Hydraulica*. Now except for a brief criticism of Newton's cataract hypothesis in 1716, there is no record that Johann had written anything whatever on the subject of hydraulics until some months after his son's treatise was off the press, when he stated in a letter to Euler that he was preparing a manuscript on hydraulics which was already well along. Nevertheless, he indicated in the first part that it was written in 1732, a full year ahead of the Russian version of his son's! The second part followed the first to St. Petersburg in 1740, and the two were eventually published, as Johann had requested, in the *Memoirs* of the Imperial Academy of Science for 1737 and 1738 (which were printed, respectively, in 1744 and 1747). They actually first appeared in his collected works, published in Switzerland in 1743.

As surprising as Johann's obvious attempt at seeming to predate his son in publication was Euler's delay in acknowledging Daniel's book until such time as he could do the same for Johann's. In fact, of the two letters of acknowledgment and praise, written on the same day, that to Johann was far more flattering. A subsequent letter that he wrote Johann about the book was quoted in part in the foreword to the Swiss version of *Hydraulica* as translated herein (see p. 347). Johann's reason for using only the first paragraph of Euler's letter is evident from the following translation of the second:

> Truly, regarding the force by which vessels are driven backward, I certainly do not have the least doubt concerning that very method you use for determining this; but when, for pipes attached horizontally to a vessel, you find that the pressure driving the vessel backward is different from that which agrees with the hypothesis of Your son, that force as it is determined by Your Illustrious Son seems to me certainly to be more suited to the truth than Yours; may I have said this without offense to You. Indeed, from the formula which you present for retroaction in this case it follows that the retroaction can be indefinitely great, even if the orifice is very small and the motion very slow, and the expression given by Your Son does not contain this inconsistency; but I am convinced that if you will deem it worthy to subject this part to examination once

more, Your theory will agree most perfectly with Your Son's idea; indeed, I suspect that fractions have to be inverted, and, with this done, it will agree most perfectly with the truth and with Your Son's expression.

Apparently Daniel first saw his father's treatise when it appeared in 1743 in the collected works, for he thereafter wrote Euler:

> I beg your Excellence to tell me in sincere friendship and confidence your opinion of my father's Opera, particularly of the last volume. I for my part have reason of the highest degree to complain about it: The new mechanics problems stem mostly from me, and my father had even seen my solutions before he solved them in his own manner, and nevertheless I am not acknowledged with even a word, which I find the more annoying as my solution is not yet published. My first solution of rotation around an instantaneous center, found from the nature of the least inertia, he has questioned and also contemned for a long time, and finally he published it as his own. But since by a miraculous hazard I obtained a page from his manuscript in which this, his pretended solution, was written, and I complained about it through my brother, he barely let me pass as a second inventor. The matter is roughly the same with the remaining new problems in mechanics. Of my entire *Hydrodynamics*, of which indeed I in truth need not credit one iota to my father, I am robbed all of a sudden, and therefore in one hour I lose the fruits of a work of ten years. All propositions are taken from my *Hydrodynamics*; nevertheless, my father calls his writings *Hydraulics, now first discovered, anno 1732*, since my *Hydrodynamics* was printed only in 1738. Meanwhile my father has gotten everything from me, except that he thought of a different general method to determine the increment of velocity, which invention consists of some few pages. What my father does not claim completely for himself he contemns, and finally, as the height of my misfortune, he inserts the letter of your Excellence in which you, too, diminish my inventions in a field of which I am fully the first, even the only, author and which I claim to have exhausted completely. Your Excellence says that I have determined the pressure of fluids flowing through a conduit in no other way than for the steady state, whereas I show immediately on page 259 toward the bottom that generally the pressure is $\frac{a - vv}{2c}$; and what, on the other hand, has my father done in this important new field? The invention of the argument comes from me; the idea to consider the conduit as cut at the point where the pressure is required comes from me; that one should require the acceleration of the last particle at the first instant of interruption is my idea; finally, that from this very acceleration, hindered either partly or completely, the pressure of the elemental volume determines the pressure of the

water in the conduit—this also comes from me; and my father has done absolutely nothing else than determine the velocity in his own manner and by repeated reasoning, which is his only invention in the entire work. The argument about the reaction of fluids my father does not yet understand today; nevertheless, he refutes me in the corollary on page 488 [page 336 of present translation; see also foregoing paragraph from Euler's letter]. All of this is still the least about which I can complain. In the beginning it seemed almost unbearable to me; but finally I took everything with resignation; yet I also developed disgust and contempt for my previous studies, so that I would rather have learned the shoemaker's trade than mathematics. Also, I have no longer been able to persuade myself since then to work out anything mathematical. My entire remaining pleasure is to work some projects on the blackboard now and then for future oblivion. I could not accept with a clear conscience the call to Berlin, even if the King should give me the honor to send me one, and I beg you therefore not to think of me any more with respect to this matter. However, I am strongly obliged to your Excellence for your kind services; your most valuable friendship presents me with an innermost and true pleasure, and I esteem such [friendship] much higher in itself than in the profit which could arise to me from it. I could not abstain from complaining to your Excellence, as my best friend, seeing that the occasion might well arise that you vindicate me of the unjust suspicion of plagiarism without doing wrong to my father, and also bring it about that the truth, as far as the controversial points between my father and me are concerned, does not suffer any injury. It does not seem proper to me to defend myself.

Far from being guilty of plagiarism, Daniel had based his treatise on material that was not only original but lasting in its interest. The reader will find in the following pages of translation the initial appearance of topics that are still prominent in the literature even today—from the kinetic theory of gases to the principle of jet propulsion. Daniel was also the first to connect manometers to piezometric openings in the walls of vessels, to consider the establishment with time of flow in a long conduit, and to attempt to predict conduit pressure in terms of the velocity. However, his derivation of what has come to be known as the Bernoulli theorem will hardly satisfy any reader but the most casual. There is no doubt that Daniel understood the theorem in its two-term (velocity head and piezometric head) form. However, the simplicity of the relationship in comparison with the cumbersomeness of his analysis leads one to suspect that—despite his claim of invariably reasoning first and experimenting thereafter—he actually knew the answer in advance. The artifice of cutting the conduit and relating the pressure to the assumed acceleration seems forced at best.

According to Truesdell [8], most of Daniel's difficulty lay in his imperfect understanding of fluid pressure. Straub, on the contrary, points out that a paper published by Daniel in the St. Petersburg *Commentarii* of 1729 had already contained "an essentially correct formula for the so-called hydrodynamic pressure" [7]. In any event, though his *Hydrodynamica* divided a moving fluid for convenience into a series of slices normal to the direction of motion, in no case was action stated to occur between them. Pressure was treated rather as a condition existing at the conduit wall which would produce a jet (or manometric column) equal in height to the piezometric head if the wall were pierced. Now Johann also assumed the fluid to move in normal slices, but the concept of pressure as a mutual interaction at their surface of contact was essential in his analysis. Moreover, in order to avoid the anomaly of a discontinuity in pressure and velocity at an abrupt conduit contraction or expansion, he imagined the actual flow section to change gradually before the contraction or after the expansion, in a manner reminiscent of Newton's cataract. He called this transition "gurges," which Truesdell insists should be translated as "eddy," despite the fact that Johann referred to the flow within the imaginary throat-like passage rather than to that around it. In fact, far from visualizing circulatory motion in the zone of separation, he specifically considered it to be occupied by stagnant fluid: "And, accordingly, there is formed along the indefinitely small length HG something like a *throat*, IFGH, contracting from the wide into the narrow, through which the liquid must pass, the acceleration being continuous but nevertheless augmented gradually, with a rather small portion of the liquid (which fills the small space IFD) remaining at perpetual rest" [page 357 of the present volume]. The word "gurges" has hence invariably been translated herein as "throat."

Johann was obviously a stride beyond his son in his analysis of the Bernoulli relationship. Whereas Daniel had treated pressure primarily in terms of height of a manometer column or jet, Johann visualized it as a force, albeit one acting over the area of the slice as a whole. It remained to Euler—under the considerable stimulus of Johann's analysis rather than Daniel's—to originate the concept of pressure at a point and to incorporate the pressure gradient into his equations of acceleration. These he finally integrated for specific conditions [9], thereby first deriving in a rigorous manner what is now known as the Bernoulli equation. As in the case of many another aspect of fluid motion for which he has not received proper credit, it is thus the name of Euler which should be most often upon the hydraulician's lips.

Although Daniel's lengthier treatise is much easier to read than Johann's, each will give the reader some difficulty in accustoming himself to the mathematical and the physical style that the respective author used. He must hence recall that the calculus was still a relatively new concept; that the energy relationship was not yet correctly written and even less well understood than the momentum relationship, with which it was thought by many to be in conflict; and that the theory of dimensions was wholly in its infancy. So long as the only dimension was length, little difficulty was encountered, as in Daniel's interrelationship of velocity and piezometric heads. Kinematics was somewhat more complex, as witness his expression of velocity through either its head or a proportionality with an observed time of fall; in fact, it was Johann who first introduced the coefficient of proportionality g. Dynamics was the most confusing, for force had not yet been universally defined in terms of length, time, and mass through the Newtonian equation, and their interrelationship was still arbitrary; even Euler, whose equations were in fact dimensionally homogeneous, still defined mass and weight as he saw fit.

Except for calling attention to certain of the anomalies or actual errors, the translators have in large part left the assessment of the authors' analyses to the reader himself. To be sure, many mannerisms of the times have been eliminated, such as the excessive use of abbreviations and symbols like the ampersand and equality sign as parts of speech, and the mathematical notation has been very slightly modernized. Translators' insertions are invariably placed within brackets, and the bracketed numbers refer to the lists of cited works at the ends of the two books. For the convenience of the reader, each illustration has been introduced into the text at its first point of reference. The original figures have been reproduced in all cases.

This Preface would not be complete without reference to a contemporary undertaking [10] begun in the Bernoulli city of Basel some two decades ago and already assuming tangible form: publication of the collected writings of the Bernoulli family, in particular the members referred to herein. The project is under the general editorship of Dr. Otto Spiess, and the first of well over a dozen projected volumes (the early letters of Johann) recently appeared under his direct guidance [11]. Another, containing Daniel's correspondence as edited by Dr. Hans Straub, is now in preparation. The scientific world stands greatly in the debt of Dr. Spiess and his colleagues for the impetus that they have given to the tremendous task of making this material generally available. The writer acknowledges his particular gratitude for stimulating conferences with Drs. Spiess and Straub, for their willingness to review the foregoing pages, and their kindness in providing portraits of Daniel and Johann from approximately the times their books were published. A final touch was Dr. Spiess' indication, during a visit to his home by the writer, of the impending appearance of Daniel's *Hydrodynamica* as translated into his native tongue [12].

Iowa City, Iowa HUNTER ROUSE
January, 1967

REFERENCES

[1] ROUSE, H., and INCE, S., *History of Hydraulics*, Iowa Institute of Hydraulic Research, 1957. Dover reprint, 1963.
[2] BERNOULLI, D., *Hydrodynamica, sive de viribus et motibus fluidorum commentarii*, Dulsecker, Strasbourg, 1738.
[3] BERNOULLI, J., *Hydraulica nunc primum detecta ac demonstrata directe ex fundamentis pure mechanicis. Anno 1732. Opera Omnia*, Vol. 4, Bousquet, Lausanne and Geneva, 1743.
[4] TRUESDELL, C., Editor's Introduction to Vol. II 12 of Euler's *Opera Omnia*, Füssli, Zurich, 1954.
[5] SPIESS, O., "Die Basler Mathematiker Bernoulli," *Bulletin de l'Association Suisse des Électriciens*, Vol. 43, No. 8, 1952.
[6] BERNOULLI, D., *Gidrodinamica ili Zapiski o Silakh i Dvizheniyakh Zhidkostei*, translated by V. S. Gokhman, A. I. Nekrasov, K. K. Vaumgart, and V. I. Smirnov, *Izdatel'stvo Akademii Nauk SSSR*, 1959.
[7] STRAUB, H., private correspondence with the writer, 1964–65.
[8] TRUESDELL, C., "Zur Geschichte des Begriffes 'innerer Druck,'" *Physikalische Blätter*, Vol. 12, No. 7, 1956.
[9] EULER, L., "Principes généraux de l'état d'équilibre des fluides"; "Principes généraux du mouvement des fluides"; "Continuation des recherches sur la théorie du mouvement des fluides"; *Histoire de l'Académie de Berlin*, 1953–55.
[10] TRUESDELL, C., "The New Bernoulli Edition," *Isis*, Vol. 49, Pt. 1, No. 155, 1958.
[11] SPIESS, O., Ed., *Der Briefwechsel von Johann Bernoulli*, Birkhäuser, Basel, 1955.
[12] FLIERL, K., "Des Daniel Bernoulli, Hydrodynamik," *Veröffentlichungen des Forschungsinstituts des Deutschen Museums für die Geschichte der Naturwissenschaften und der Technik*, Reihe C, Nr. 1a, Munich, 1965.

DANIELIS BERNOULLI Joh. Fil.
Med. Prof. Basil.
ACAD. SCIENT. IMPER. PETROPOLITANÆ, PRIUS MATHESEOS
SUBLIMIORIS PROF. ORD. NUNC MEMBRI ET PROF. HONOR.
HYDRODYNAMICA,
SIVE
DE VIRIBUS ET MOTIBUS FLUIDORUM COMMENTARII.
OPUS ACADEMICUM
AB AUCTORE, DUM PETROPOLI AGERET, CONGESTUM.

ARGENTORATI,
Sumptibus JOHANNIS REINHOLDI DULSECKERI,
Anno M D CC XXXVIII.

Typis Joh. Henr. Deckeri, Typographi Basiliensis.

BY
DANIEL BERNOULLI, Son of Johann,
MEDICAL PROFESSOR AT BASEL
*Formerly Professor Ordinarius of Higher Mathematics
Now a Member and Honorary Professor of the
Imperial Academy of Science of St. Petersburg*

HYDRODYNAMICS

OR

Commentaries on Forces and Motions of Fluids.
AN ACADEMIC WORK
Written by the Author While He was Engaged at
St. Petersburg

STRASSBURG,
PUBLISHED BY
JOHANN REINHOLD DULSECKER
Anno MDCCXXXVIII
TYPE BY JOH. HENR. DECKER, BASILIEN TYPOGRAPHER

---

*(Translation of original title page,
reproduced on preceding page.)*

TO THE
MOST ELEVATED
and
MOST SERENE
Prince and Lord
LORD ERNST JOHANN

Duke, by the Grace of God, of
CURLANDIA
and
SEMGALLIA
IN LIVONIA

Most Elevated and Most Serene
PRINCE
Most Beloved Lord

I would not have dared to dedicate this *Hydrodynamics* to Your Most Serene Name if it had not been written by me with the advice and help of the Academy of Sciences which is flourishing under your protection at St. Petersburg. We know how much we owe You, Most Serene Prince, Magnanimous Protector of the Academy, second to that August Athene of the northern world, and with pious and undying memory we shall reflect upon this with the entire learned world, which is inspired henceforth by the excellent Academy enlightened by the pleasant rays of your benevolence. Flourishing in a shrine of eternity will be: among the Russian people, the greatness of Your services to them; among the Curlands, the memory of the fortunate predictions which divine providence has destined for them under Your Sceptre; and finally, among universal peoples, the perpetual admiration of Your most glorious life. The illustrious destinies of the present times show us how dear to the gods is the Majesty of the Russian Scepter and the happiness of Your people. May these [gods] grant the prosperous occurrence of great deeds; may they grant a great length of life and Reign to You; may they grant successors of your blood, rivaling your virtues in a long series to the end of all time, with the universal world approving. Thus dedicates,

Most Serene and Most Elevated Prince,

MOST GRACIOUS LORD,

Written at Basel
10 March 1738

To Your Highness
A Most Humble & Most Obsequious Servant
DANIEL BERNOULLI

PREFACE

Finally our *Hydrodynamics* is published after all the obstacles which delayed its printing for almost eight years have been overcome; perhaps it would never have seen the light if all that labor had been mine alone. With pleasure indeed I acknowledge that the principal portions of this work are due to the guidance, counsel, and support of the Academy of Science of St. Petersburg. The opportunity for the book arose from the very purpose of the latter, according to which the first Professors who had gathered to form it were retained to write a Diatribe on some useful and, as much as possible, new subject, and they were advised, certainly, that the Theory of forces and motions of fluids, unless it would be undertaken with an unwilling Minerva, is neither a useless nor a trite matter, as anyone will easily concede. In addition, in order to suppress the Reader's boredom, from the outset I paid attention to a variety of things, especially in the last five Sections, and I inserted analytical, physical, and mechanical examples, theoretical as well as practical, some geometric, nautical, astronomical, and others, the understanding of which, nevertheless, did not seem to support as much as to postulate the exposition of the undertaken work. The calm Reader, understanding these matters, will easily correct whatever mistakes have escaped [this] hasty person. The intent of this writing is unique: that I may be useful to the Academy, all labors of which are organized in such a way as to promote public convenience and an increase of good literature.

CONTENTS

Page

FIRST CHAPTER
Which Is an Introduction, and which Contains Several Matters To Be Noted in Advance   1

SECOND CHAPTER
Which Deals with Standing Fluids and Their Equilibrium, Either Between One Another or Related to Other Forces   18

THIRD CHAPTER
Concerning the Velocities of Fluids Flowing out of a Vessel Formed in any Way Whatsoever through any Kind of Opening Whatever   35

FOURTH CHAPTER
Concerning the Various Times which can be Expected in the Efflux of Water   71

FIFTH CHAPTER
Concerning the Motion of Water from Constantly Full Vessels   101

SIXTH CHAPTER
Concerning Fluids not Flowing out but Moving within the Walls of Vessels   124

SEVENTH CHAPTER
Concerning the Motion of Water through Submerged Vessels, where it is Shown by Examples how Significantly Useful is

CONTENTS

the Principle of the Conservation of Live Forces, even in those Cases in which Continually some Part of Them is to be Considered Lost   139

EIGHTH CHAPTER

Concerning the Motion of Homogeneous as well as Heterogeneous Fluids through Vessels of Irregular and Abrupt Shape, where from the Theory of Live Forces, a Part of Which is Continually Absorbed, are explained Excellently Singular Phenomena of Fluids driven through Several Orifices, after General Rules have been Set Forth for Defining the Motions of Fluids Anywhere   159

NINTH CHAPTER

Concerning the Motion of Fluids that are Pushed forth not by their own Weight but by an Outside Force, and particularly concerning Hydraulic Machines and their Ultimate Grade of Perfection that can be Attained, and how this could be Perfected further through the Mechanics of Solids as well as of Fluids   183

TENTH CHAPTER

Concerning Properties and Motions of Elastic Fluids, but especially of Air   226

ELEVENTH CHAPTER

Concerning Fluids acting in a Vortex, while also Concerning Those which are Contained in Moving Vessels   275

TWELFTH CHAPTER

Which shows the Statics of Moving Fluids, which I call Hydraulico-Statics   289

THIRTEENTH CHAPTER

Concerning the Reaction of Fluids flowing out of Vessels and the Impetus of the Same, after They have Flowed out, on the Planes against which They Strike   315

HYDRODYNAMICS
BY
DANIEL BERNOULLI

HYDRODYNAMICÆ
SECTIO PRIMA.

Quæ introitus est, variaque continet prænotanda.

§. I.

FIRST CHAPTER

Which Is an Introduction, and which Contains Several Matters To Be Noted in Advance

§1. Since the Theory of Fluids is twofold, of which the one, Hydrostatics, considering the pressures and various equilibria of stagnant liquids, and the other, Hydraulics, considering the motion of fluids, have been treated separately by writers, and since I indeed understood both of them to be interrelated by so close a link that the one is in very great need of the other, by no means did I hesitate to combine them, inasmuch as the order of things seemed to require it, and to describe them both under the common and more general name of Hydrodynamics. However, although from the most ancient times the Theory of fluids has been continuously refined, nevertheless it did not gain very noteworthy additions. Certainly the knowledge of the ancient Mathematicians was terminated by this, that they understood the common equilibrium of standing fluids or also of bodies together with the fluids within which they lie, about which Archimedes wrote. And since in addition it is self-evident that, where equilibrium does not exist, motion occurs toward the region of lesser pressure, hence they were able to contrive various games and hydraulic machines, serving excellently partly for pleasure and partly for public interests, in which matter certainly they showed themselves to be very ingenious. They also perceived, but rather as through a veil, those motions which are due to the pressure of the air. But they were clearly ignorant of the true reasons and accurate measures in matters of Hydraulics, and thus they were merely standing on the threshold.

§2. The efflux of water from a vessel through a very small orifice serves excellently for defining the motion of fluids. But although it was not wholly unknown to Frontinus and others, as some believe, that the velocity of water flowing out of a vessel or container increases because of an increased height of water above the point of efflux, it nevertheless must be known that indeed the same Frontinus, in computing small amounts of water, or, rather, the water to be expended, committed disgraceful and improper errors. Benedetto Castelli first pondered about the relation between velocities and heights, but he believed a false law, thinking that the two follow the same proportion. After this Torricelli finally observed that the velocities increase in the ratio of the square roots of the heights, which everyone followed. But even though they were not agreeing on the absolute measure of velocity, they nevertheless began experiments by which they estimated that measure to be defined, from which it is customary to acknowledge especially that which was performed by Guglielmini and repeated eight times, although it departs, of course, from other experiments performed at that time. However, it is usual that all [experiments] performed under different circumstances differ from each other, and it is not always safe—as we shall say about many in the appropriate place—to pass judgment concerning the velocity of the water according to the quantity of the same flowing in a definite time through a definite orifice. Thus, when we call to account the Guglielminian experiment of which we just made mention, it should be concluded that the velocity [obtained] from the quantity of water which flowed through a given orifice in a given time was not greater than that which is due to the fourth part of the height of the surface of the water above the orifice. And there are other experiments by the same author which are enumerated in Book 2, Prop. 1, *aquarum fluentium mensura*, by virtue of which the water flowing out can ascend by its own velocity to two-thirds of that height. Among the works of Mariotte and others there stand out those [experiments] which decide for half the height; this diversity of the velocities so estimated notwithstanding, I persuade myself that the true velocities hardly differed from each other, that they had been in the usual ratio to the heights of the water, and that they were everywhere approximately such as those which are due to the entire height. But those [experiments] which were last mentioned, which at first glance seem to militate for half the height, by number many among the works of the authors, without doubt moved Newton, a Man immortal for his merits, to speak somewhat more boldly about the Theory by which he had found that water springing vertically upwards from a vessel through a very small opening can ascend to half the height of the water standing in the vessel, although he contradicts that assertion in all the experiments which were conducted concerning these heights directly. He published the Theory in the first edition of *Principia Mathematica Philosophiae Naturalis* and attacked it from the pressure by which the water situated in front of the orifice and just about to flow out is driven into motion. But since the nature of the matter by no means seems always to permit that the force exciting the water to flow out be defined a *priori*, and since, rather, concerning this it is hardly pleasing to resolve it otherwise than from the phenomena of motion, that is, a *posteriori*, which I often found, the thinking relied upon for that principle must be mistrusted. Hence the Man just praised changed his statement in the second edition of his Work, and he changed it back somewhat in the third, affirming that the water rises indeed to the total height but that the stream which it forms is contracted or made slender in front of the orifice, thus giving satisfaction to both the phenomenon of the velocity and that of the quantity flowing out in a given time, which seemed to contradict each other. But although it is not to be denied that the contraction of the aqueous filament is the true reason on account of which the velocity of the water flowing out cannot be estimated from the quantity, nevertheless I consider that the Theory is not to be overemphasized, because it is *accidental* and not everywhere faithful to itself even while the velocity does not vary, let alone different reasons such as friction, viscosity of the water, and other similar things. Thus when the water flows out not through a simple orifice, but through a small cylindrical pipe, the stream is not notably contracted, the velocity being preserved, after that has been excepted which is lost by it because of friction. But if, this notwithstanding, anyone supposes that the flow of water can be deduced correctly and wholly from the pressure, I may have asked this, that he pay attention to the more composite cases, for example to the flow of water, which Mariotte calls extraordinary, from a vessel which some diaphragm perforated by an orifice separates into two cavities to be filled with water, so that the water is forced to flow through the two orifices. Mariotte speaks about this motion in his excellent *Traité du mouvement des eaux*, Part IV, p.m. 442.

§3. Since these things are so, anyone will decide easily for himself how little hope there is that somehow the Laws of motions for fluids will be reduced to the rules of pure Geometry without any physical hypothesis, since certainly on the threshold itself they may have gotten away from the clearsightedness of this Man superior and incomparable in ability; nor do I believe that these things which I am about to present in this work can endure all mathematical rigor. The principles of the Theory are physical and are to be accepted, not without generosity, as approximately true. But yet, after the principles have been accepted, all will be geometric, subject to no restrictions, and connected to each other by a necessary interrelationship.

Nevertheless, I cannot but feel well concerning those physical principles with which I became strongly involved, since indeed they led me by the hand to exposing many new properties concerning both the equilibrium and the motion of fluids, which, unless the love of the undertaken work deceives me, will some day promote Hydrodynamics significantly, if they are refined more than I was allowed [by circumstances] to do. At this point it may be suitable to admonish—since to many anything which is new is customarily suspect—that I conceived the whole Theory in my mind, wrote the treatise, communicated most of it privately among friends, even sketched some things in the presence of our Society, before I undertook any experiment, lest I should be liable to be deceived by preconceived measures through a false opinion which is nevertheless approximately satisfactory to those [measures], and that even at some time Men most perspicacious in known theorems confessed openly that they cannot persuade themselves so, nor do they consider that [the theorems] are about to be confirmed by experiments; and after all that had happened, at last the experiments were made before friends, and they agreed with the Theory as much as I myself could barely hope. But now let us return to that from which we digressed.

§4. After Authors were certain concerning the variety of velocities caused by changing the heights, they began to consider more composite vessels, namely those furnished with variously inclined and unequally large ducts. But Frontinus already knew in his own time the nature of these ducts to some degree, knowing that a quantity of discharge is increased by the slope or depression of a pipe, that is, of the designated duct which has been attached to a reservoir or even sometimes to a small river; whence also he ordered that the pipes be arranged in line, as he says, and put at the same height. And indeed Frontinus is unjustly accused in this regard by some that he had no understanding of velocity; however, when he makes a calculation of all the water received and compares the latter with that about to be expended, I do not see how he can be excused. By experience, also, he had been thoroughly informed, which deserves to be mentioned, that more water is expended than should be through a pipe of both the proper size and position to which pipes of larger size are immediately attached. I will show in the appropriate place that this is so, and that it had been indicated correctly by Fabretti, although otherwise very skilled Men indicated that it was not evident to them, or rather that they were in doubt concerning it.

§5. However, what the ancients observed obscurely and without true measurements, that at last Mr. Guglielmini grasped in his treatise *aquarum fluentium mensura* by means of the following more accurate and more general proposition, saying that the *velocity of the water flowing through an inclined conduit is the same as if it would have flowed from a vessel through an orifice similar and equal in cross section, just as far below the surface of the water as the section is below the horizontal* [passing] *through the beginning of the canal*, which proposition Denis Papin attacked, he himself diverging greatly from the truth. But since we are at it, in order that we may review the principal comments of both Hydrostatics and Hydraulics, in this place the following remark is also to be listed about investigating the pressure of fluids from an impetus, namely that the *force of a fluid dashing against a perpendicular plane at a given velocity is equal to the weight of the cylinder of fluid erected above that plane, of which the altitude is such that from it something movable, by falling freely from rest, would acquire the velocity of the fluid*. With the help of this most useful Problem one may estimate the force of fluids driving machines or (which is the nature of the wind) propelling ships, the motion of solid bodies in resistant media, and many other things. About Hydrostatics, however, which is particularly concerned with very slender tubes or capillaries, I say nothing, because thus far it could not be reduced to the general Laws common to all fluids. Besides, it is uncertain as to which Author will have first observed the nature of these small pipes; nevertheless it is agreed that the observation is recent, because concerning it there is nothing to be seen in books published before these last seventy or eighty years.

§6. In addition to those cited, Authors from the times of Galilei rather celebrated in aquarian matters are Torricelli, Borelli, Viviani, Pascal, Boyle; and of a more recent age are Varignon, Newton, Poleni, Hermann, and Jakob and Johann Bernoulli, the discoveries of whom are found in the *Commentaries of the Royal Academy of Science of Paris, Principia Mathematica Philosophiae Naturalis*, the treatise *de Castellis* and notes pertaining to Frontinus, *Phoronomia*, the *Acta Eruditorum* of Leipzig, and various other works. But these discoveries about the curvatures generated from the pressure of a fluid, and others of this sort, were presented by Geometers because they are easily reduced to pure Geometry; however, concerning the rest, I pass in silence over things worthy of all praise.

Since these things have been presented which pertain to the works of others, I feel that it is reasonable that I advise sincerely, the opinion of my colleagues also having been considered, whether any and how many additions to Hydrodynamics can be or must be hoped for from the former. Briefly therefore, as much as I will be able, I will indicate the important points of the undertaken work.

§7. First in order are shown the outstanding Theorems which pertain to the equilibrium of standing fluids: the custom of the practice has seemed to me to demand it, although I acknowledge rather freely that no new propositions have been added by me. Indeed the method of demonstrating, as far as I know, is original with me, but since it is easy to compose for oneself innumerable demonstrations, there is little also in this portion that I claim for myself. In addition, some phenomena of capillary tubes are reviewed in passing, and finally on occasion of the pressure which fluids exert on the sides of vessels, many new and different theorems are added about the shape of bladders full of liquid, about their capabilities for elevating loads, about the construction and strength of aqueducts, and about other associated things.

§8. Thereafter the motion of fluids flowing out of vessels is treated, and since all who were engaged in this matter up to now will have considered in their own Theories the unique and most obvious case in which the orifice is taken as infinitely small with respect to the internal area of the vessel, ours is recommended not a little for its own breadth, for it extends itself to the situation of an orifice of any size whatever, and indeed to vessels of any shape whatever. For although consideration of the internal shape of the vessel is least required when the orifice can be considered as infinitely small, nevertheless without it the motion of the water cannot be defined when [the orifice] is of notable magnitude. General corollaries are deduced from the Theory which illustrate splendidly the variable motion of water and the disposition of the same, and they confirm whatever either experience has shown or the attributes of the matter indicate manifestly through themselves. Certainly, when the internal areas are, for example, moderately greater than the area of the orifice, the Theory shows that the error that follows from the consideration of the orifice as infinitely small is unnoticeable, and accordingly our several additions will perhaps seem rather useless. But let me wish that individuals, even if only those in the future, consider for themselves that—apart from the fact that I am writing not only for [scholars of] hydraulics but also for Geometers who are likewise delighted by bare truths—the use of our meditations is very great in other affairs, which [use] they will understand more when they have considered that the motion begins from rest and passes through infinite changes before it attains a certain speed, and that the greatest changes occur often in so short a period of time that they can in no way be clearly perceived by the senses, nevertheless that they are to be determined at the individual points both so that the motion be understood correctly in one's mind and because then various Theorems can be deduced. So I noticed (let this example, for the sake of the momentum of the discussion, be representative of all) that it cannot occur that the pressure of water flowing through a conduit at a given velocity be defined along the sides of it unless those changes, which let me call instantaneous, be understood correctly in one's mind, no matter how imperceptible [they are] to the senses. Concerning these things, as it was I who first thought about them, I thus added a new portion to the Theory of water with the most pleasing success, which, because it considers both the motion and the pressure of fluids together, seemed most suitably called *hydraulico-statics*. After these specimens of the general Theory are shown concerning cylindrical vessels, both the simple and those which are furnished with pipes, there are also determined, especially in these latter ones, the changes which arise at the beginning of flow while the given grade of velocity is being reached, and this certainly under the hypothesis of very wide vessels. But it is to be noted that these changes are quite perceptible, even if the vessels are of infinite size, and that they can be demonstrated by experiments, until the water flowing out of a very wide vessel through a simple orifice immediately at the first instant of time has the entire velocity which it can attain. The previously mentioned changes depend upon both the length and the shape of the pipe. Finally also analytical calculations are included for finding times of a different kind together with the physical notation pertaining thereto. And finally, as the theory indicates that it cannot occur that water ascends much beyond the uppermost surface of a bubbling spring, it is shown at the end of the section that there does not pertain to our hypotheses the singular phenomenon which I myself have observed rather frequently and can imitate at will, and of which mention is made in the *History of the Royal Academy of Science of Paris*, for the year 1702, where it is said that it happens sometimes that water in leaping fountains rises to three or four times that height which corresponds to the uppermost surface of the water, but that soon nevertheless the enormous thrusting of water is depressed to the ordinary height, and afterwards the genuine understanding of that phenomenon is conveyed with the true measures sought from our Theory, and the method of producing that unusual surge and finally of increasing it at will is indicated.

§9. Further, the Theory is extended to the examination of motions from constantly full vessels, to which certainly as much water is continuously supplied as flows out of them. The nature of these consists most of all in this, that emanating fluids approach more and more that level of velocity which is due to the full height of the surface of the fluid above the orifice, but that they never wholly attain it, except after an infinite time. Nevertheless the water is shown to converge so quickly to that velocity that after an unnoticeably short time it very nearly acquires the full value, except when it is carried through very long streams or aqueducts and discharged through a great orifice. Then indeed the accelerations are not so rapid that they cannot be perceived, which is confirmed by a singular example taken from the book *du mouvement des eaux* by Mr. Mariotte. But since the motion begins from rest and increases forever, formulas are given by the aid of which either from the time of flow or from the quantity of water discharged the velocity at individual points of time can be defined, and vice versa.

§10. Following that, fluids are considered which are being moved within vessels, where chiefly the reciprocal or oscillatory motions of the fluids are submitted to measurements and their relations are indicated. However, Newton gave a similar Theorem for the oscillations of a fluid in a pipe of uniform cross section (the two extreme legs of which are vertical, the intermediate part horizontal), which Theorem my Father rendered more generally in the *Commentaries of the Imperial Academy of Science of St. Petersburg*, Book II, p. 201, for any given inclination whatever of the extreme legs with respect to the horizon. Our Theory explains the entire matter without any restriction, considering that at individual places the pipes are variable at will with respect to direction, position, or area. Next it is shown in which cases it may occur that the different oscillations of a swinging object are Isochronous, under which conditions the length of a simple Isochronous pendulum is determined most generally. But in addition to this type of oscillation certain others are subjected to examination in the subsequent section, such as those which occur in pipes immersed in infinite or even confined water, in which there is a need for singular caution, insofar as all the applied phenomena are responsible for the departure in the calculation; but if the same things have been neglected, the difference between them becomes as great as it is between the laws of motion which are valid for perfectly elastic bodies and those for pliable bodies.

§11. After this I progress to other more composite subjects, considering certainly the motion of either homogeneous or heterogeneous fluids which are forced to flow through one or more orifices before they are discharged into the air, where that rule commonly accepted concerning the surge of the water to the uppermost level of the surface fails decidedly, with even the ordinary laws of pressure ceasing to hold. However, of all these things not even a vestige is found among the

SEVERAL MATTERS TO BE NOTED IN ADVANCE

works of the Authors except that which Mariotte has at the place cited above, Part IV, p.m. 442, *du mouvement des eaux*, where certainly he shows that he had been taught by experience that the flow of water is retarded; however, at the same time it is clear how far off he had been from the true Theory of these motions, and this Theory seems indeed to avoid the influence of almost all the principles customarily applied up to this point in similar cases, so that there is nothing which further confirms the superiority of our [principles]. Certainly the experiments performed do not allow me to doubt the truth of these any further. However, these things which have been considered do not lack their usefulness whenever they can be of great importance in improving hydraulic machines.

§12. Comments follow about hydraulic machines, about which it is shown principally that there is some definite termination of perfection beyond which one may not be able to proceed. But the falling short of this ultimate level of perfection in many widely accepted machines is subjected to numerical calculation with added rules or concepts, to which in constructing new machinery one should pay attention. In place of an example is mentioned the *machine de Marly*,* very well known throughout the world, about which it is shown, if only the descriptions are to be trusted, that it supplies not more than about one fifty-sixth part of that quantity of water which, the remaining things being equal, a theoretically perfect machine can supply. Also special examination is made of a machine most familiar from very ancient times right up to our age, namely the Cochlea [waterscrew] of Archimedes, not unworthy of the attention of Geometers as much because of the understanding of those things which pertain to pure Geometry as of those which pertain to Hydraulics.

§13. There follow some specimens of the motion of elastic fluids, such as air and exploded gunpowder, these things having been set forth which pertain to the nature of these fluids; but I myself consider these not differently from physical hypotheses, about which I will affirm nothing confidently. The Propositions and Problems of this section are new and selected with the intention that they can give occasion for illustrating or even solving many physical questions. Certain things are added about the estimation of live forces innate to elastic fluids, which at some time probably will be of frequent use in mechanical practice. Indeed, it is shown that the effect of, let us say, one pound of inflamed gunpowder can be greater in elevating weights

---

* [Huge river-driven pumping works, built 1684 in Marly-le-Roi, a suburb of Paris, to supply the fountains of Versailles—*Trans.*]

than that which one hundred very robust men can accomplish by continuous labor within one day's span.

§14. Further, the circular motion of fluids is treated, and as well fluids which are standing in vessels having been set in motion; several other matters are intermixed. But the [statements] which are proffered on circular motion can serve in a certain way in explaining the phenomena of gravity through vortices; the remaining things may be applied as far as possible.

§15. The previously discussed Theory of motions is brought right back again to the equilibrium of fluids, but of moving fluids, of which the laws have not yet been shown. It is amazing, since motion is defined elsewhere from pressure, that here by the inverse method the pressure is sought from the motion by defining it beforehand from the environment. Nor should I have believed that another way could be begun safely apart from that which I followed. However, I considered that the conduit through which the water flows is shortened in that place and at that point of time which comply with the question; and afterwards, through our previously mentioned rules, I investigated the acceleration of a particle of water just on the verge of flowing out. From that acceleration one was able to understand the pressure upon that aqueous particle, which compression, by the nature of fluids, is equal to the pressure on the sides of the conduit. After this pressure has been determined, it is evident what should happen if the conduit would have been perforated in just this place and a small pipe were in place of the orifice; indeed it will happen that the water in it ascends up to a certain standing level in the little pipe, sustained by the water below flowing through the conduit, so that here equilibrium is present between flowing and standing water: but under this name I considered that this Theory could be conveniently called *hydraulico-statics*. Further, it may merit being noted that this same Theory is in turn a basis and a source of other previously unknown motions. The Theorems which are presented are not only new, but also the majority are unexpected, of the truth of all of which I was not able to convince myself clearly until I had conducted experiments which removed all my doubt. But they have a significant use whenever the true estimation of the pressure of water flowing through aqueducts or streams is based upon them, and hence for deducing the required strengths of pipes. From this also depend the accurate measures of water to be expended through small water meters inserted laterally in a stream. In Physiology those things which pertain to the motion of liquids in an animal body are already better understood, and there are others.

§**16.** Finally, I progress to explaining certain other methods by which water can cause pressure: namely, while it is flowing out of an orifice, water thus presses a vessel in the opposite direction no differently than a cannon ball pushes back the cannon from which it is driven out. Many new properties of that repulsion are discovered which illustrate the nature of pressures splendidly, and their general laws in mechanics will indicate this matter to those thinking seriously about it. I performed these investigations because it seemed to me that they can at some time offer the occasion for discovering something new for navigation without oars or the help of the wind, about which matter I may in the proper place convey a little information, although I know that the origins of all things of this sort in themselves seem ridiculous to most people. Finally certain Theorems are also included on the force of water from impulse and back-pressure thence developed that bodies encounter while moving in fluids.

§**17.** And certainly these are things which seemed to me to allow geometric deduction from accepted principles. But since there is nothing in Theory so rigorously proven that it does not require some restriction in its application to solid bodies, therefore it is readily evident that no Theory regarding fluids should be anticipated that satisfies most fully all measurements ascertained by experience; I want those who will try to confirm our Theorems by experiments to be mindful of this matter. Certainly they discover some, but not perfect, agreement everywhere, which is either more strict or more lax depending on the circumstances of things. But whenever I myself performed some experiment, first of all I pondered until the principles of the Theory agreed with the proposed case, and thus the experiment never or very rarely failed me. Indeed, not only was I accustomed to discern beforehand in which region the difference would be, if it was to be noticeable, but also how large it would be; thus it is clear enough in itself, if I judge correctly, that fluids certainly follow the laws which we assert to be prescribed for them, although they encounter everywhere now greater, now lesser obstacles. In addition, I performed not a few experiments, of which I placed individual ones at the end of the section to which they pertain; but I was especially anxious to confirm propositions previously unknown and for the most part paradoxical enough. Concerning belief in the experiments, there is nothing which anyone may doubt, since I performed the important ones in the presence of Friends after the Theory had been made public, nevertheless leaving a large portion of the experiments which I conceived in my mind to be performed by others, since it is not pleasing to go through them individually. After our propositions have been read thoroughly, anyone may propose innumerable others for himself, and I judged that it was not my task to explain all that are desired by me; nevertheless, I explained some.

§**18.** But now, finally, the understanding of the principles that we have mentioned so often is to be rendered. The primary one is the *conservation of live forces*, or, as I say, *the equality between actual descent and potential ascent*. I shall make use of this latter term, because that which the other one signifies finds perhaps a more liberal usage among some Philosophers, who indeed are inclined to the name *vis viva* only. I plan, here and in our work to come, to speak of this matter a little more fully.

§**19.** After Galilei had shown that a body descending either vertically or on some curved plane acquires the same velocity as long as the height of fall is the same, which can be shown from the nature of pressures, Huygens made use of this same proposition, but fortunately for a more general hypothesis, in bringing out the laws of motion of elastic bodies resulting from percussion, and, certainly, in stabilizing the center of oscillation of a compound pendulum; indeed, he brought forth this axiom of his own in the following words: *If any number of weights begin to be moved in some way by the force of their own gravity, and the individual weights return to rest of their own accord, the center of gravity of the same group of bodies will return to the original height*, where by the phrase *in some way* he understands *either they may strike each other during descent, or they may press, or the bodies may act on each other in any other way*. From that axiom at once follows the principle of the conservation of *live forces*, which Huygens himself showed also, and in which it is assumed: *If any number of weights begin to be moved in some way by the force of their own gravity, the velocities of the individual weights everywhere will be such that the products gathered from the squares of these* [velocities] *multiplied by their appropriate masses are proportional to the vertical height through which the center of gravity of the composite of the bodies descends multiplied by the masses of all of them*. It is amazing how much utility this hypothesis may have in mechanical Philosophy; indeed, my Father, if anyone, noticed this correctly; he showed it vaguely, but among the foremost, in the Parisian edition of his *Discours sur les lois du mouvement* and in Book II of the *Commentaries of the Imperial Academy of Science of St. Petersburg*, and it is the same that I employed for investigating in fluids the laws of motion arising from their own gravity; for I set the velocities of the particles constantly to be such that, after the individual particles were moved vertically upward to the state of rest, their common center of gravity ascended to the original height. However, I preferred to adopt this hypothesis, on account of the reason mentioned above, with

Huygenian rather than Paternal words, and to mark it with the name of *the equality between actual descent and potential ascent*, rather than by that other of *conservation of live forces*, which some even yet dislike, chiefly in England, I know not by what misfortune. Certainly it seems to me that in the entire Leibnitian doctrine about *live forces* there is nothing concerning which not everyone, in his own manner of speaking, would agree, which, unless I am in error, I showed clearly in the *Commentaries of the Imperial Academy of Science of St. Petersburg*, Book I, p. 131ff., and to which I wished to commission a place here, lest any of the Readers be offended by these words, and so that he knows that nothing is accepted by me which is not received in Mechanics by all, and which does not join by a necessary link with what Galilei already showed when he established that the increments of velocities follow a proportion composed of pressures and instants of time.

§20. Although, concerning the remainder, the principle mentioned above is universal, nevertheless it is not to be treated without circumspection, because it often occurs that the motion carries over into another material. So, for instance, the position of the former is valid for determining the rules of motions from percussion, if only the bodies be perfectly elastic; but when they are not so, it is easy to see that a portion of the *live forces*, or of the *potential ascent*, expended in the compression of bodies is not restored to the bodies, but remains impressed in the certain fine material to which it has transferred. If, nevertheless, the matter is correctly considered, whenever the ratio is known which exists between the portion residual to the bodies and that which transfers to the fine material, it will appear that that inconvenience can be obviated easily, and thus the laws of motions can be defined properly for pliable bodies. Something similar occurs in calculating the motion of water, where sometimes it is clear that a portion of the *potential ascent* is lost continuously; this matter should be taken into consideration in any case in the calculations to be performed; having paid proper attention to this, I came to detect many new Theorems about the flow of water, which [fact] is to be seen in Chapters VI and VII, and about which I do not yet see whether they could be proven, much less invented, by any other method.

§21. So, therefore, I have not used our principle recklessly, and in this way much presents itself that was previously unknown, not only about the motion of water but also, as one can see, surprisingly, about its pressure, which, with no Analysis yet performed, no one will have easily foreseen or expected. But when it happens that neither can all of the *potential ascent* be conserved because of the nature of the situation, nor can it be foreseen how much may be absorbed, the motion of fluids cannot be determined accurately enough, nor do I estimate that it can be done by any other method. Therefore, let me wish that the Reader be cautious in deducing corollaries from our Theory which often, on account of changed conditions, may not accurately agree with experiments.

§22. From the previous statements it is already sufficiently evident that it is required from our theory that the velocity of the individual particles of fluid be defined from an assumed velocity which exists at some place such as the point of efflux. In the same manner, it was necessary to add another hypothesis, which is this: After we understand, in the mind of course, that the fluid was divided into layers perpendicular to the direction of motion, let us consider that the particles of fluid of the same layer are moved at the same velocity, so that everywhere the velocity of the fluid is reciprocally proportional to the corresponding area of the vessel. This hypothesis is familiar, although it is known further that fluid moves slightly more slowly at the sides of a vessel but faster in the middle, which happens on account of the attrition, and that other exceptions as well are frequently to be made. Nevertheless, a noticeable error can very rarely arise from defects of this sort.

§23. Let me terminate here these warnings about our hypotheses with an enumeration of the phenomena which can both illustrate and confirm somewhat the conservation of *live forces* in the motion of fluids. Indeed, many of them occur in this very text, which, however, I will not treat on account of the calculations they require. However, that which is observed concerning a drop having fallen into standing water is trivial and obvious: certainly it creates rings in the surface of the standing water, and either the larger the drop or the higher the fall, the more of them it creates, and there is no doubt but that these rings would propagate themselves without end unless the viscosity of the fluid and other similar things were [acting] as a hindrance. Also, whenever it is pleasing to observe another effect from drops of this sort while many smaller drops are being projected upward from the surface of the water below, then what pertains here especially appears constantly: that the higher the drops surge, the fewer they are in number and the smaller in volume; and when the height of the fall was two feet, rather often the smaller drops ascended beyond the height of fall, particularly when the water was passing through a large orifice. Also it is worthwhile to note here what is observed regarding a particle of water in a narrow conduit which is horizontal and covered by a perforated cap at the end towards which the water flows. It is certain that at the instant at which the water reaches right up to the cap, a few drops spring forth with a great impetus, and soon the motion of all the water is established. But anyone could easily believe that the water next to the orifice continues to be moved at the proper velocity, and that the remainder of the flow is unchanged; this, however, would represent the conservation of live forces very poorly; that violent *instantaneous* efflux of water, just like an explosion, represents it splendidly; more about these things elsewhere.

§**24.** These are the things which I wished to point out in advance concerning our hypotheses and both their eminence and deficiency. It remains that I say something about the nature of the fluids toward which our extensive efforts will certainly be directed, not because I consider that I have more understanding of them than others, but because I believe it is a sin to depart from this custom common to all writers. And first, certainly, all usually agree that internal motion exists in all fluids whatever, without which indeed no one understands correctly such fluidity, effervescences of different fluids, dissolutions of solids submerged in fluids, evaporations, and infinitely many other phenomena. Hence a great part of the most solid matter liquefies with sufficient heat, which impels all things into motion; but that internal motion causes the particles not to remain adjacent to one another, but rather to move to and fro, by which it occurs that without friction they withdraw from a spot upon receiving a very small impulse, which would certainly not happen if the same particles were placed adjacent to one another as in a pile of sand. So it is easy to understand that the dust from egg shells held in a pan over a fire is said to imitate boiling milk. But the greater the heat is, the more violent is the motion of all the particles, and these are dispersed at a greater interval from one another; this agrees with the dilatation of all fluids from added heat, and their contraction from cold, to which law even water itself, not yet frozen, is subject; but that it is of a different nature when it is frozen seems to be deducible from another cause occurring by chance, namely this: that water supports air particles in its interstices which thus do not increase the volume of the water, just as sugar dissolved in water does not increase its volume; that at the time of impending freezing the motion of the aqueous particles is lessened; that thus the same particles approach each other more; and that thus they drive the air particles from their own interstices, which, then less suitably arranged in a different place, can increase the volume, just as sugar not yet dissolved can increase the volume of the water with which it is mixed. Hence the reason is easily deduced as to why ice made of water well purged of air before freezing becomes not specifically lighter, but rather somewhat heavier. But Mariotte instituted excellent experiments about the true solution of air in water right up to the point of saturation, and he enumerated these in his *Traité du mouvement des eaux*. Therefore, there is an opportunity for doubt that fluids (as I said) are frozen when the intestine motion ceases or is greatly diminished, when, namely, the particles collapse on each other and become adjacent, and at the same time expel the heterogeneous particles from the interstices, if any linger there; nevertheless, the hardness of frozen bodies is hence not understood more clearly, since it seems that with that motion stopping, a body is formed with a nature halfway between a fluid and a solid, unless something else occurs, and it should be compared with a pile of sand; but whatever occurs to the matter, lest certainly I follow conjectures, in the meantime it will be pleasing to assume: that as many particles as you please gravitate to each other, or, in order that I use the word familiar to the English, attract each other, and that the attraction increases significantly when the particles approach one another; that it is of different strength in different bodies, for example less in oil than in water, the ice of which is harder; that fluids, the particles of which either attract each other more strongly or are moving in a slower motion, freeze more quickly and more easily. Then it would be pleasing to conjecture that water impregnated with sugar or salt freezes more slowly, because the particles of sugar or salt placed between the aqueous particles diminish the attraction of the latter, so that these cannot be joined and frozen solid unless the heterogeneous particles are driven off; and certainly in all fluids which are impregnated with heterogeneous particles, at the time of freezing a certain expulsion or secretion or precipitation of portions from the pores occurs. There are an infinite number of other phenomena of either solid or fluid bodies which agree altogether wonderfully with the principle of mutual gravitation, such that it is a pity that the principle itself is established so far above the human mind that I consider that there is no one who can understand it in any way.

§25. At last let one admit to having been warned that this treatise is considered by me as Physical rather than Mathematical, and that I did not conduct this plan to obtain a Geometric method in hypotheses, definitions, and other devices to be set down in advance beyond measure, and I follow everywhere the order and discourse of Geometers who customarily start from the beginning, having explained the propositions, and that I treat all in such an order that the individual things are deduced properly from the first premises, and they leave nothing unproven behind them, however much this may already have been proven by so many others. I did not have any trouble in understanding those things which have been handed down by others, whether they were definitions and axioms, or even theorems, nevertheless I did not omit their proofs, which are new, and, finally, even in the first section are placed the proofs of the Theorems demonstrated at random by others; and since certain terms occur that are neither explained nor used by others, I shall give their definitions in the text itself. I shall propose the remainder sometimes in the form of Propositions, Theorems, Problems, Corollaries, and Scholia, according to the custom of the Geometers, and sometimes I shall give it in a continuously explained discourse.

One thing remains about which I wish the Reader to be especially warned in advance: that I was not able to apply to this work that diligence or attention which I should have, and which I myself desired. And therefore I have no doubt but that some errors will have crept in while I was doing the calculations, which I hope no one will employ wrongly; others which met my eye while I lightly read over the treatise I myself corrected; nevertheless I am convinced that still others remain even yet.

SECOND CHAPTER

Which Deals with Standing Fluids and Their Equilibrium, Either Between One Another or Related to Other Forces

THEOREM I

§1. The surface of a standing fluid is parallel to the horizon.

PROOF. Let the vessel *ABCD* (Fig. 1) contain the fluid *EBCF*, the

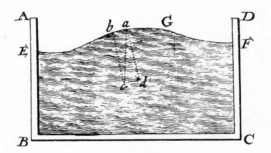

FIGURE I

surface *EGF* of which, if it can be done, is made not parallel to the horizon. Let an elemental volume in the rather elevated position *a* be considered, which by its own gravity is driven vertically downward by a force represented by *ac*. Let this force be resolved into the two components *ad* and *ab*, one perpendicular to the surface and the other tangent to it. But since there is nothing present which resists this latter force, this cannot but spread its own effect and therefore draw the elemental volume itself toward *E*, which would be contrary to the hypothesis of stagnation, or of the permanent state. Therefore, it is necessary that the tangential force *ab* be everywhere zero, which does not occur otherwise when the entire surface is parallel to the horizon. Q.E.D.

STANDING FLUIDS AND THEIR EQUILIBRIUM

§2. COROLLARY. Hence the truth of the general propositions is understood: that certainly the surface of a fluid, the portions of which are acted upon by any type of force whatever, always composes itself so that any volume element whatever, placed on the surface, is drawn in the direction perpendicular to the surface.

THEOREM 2

§3. A homogeneous fluid contained in two connecting pipes formed in any way whatever is in a state of equilibrium when both surfaces are placed at the same level, that is, [when] they maintain an equal vertical distance from the lowest point of the vessel.

PROOF. Let fluid be contained in the vessel ABC (Fig. 2) composed of two connecting legs or pipes, and let it be placed in each leg to the

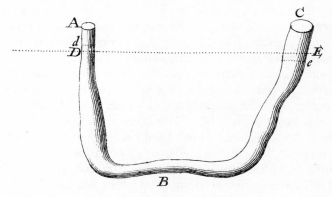

FIGURE 2

same height. I say that this position cannot be changed without some heavy body betaking itself from a lower position to a higher, which would be contrary to the nature of weights. For if the surface E descends to e, and in the other part D is elevated from D to d, then, since the remaining part of the vessel is full of the same fluid before and after the position has been changed, the effect of all this change is manifested in this: that the particle Ee will have ascended to Dd.

Besides, the same is evident as well from the first Theorem, since a pipe in standing water can be assumed to be formed in any way whatever, in which certainly the water will maintain the position which it had before, since it is just the same whether the water enclosed in the pipe is pressed by the sides of the pipe or by the surrounding water.

Scholium 1

§4. If in the first proof of the preceding paragraph the total mass at DBE is considered to have exchanged its position with the position dBe, it is easily shown that the center of gravity of the entire mass has ascended to a higher position, which is no less absurd. But since in our proof there is no particle at Ee which will not ascend after the position has been changed, I considered that the proof will be more precise and more clear if there be no consideration of the center of gravity.

Scholium 2

§5. We have some individual phenomena concerning capillary tubes. For instance, water ascends above the level in a rather narrow tube, the other extremity of which is submerged in the water, while Mercury does not reach that level. Truly, since I considered this very carefully at one time, I came to the same conclusion more or less which my uncle, Jakob Bernoulli, of blessed memory, once gave in his *dissertatio de gravitate aetheris*, namely that the water in a rather narrow tube ascends there beyond that level because the number of aëreo-aethereal particles at the base of the column which lies above the water in the tube is less than the number of the particles at a similar base beyond the tube. This is certainly understood from the following: after the globules have been placed next to each other on a horizontal table, if a circle is made with a compass, some particles are necessarily excluded because they cannot be divided; but the pressures of the aëreo-aethereal columns (of which one base is in the tube, the other outside the tube) are in proportion to the bases, that is, to the numbers of particles in the bases; hence, if the number of particles in the first base equals a, in the second base equals $a + b$, and the pressure in the first column equals g, the pressure in the other column will be $\frac{a+b}{a} g$, thus the difference of the pressures equals $\frac{b}{a} g$, to which the height of the water above the level must be equated. In order that these things be understood more clearly, it will have to be considered that g is proprotional to the square of the diameter which corresponds to the surface of the fluid contained in the tube, and that a is also proportional to the same square because of the extreme smallness of the particles, such that the ratio of g to a is to be considered constant, and accordingly that the height of the water above the level must follow the proportion of b itself. But, which is intrinsically evident, b is in proportion to the periphery of the surface of

STANDING FLUIDS AND THEIR EQUILIBRIUM

the fluid contained in the tube; therefore, the height above the level will be in proportion to that same periphery, which experience has confirmed for a long time now. If further we should now consider different fluids, we will see that the previously mentioned periphery is the more complicated and accordingly the larger, the greater are the fluid particles, and since the height of the fluid above the level depends upon the magnitude of this periphery, we understand why this height does not follow the inverse ratio of the specific gravity in the same tube. Thus if the same tube is immersed in spirits of wine and water, the former will ascend less than the latter, although, nevertheless, the spirits should ascend more on account of the lesser gravity. But this indicates, if I have followed the matter correctly, that the particles of the spirits of wine are less [in number] than the particles of water. Nevertheless, never in my judgment can the ascent above the level in any fluid be changed into descent, and I should believe that all fluids are of the same nature in this respect, unless some other reason not yet considered so far appears in addition, and if we should argue according to our hypothesis, it should be said that Mercury also would have ascended above the level if only its particles were not attracted mutually to each other by a greater force than are the particles of water. Indeed, to this attraction I attribute all those things which make Mercury go in a different way. At the end of this section let me show the experiments which led me to this thinking.

Lemma

§6. Let the cylindrical pipe *ABDC* (Fig. 3), the base of which is perpendicular to the sides of the pipe, be inclined to the horizon in any

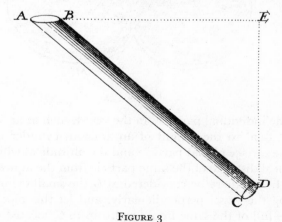

FIGURE 3

way whatever, and let it be considered filled with water right up to *AB*. I say that the pressure of all the water on the base *CD* is equal to the weight of the aqueous cylinder the base of which is *CD* and the height of which is the vertical *DE* terminated by the horizontal *BE*.

PROOF. When the shape of the pipe is cylindrical, and in addition the base is perpendicular to the sides of the pipe, anyone sees that the action of the fluid on the base is the same as if there were a solid cylinder of the same weight above the inclined plane; but it is established from mechanics that the pressure of the solid cylinder on the base is that which is defined in the proposition, and therefore the action of the fluid will be such, if only one does not regard the adhesion of the fluid to the sides of the pipe and also the behavior of the same with regard to capillary tubes, from which we diverted our thinking. Q.E.D.

THEOREM 3

§7. Now if, generally, a vessel *AHMB* (Fig. 4) is formed in any way whatever and filled with water right up to *DE*, the pressure of the

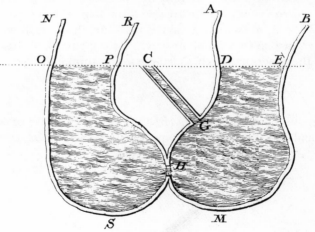

FIGURE 4

water on the individual particles in the vessel, such as at *G* or *H*, will always be equal to the weight of an aqueous cylinder the base of which is the surface of that particle and the altitude of which is equal to the vertical distance of the same particle from the aqueous surface.

PROOF. I. Let there be considered at *G* the small cylindrical pipe *CG* entering the vessel perpendicularly, and let this pipe be understood to be full of the same liquid right up to *C*, located on the line

ED. If now the vessel is considered to be perforated at *G*, the fluid in both places will be in equilibrium (through §3); therefore, the fluid in the tube *CG* presses against the interior just as much as the fluid of the vessel presses against the exterior. But the former pressure agrees with the proposition (through §6), and therefore the other [does] also.

II. But if, in place of the point *G*, another point *H* is assumed so that a line which enters the vessel perpendicularly at that point lies within the vessel, then the whole vessel *RHSON* can be considered united with the former at *H*, full of water right up to *PO*. Thus, indeed, it appears, if the particle at *H* which is common to both vessels is perforated, that the fluid thus will be in equilibrium (§3), and therefore that the pressure of each at *H* is equal. But the pressure of the fluid in *RSN* is that which is indicated in the proposition (according to the first part of this proof), and therefore it is the pressure of the fluid which is in the vessel *AMB*. Q.E.D.

Scholium

§8. The equilibria of standing fluids in more composite cases are easily deduced from these propositions. However, I, being content with the proofs which I just gave of the *fundamental* propositions in hydrostatics, do not wish to follow all of them, for the understanding of our practice does not require it. But those things which are important to the pressures of fluids not being at rest surely require a more profound investigation. Nor yet has the pressure of fluids flowing through conduits or pipes at a given rate of speed been properly determined by anyone, although this type of argument may be very useful in hydraulic matters as well as in many others. But it is not advisable to deal with these latter things before we have commented on the motion of fluids.

§9. From the preceding an understanding is evident of the powers of bladders by which immense weights can be supported. Hence even the force is known which is sustained by the walls of a pipe in which water is standing. Since it is customarily handled by writers on hydrostatics, we will now treat this argument, particularly since many other things are supported by it which we will have to discuss.

At first let there be the bladder *onmp* (Fig. 5), placed between a hard floor and the weight *B*, into which water is poured through the pipe *FRo*, the vertical leg of which, for the sake of brevity, we will make incomparably longer than the diameter of the bladder. The weight *B* will not be elevated immediately. But if water is poured in further right up to F, for instance, finally the weight will become elevated; however, there will be equilibrium, since the region of contact cd remains with respect to the orifice o just as the weight B is with respect to the weight of the aqueous cylinder of height FR standing

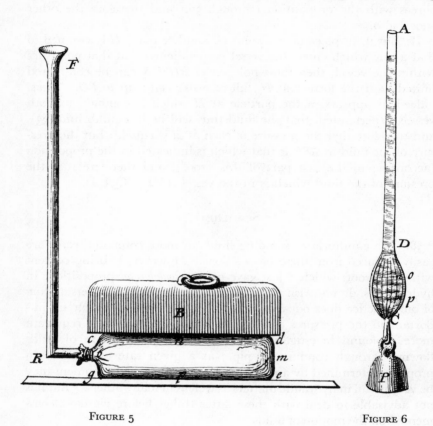

FIGURE 5   FIGURE 6

above the base o. And so the absolute determination of the elevation depends upon the structure of the bladder; if, for example, it would have been composed of perfectly flexible filaments admitting no extension, and if it also had a natural Spherical shape, it is readily apparent that the regions of contact cnd and gpe will be equal and folded, and that the remaining expanded portion will have the shape of a spherical Zone. And so through Geometry a measure of the elevation np is deduced, which will be zero as long as the greatest circle of that bladder will have a smaller ratio to the orifice o than that which exists between the weight B and the weight of the previously mentioned aqueous cylinder, and the whole bladder will not be unfolded before the height will be infinite, that is, never. But if the fibers are of a different nature, the situation will be otherwise; many persons who have discussed the shape of an inflated bladder and who wanted to apply it to the muscular caverns in the classifications of animals have not considered this sufficiently. I now wish to treat this matter a little more fully.

§10. Let there be a bladder DC (Fig. 6) and a weight P hung from the same, and at the same time [let the bladder be] attached to the tube DA, the length of which, in turn, we may consider for our own benefit to be incomparably greater than the length DC. After these things have been established, certainly anyone easily ascertains that, after the vessel and the tube have been filled, the former will become inflated and will lift the appended weight P. But no one will know the state of equilibrium and the shape of the bag unless the structure of the bladder and of its fibers is clearly understood; since these matters are so, we will examine some individual cases which can occur rather frequently.

Case I

§11. Let the bladder be composed of the longitudinal fibers DpC, DmC, etc., in the form of meridians, concurring uniformly at the points D and C, or the Poles, perfectly flexible and uniform, the individual ones of which are connected mutually to the next by minute transverse fibers, which are so lax that they admit sufficient extension under a minimum or practically null force. Thus any fiber DpC whatever will be curved in the shape of an elastic, and the whole vessel will assume the form of a solid which is generated from the revolution of this curve about the axis DC. If, further, the height AD is infinite, the elastic DpC becomes a rectangle, and then the maximum breadth of the bladder is to the length of the axis DC as 25 is to 11, more or less, and the length of the arc DpC is to the same axis approximately as 5 is to 2, so that the vessel will be shortened by three fifths at the maximum elevation of the weight.

Case II

§12. If, after the remaining things have been established as before, the minute transverse filaments no, mp, etc., which are perpendicular to the longitudinal fibers are resistant to extension, it appears that the shape of the fiber $DopC$ cannot be determined unless two kinds of forces are considered to be applied at any one point, one of which acts perpendicularly to the curve and presses the fiber outward, and the other is perpendicular to the axis DC of the curve and draws inward. Thus it is easily seen that an infinite number of laws for these pressures can be devised so that the fiber $DopC$ conforms to any given curve whatever, and thus, for instance, even to a circular one, which shape is attributed by most Physiologists to fibers which pertain to small muscular mechanisms. But there is still another way by which the longitudinal fiber $DopC$ can acquire the shape of a circular arc, namely, when the transverse fibers np, mp, etc., are completely absent. Thus, indeed, while the bladder is inflated, an opening is made between two adjacent longitudinal fibers $DopC$ and $DnmC$ through which the fluid escapes, but at the same time, since it cannot flow out fast enough, it extends the fibers and composes them to the circular shape. And in this case the greatest shortening of the bladder, which in the first case was $\frac{3}{4}$ of the total length of the uninflated bladder, now is only approximately $\frac{4}{11}$.

§**13.** It follows from the above that it is difficult to determine correctly the shape of an inflated bladder to which a weight is appended since, indeed, there is no one who can know perfectly the nature of the minute fibers. Nevertheless, here I will transcribe certain examples, which seem to be especially plausible, from my notes, without proof, which, if anyone desires, he will find in Book III, *Commentaries of the Academy of Science of St. Petersburg*. But first of all I will give the equation for the curve which is formed from the two kinds of forces, as I mentioned in the previous paragraph, with these following any law whatever.

§**14.** Thus, let the thread AEG (Fig. 7) be fixed at the two points A and G. Let the straight line AG be drawn, and let there be two infinitely close points D and E on the thread from which the perpen-

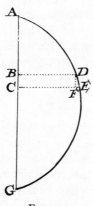

Figure 7

diculars DB and EC are drawn to AG; moreover, let the small line DF be parallel to the line AG. It is known that at the individual points D or E two forces variable in any way whatever are applied, one of which is everywhere perpendicular to the curve, the other everywhere perpendicular to AG; we will set the first one equal to A at the point D and equal to $A + dA$ at the point E, the other $= C$ at the point D, and $= C + dC$ at the point E. Further, let $AB = x$, $BD = y$, $AD = s$, $BC = dx$, $FE = dy$, $DE = ds$, since the element of the curve is assumed to be of constant magnitude. The radius of the Osculating circle at point D is R, at point E is $R + dR$. I say that the following will be the equation pertaining to the curve: $-A\,dR - R\,dA = (R\,dC\,dx + 2C\,dy\,ds + C\,dx\,dR)\,ds$, or, $CR\,ddx$ having been substituted for $C\,dy\,ds$ $\left(\text{for } R \text{ is } = \dfrac{dy\,ds}{ddx}\right)$, one will have $-A\,dR - R\,dA = (R\,dC\,dx + CR\,dds + C\,dy\,ds + C\,dx\,dR)/ds$, or $\dfrac{-AR\,ds - RC\,dx}{ds} = \int C\,dy$.

§15. It is seen from the preceding equation that, when the forces which are perpendicular to the curve act alone, AR becomes a constant quantity, because certainly thus C becomes 0. Therefore, then, the radius of the osculating circle everywhere follows an inverse ratio to the corresponding force. But if the forces perpendicular to the axis are present alone, then, with the letter A vanishing, there results $-\dfrac{RC\,dx}{ds} = \int C\,dy$. But this equation can be integrated and reduced to the form $RC(dx)^2 = $ a constant quantity; from this it appears that the force multiplied by the radius of the osculating circle is everywhere in inverse proportion to the square of the sine which the ordinate makes with the curve. Similarly, the canonic equation admits integration when the forces which are perpendicular to the axis are all equal to one another or proportional to the element ds of the curve. So, indeed, after dC has been set $= 0$, one obtains $-A\,dR - R\,dA = 2n\,dy\,ds + n\,dx\,dR$, by considering n as a constant quantity, with which, after the equation has been properly treated, there results $ny\,dy + mm\,dy - ns\,ds = \int A\,dx$, where m is the constant arising from the integration.

If, in addition, the forces normal to the curve are assumed proportional to the ordinates y, the last equation can be reduced further to this:

$$-dx = \left(2ff - \frac{gyy}{h}\right) dy \bigg/ \sqrt{(2ny + 2mm)^2 - \left(2ff - \frac{gyy}{h}\right)^2},$$

the constants f and m of which will have to be applied to particular cases, while n and g depend upon the relationship of the forces at some particular point; whence, if $g = 0$, the catenary appears, and if $n = 0$, the elastic curve appears; but generally the equation serves for determining the curvature of a uniformly heavy cloth over which fluid is lying. The most simple case in this matter is that when it is supposed that $f = m = 0$, for then, indeed, there results

$$-dx = \frac{-gy\,dy}{\sqrt{4nnhh - ggyy}},$$

or, after the integration has been performed, with the addition of the required constant, $x = -\sqrt{\dfrac{4nnhh}{gg} - yy} + \dfrac{2nh}{g}$, which is the equation of a semicircle to which certainly the cloth will adjust itself in the following hypothesis: let a rope of heavy cloth AEG (Fig. 8) be curved

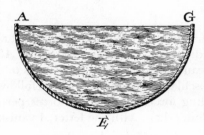

FIGURE 8

in a semicircle, the diameter AG of which is set level, and let fluid lie over the rope right up to AG. If the weight of the fluid is equal to the weight of the rope, I say that a perfectly flexible rope of uniform thickness will preserve the semicircular shape. But in what manner it is to be effected that the weights of the rope and the fluid become equal is well known from the elements of Geometry. Finally, if it is stated that the forces A as well as C are everywhere proportional to the corresponding ordinates y (which hypothesis seems clearly to agree most closely with the true shape of the bladder in Fig. 6), then again the canonic equation which contains differentials of the third Order can be reduced simply to a differential equation, and this should be solved easily through quadratures. If, indeed, $A = my$ and $C = ny$, I say that the nature of the curve ADG in Fig. 7 is expressed by this equation:

$$dx = (g^3 + \tfrac{1}{2}myy)\,dy / \sqrt{(f^3 + \tfrac{1}{2}nyy)^2 - (g^3 + \tfrac{1}{2}myy)^2}$$

in which the letters f and g of constant magnitude appear again from the integrations; but the value of the letter n becomes negative when the equation is applied to determining the shape of the inflated bladder.

§16. I did not wish to pursue these things too much, because they do not pertain very closely to Hydrodynamics. Indeed, I include nothing about elastic fluids because I arranged to treat the theory of them separately; but, nevertheless, because it pertains to the pressures of elastic fluids, the former can easily be deduced and proven from the nature of simply heavy fluids shown above by assuming that the fluid is destitute of elasticity and that a cylinder of the same fluid of infinite or almost infinite altitude is lying above it; but we will mention how these things are to be understood in the proper place. Now, indeed, I continue to that which is customarily sought above all in aquatic matters, namely, how great the strength of conduits must be in order to resist the pressure of water, where chiefly conduits are considered which carry water to fountains, about which I will also say a few things.

§17. The pressure of water standing in conduits ought to be properly distinguished from the pressure of flowing water, although no one, as far as I know, has paid any attention to it up to this time; hence it is that the rules presented by others are valid only for standing water even though they use words which can persuade equally that these apply to flowing water. But in order that the distinction of either Theory may appear in its own light, I will give a certain example the proof of which will be evident from what is below. In place of a reservoir let a very wide vessel $ABCD$ (Fig. 9) be full of water right up to EF, and be connected in its lower part to a horizontal cylindrical pipe $MOmo$ through which it is understood that

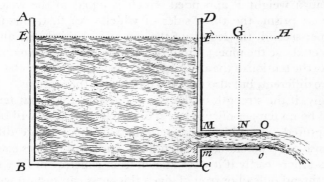

Figure 9

water can flow without impediment. Let the vertical NG be drawn, terminated by the horizontal EH. After these things have been so prepared, I say that if the entire orifice Oo is obstructed by one's finger, the point N is pressed outward according to the total height NG; that if half the orifice is obstructed, this pressure is diminished by a fourth part of the original, and if, finally, after the finger is removed the water is allowed to flow very freely, that all the pressure vanishes, in the same way that the whole is customarily confused with the part or even with nothing by the Authors. But I will demonstrate that the pressure can even be made negative and thus be changed into suction. But since I cannot treat this before I have treated the whole theory of flowing water, I shall now consider standing water only, just as if the entire orifice Oo were closed.

§**18.** Moreover, it is definite from Mechanics that the walls of the pipe $MOmo$ (the diameter of which we will consider to be incomparably less than the height NG) are not extended differently than if they were arranged in the rectangular shape $MOmo$ (Fig. 10), and

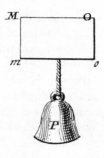

FIGURE 10

if they had a weight P appended which is equal to the weight of the aqueous prism, the three sides of which are: first, the radius of the pipe; second, the length of the same; and third, the altitude of the water above the pipe. From this proposition is known not only the ratio of the tensions if the altitudes of the water or the diameter of the pipes were different, but also the very measure of the tensions. Thus, accordingly, if the strength of the pipes is greater than that tension, there will be no danger of rupture; if otherwise, the pipe will be certainly ruptured. In addition, experiments of this sort are difficult and expensive. Therefore, the strength of lead or iron pipes could be understood more easily if it were known from experiment how much weight a thread of lead or iron of given thickness can sustain without danger of rupture. At the end of the section I will add a similar experiment performed by me to show how from this the strength of a pipe of given thickness and diameter can be deduced.

EXPERIMENTS WHICH PERTAIN TO CHAPTER II

Pertaining to §5

Concerning capillary tubes: Innumerable experiments concerning the nature of these tubes have been undertaken by many, among whom Georg Bernhard Bilfinger stands out, who not only collected the important ones but also added many of his own; see *Commentaries of the Imperial Academy of Science of St. Petersburg*, Book 2, p. 233ff.

I. In order that it might appear properly to the eye how contrary in Character mercury and the rest of the fluids are in this area, I ordered a glass vessel ABD (Fig. 11) to be made composed of two vertical legs,

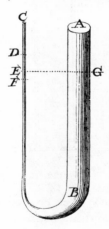

Figure 11

of which the one AB had a diameter of three or four lines and the other BC of hardly a third part of a line. When the vessel was filled with any liquid whatever, the surface was higher in the narrower leg than in the wider, as at D and G; however, mercury alone was more depressed in the narrower than in the wider, as at F and G.

II. In order to show that mercury differs from the nature of the other fluids for no other reason than on account of the stronger mutual attraction of its own particles, I reflected on these experiments: indeed, I filled a slender pipe with mercury by suction and erected it slowly from its horizontal position. Accordingly, the mercury, although not all of it, flowed out, and the vertical height of the mercury remaining in the pipe was consistent with itself in every position. However, when the mercury is suspended in this way in the pipe, and if then the extremity of the pipe is brought in touch with the mercury standing in the vessel, it all flows out directly. The prior Phenomena, unless I am mistaken, indicate that the same thing occurs with mercury and the other fluids when there is no opportunity for an attractive force; but the last phenomenon shows that mercury attracts itself very strongly.

III. Let there be assumed a cylindrical glass pipe of a diameter of three or four lines, furnished with a base of delicate Paper or of a very thin plate of iron prepared and perforated in the middle by a tiny little orifice, as Fig. 12 shows. Let the pipe $ACDB$ be inclined and

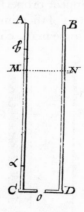

FIGURE 12

filled entirely with mercury, then erected little by little; what happened before will occur, and although the pipe is very wide, nevertheless, not all the mercury flows out, but part of it will remain suspended, as for example $MCDN$, and the smaller is its little orifice o, the greater will be this amount. When then the base is submerged in the mercury in some associated vessel just a little bit, so that the submerged part of the pipe is $C\alpha$, not only does the mercury not ascend in the pipe right up to β (it having been assumed that $C\alpha = M\beta$ of course) but also almost all of it flows out, until the surface MN reaches α. Next I submerged the empty pipe $ACDB$ reasonably deeply in the mercury which was in the other vessel, and nevertheless nothing of it began to flow from the vessel into the pipe before it had been submerged to the height CM; and then suddenly it flowed right up until it had reached a level in each part, namely right up to MN, if it was submerged up to that point. All these things are deduced easily from the mutual attraction of the mercurial particles. In addition, I performed a test to investigate the relationship which exists between the height MC and the area o of the little orifice; at any rate, it is probable that that diameter is in a reciprocal ratio to the diameter pertaining to the little orifice; nevertheless, I was not able to confirm the idea sufficiently by experiment, sometimes because of the impurity of the mercury which I used, which caused the height of the suspended mercury to be not completely consistent with itself when the orifice had not been varied in repeated experiments, or sometimes also because it is difficult to measure very small orifices accurately. Indeed, the orifices must be a minimum, since the height of the suspended mercury is barely six or eight lines when the diameter of the orifice equals the sixth part of a line; nevertheless, let me tell the method which I have used. Indeed, by means of copper wires of different thickness which are used in musical instruments, the very small diameters of which I found very correctly from the length and the weight of them, I perforated the little paper CD; but in this way shreds usually appear around the walls of the orifice which impede the efflux, and thus it easily happens that the orifice is greater than is the thickness of the wire.

Pertaining to §18

Concerning the strength of pipes. A round copper wire the diameter of which was $2/11$ of a Paris line, to which successively continuously greater weights were added, did not break until the weight exceeded 18 Nuremberg pounds. Then I observed that a very thin lead plate, which was of rectangular shape, $5/4$ of a line wide, $1/131$ line thick, was broken when to it was appended a weight of three and a half ounces. From these two observations it followed, with all the remaining things being equal, that the copper wire is more than 28 times as strong as the lead wire. From the previous experiment it is also deduced that, if a copper pipe should have a diameter of 1 foot and the thickness of the walls were $2/11$ line, it can sustain water to a height of 518 feet before it is broken. In this calculation I used 70 pounds for the weight of a cubic foot of water. But if the same pipe is of lead, it will sustain water to a height of 18 feet in the light of the other observation, and it can bear a height of water of 99 feet if the walls of the pipe are a whole line in thickness. This agrees with what Mariotte has in his *Traité du mouvement des eaux*, p. 472, where indeed he says that a lead pipe, the diameter of which was 1 foot and the thickness of the walls two and a half lines, has supported water without rupture to the height of a hundred feet, and that while he was observing this he shaved off the sides little by little until at last they were diminished to a thickness of one line, and that then at last the force of the water destroyed the pipe.

From the observed strength of copper wire, the strength of cannons is also determined: let there be, for instance, a cannon, the internal diameter of the barrel of which is three inches; moreover, the thickness of the walls not far from the touch-hole, where the force of the powder is greatest, is customarily more or less equal to the internal diameter, so that the total diameter is thrice the internal diameter of the barrel. Because, therefore, this thickness is not to be neglected with respect to the internal diameter of the barrel, we shall consider all the material concentrated in the middle and thus at a distance of three inches from the axis of the barrel. This having been established, the maximum height of water which a cannon can support not far from the touch-hole will be $\frac{11}{2} \cdot 12 \cdot 3 \cdot 2 \cdot 518 = 205{,}128$, which force exceeds the elasticity of natural air by about seven thousand times. But I will show in the following that ignited gunpowder can exert a force for rupturing any cannon greater indeed than that which was mentioned, but nevertheless not exceeding it much. But the cannons obtain the additional strength that they require from belts or bands which are called *plattes bandes et moulures*, apart from the fact that at the very rear of the cannon (*á l'endroit de la culasse*) the thickness is greater than that which we assumed. Nevertheless, we will not be surprised that quite a few cannons are shattered.

THIRD CHAPTER

Concerning the Velocities of Fluids Flowing out of a Vessel Formed in any Way Whatever through any Kind of Opening Whatever

§1. Before we may attempt to define the motion of water developing from its own gravity, we will look again at what we set forth in the First Chapter, §§18, 19, 20, 21 and 22, concerning the principles to be applied to the following matters.

We will recollect, certainly, that the *potential ascent* of a System, the individual portions of which are moved at any velocity whatever, indicates the vertical height to which the center of gravity of that System reaches if the individual particles, their motion having been turned upward with the proper velocity, are understood to ascend as far as they can; and that the *actual descent* denotes the vertical height through which the center of gravity descends after the individual particles have come to rest. Then as well we will be mindful that, necessarily, the *potential ascent* is equal to the *actual descent* when all the motion remains in the scattered material, and none of it goes over into unobservable or other type material not pertaining to the system, and, finally, that the motion of fluids is approximately such that everywhere the velocity is reciprocally proportional to the corresponding area of the vessel, concerning which we will add certain other things in the proper place. Now it is fitting to examine the following proposition.

PROBLEM

§2. If water flows through a conduit formed in any way whatever and its velocity is known in some place, find the *potential ascent* of all the water contained in the conduit.

SOLUTION. Let there be formed in any way whatever the conduit ST (Figs. 13 and 14), through which part *bcfg* water flows; it is assumed, if on the axis *ae* is taken some point *n* through which the plane *pm*, perpendicular to the axis, passes, that all aqueous particles

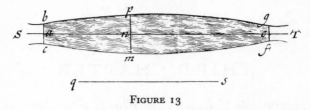

FIGURE 13

existing on that plane will flow at an equal velocity and indeed such that it is everywhere inversely proportional to the area of the section *pm*. Moreover, let the velocity of the water at *gf* be such as is due to the vertical height *gs*, that is, let the *potential ascent* of the aqueous

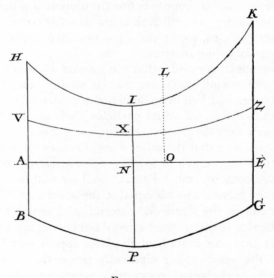

FIGURE 14

stratum at *gf* be equal to the line *qs*, and since heights of this sort are in proportion to the squares of the velocities, it follows that the *potential ascent* of the water at *pm* is equal to the fourth proportional of the square of the area *pm*, the square of the area *gf*, and the altitude *qs*, indeed equals $\frac{(gf)^2}{(pm)^2} \cdot qs$. Thus, with these things having been set forth, we will assume: that the curve *BPG* in Fig. 14 is the scale of the areas of the conduit, so that, with $AN = an$, *NP* denotes the area at *pm*; hence that the curve *HIK* is the scale of the *potential ascents*, so that

VELOCITIES OF FLUIDS FLOWING OUT OF A VES

$NI = \dfrac{(EG)^2}{(NP)^2} \cdot qs$. Now let it be assumed that the indivi(?) of the curve HIK have a weight equal to the weight of (?) ding aqueous stratum, and that the center of gravity of (?) falls at the point L, and let LO be drawn perpendicular to the axis AE; thus LO will be the desired *potential ascent* of all the water. From mechanics, moreover, it follows that if a third curve UXZ be formed, the ordinate NX of which is everywhere equal to $\dfrac{(EG)^2}{(NP)}$, LO will be equal to the fourth proportional of the area $AEGB$ and $AEZU$ and the line qs or EK. Therefore, that which is sought is evident. Q.E.I.

§3. For instance, if there is a conic conduit, the anterior and posterior surfaces gf and bc of which have diameters in proportion as m is to n, the *potential ascent* of the water will be

$$\frac{3m^3}{n(mm + mn + nn)} \cdot qs.$$

Problem

§4. Given infinitely small variations, with respect to position as well as to velocity, which correspond to the anterior surface of the water, find the variations pertaining to the *potential ascents* throughout the water.

Solution. Let the area $AEGB = M$ [Fig. 14], the area $AEZU = N$, $qs = v$; the *potential ascent* will be $\dfrac{Nv}{M}$. Truly, because the quantity of water in the conduit is considered constantly the same, the area $AEBG$ will be invariable, and thus $dM = 0$, so that the differential of the *potential ascent* is simply $\dfrac{N\,dv + v\,dN}{M}$, and also dN is obtained from the variation of the position of the water. Therefore, the proposition is evident. Q.E.I.

Scholium

§5. These propositions can serve for defining the motion of fluid moving within vessels, that is, not flowing out, as I shall show in the proper place; but certainly when the fluid flows out through an orifice, a more appropriate computation is established differently, namely, as follows.

Problem

§6. Find the difference in *potential ascent* after a volume element has flowed out through an orifice.

SOLUTION. Let us consider that water flows out of the vessel *aimb* (Fig. 15) formed in any way whatever; let the base *im* be perforated

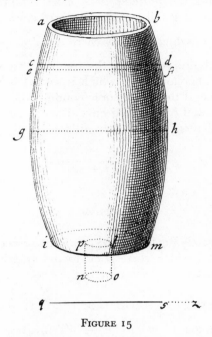

FIGURE 15

by the orifice *pl*; let the quantity of water remaining in the vessel after a given quantity of it has already flowed out be *cimd*; moreover, let the volume element *pnol* flow out in an infinitely short instant, with the surface *cd* descending to the position *ef*. Let a section *gh* be assumed in the middle of the water, parallel to the surfaces *cd* and *ef* and to the base *im* itself; and let the velocity of any one of the particles on *gh* be such that it can ascend to a height *qs* or v when the volume element has not yet flowed out, and to a height *qz* or $v + dv$ after that very volume element has flowed out. With all these things having been set forth, the increment of *potential ascent* of the water is sought after the position *cimd* is replaced with the position *eipnolmf*, that is, after the volume element has emerged.

As before, let the curve *CGI* (Fig. 16) be drawn as the scale of the areas, where, precisely, *CD* or *EF* will represent the area of the aqueous surface before or after the efflux of the volume element,

VELOCITIES OF FLUIDS FLOWING OUT OF A VESSEL

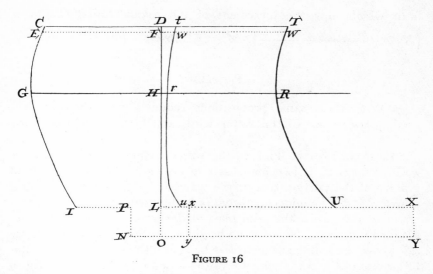

FIGURE 16

GH the proper assumed area, IL the size of the base, PL the size of the orifice, while the very small connecting parallelogram $PNOL$ corresponds to the cylindrical volume element $pnol$. Next, let there be constructed another curve TRU, the ordinates of which are again equal to the square of the line GH divided by the corresponding ordinate of the curve CGI, to which curve in the same manner is annexed the small parallelogram $LOYX$, the side LX of which, certainly, is equal to the square of the line GH divided by the line PL.

Therefore, now it is apparent that the *potential ascent* of the water before the efflux of the volume element is equal to the fourth proportional of the area $DCIPL$, the area $DTUL$, and the height qs, and that the same after the efflux of the volume element is equal to the fourth proportional of the area $FEIPNOL$, the area $FWUXYOL$, and the height qz; moreover, in both analogies the first terms (namely, the area $DCIPL$ and the area $FEIPNOL$) are equal to each other; therefore, if either one of these areas be indicated by M, the area $DTUL$ by N, the *area* $FWUXYOL$ by $N + dN$, the height qs by v, and qz by $v + dv$, the increment of *potential ascent* during the efflux of the volume element will be $\dfrac{N\,dv + v\,dN}{M}$. Thus, if now it is assumed that $LD = x$, $FD = -dx$, $DC = y$, $HG = m$, and $PL = n$, one will have $DT = \dfrac{mm}{y}$, $LX = \dfrac{mm}{n}$, $LO = \dfrac{-y\,dx}{n}$ (because the area $DFEC$ = area $LONP$), and hence $dN = LOYX - DFWT = -\dfrac{mmy\,dx}{nn} + \dfrac{mm\,dx}{y}$, from which now the increment of the *potential ascent* sought is $\left(N\,dv - \dfrac{mmvy\,dx}{nn} + \dfrac{mmv\,dx}{y}\right)\Big/M.$ Q.E.I.

Problem

§7. With these same propositions retained, find the infinitely small *actual descent* of the water while the volume element flows out.

SOLUTION. Since in Fig. 15 the water changes position *cdmi* for position *efmlonpi*, it is evident that in either position the center of gravity of the portion *efmi* of the water is in the same place, and therefore it can be understood that only the small portion *cdfe* (which equals $-y\,dx$, while the total mass of the water equals M) has descended into *lonp*. Now let the height of the small aqueous particle *cdfe* above the volume element *lonp* be x, and the height of the center of gravity of the water *efmi* above the base be b; then the height of the center of gravity of all the water in the position *cdmi* above the base will be $b - \dfrac{y\,dx}{M}\cdot(x-b)$, and in position *efmlonpi* the same height will be $\left(\dfrac{M+y\,dx}{M}\right)b$, whence the difference of the heights or the required *actual descent* equals $\dfrac{-y\,dx}{M}\,x$, which equation indicates that the volume element which flowed out is to be multiplied by the height of the water above the orifice, and the product is to be divided by the quantity of the water in order to obtain the *actual descent* which occurs when the volume element flows out. Q.E.I.

Problem

§8. Determine the motion of a homogeneous fluid flowing out of a given vessel through a given orifice.

SOLUTION. Since, through our hypothesis, *potential ascent* at individual instants is equal to *actual descent*, an increment of the former while a volume element flows out will be equal to an increment of the latter because it develops in the same short time. Therefore if, again, the surface of the water, after a given quantity of it has flowed out, is set equal to y, the area of the vessel at any place whatever, as it pleases, is assumed equal to m, the area of the orifice equal to n, the height of the water above the orifice equal to x, the quantity N is determined by that law which was indicated in §6, and v is understood to be the height due to the velocity of the water at the assumed place where, indeed, the area of the vessel is m, then, through §6, the increment of *potential ascent* will be $\left(N\,dv - \dfrac{mmvy\,dx}{nn} + \dfrac{mmv\,dx}{y}\right)\Big/M$, and the least *actual descent* will be $\dfrac{-yx\,dx}{M}$ (through the preceding §), from which there results $\left(N\,dv - \dfrac{mmvy\,dx}{nn} + \dfrac{mmv\,dx}{y}\right)\Big/M = -yx\,dx/M$, or $N\,dv - \dfrac{mmvy\,dx}{nn} + \dfrac{mmv\,dx}{y} = -yx\,dx$, which equation can generally be integrated since the terms N and y are given functions of x itself, and the term v is of only one dimension.

§9. COROLLARY 1. Since the velocities are in inverse proportion to the areas, it is evident that the height which corresponds to the velocity of the water flowing out will be $\dfrac{mm}{nn}v$; therefore, if this is called z, one will have $nnN\,dz - mmzy\,dx + \dfrac{mmnnz\,dx}{y} = mmyx\,dx$.

§10. COROLLARY 2. If the orifice is very small in proportion to the areas of the vessel, then $n = 0$, and the entire equation reduces to this: $-mmzy\,dx = -mmxy\,dx$, or $z = x$; accordingly, therefore, the water constantly flows out at that velocity by which it can ascend right up to the height of the uppermost surface, the only case that Geometers had understood correctly to this time; and this proposition is valid for all vessels, however formed. But when the orifice is not considered as infinitely small, by no means is the shape of the vessel to be neglected. Nevertheless, one can observe that, unless the orifice is very wide, it can be considered as infinitely small without any noticeable error at all.

§11. COROLLARY 3. When the fluid is not everywhere the same, the computation is to be undertaken in a similar manner: by inquiring, surely, both into the increment of *potential ascent* of the composite fluid and into the *actual descent*; and by equating these to each other. Thus if, moreover, the orifice is very small, it is also intrinsically evident, as the calculation shows, that the fluid will spring forth at a velocity due to a certain height such that, if the vessel were refilled to the same height by the same liquid which flows out, the walls of the vessel would sustain the same pressure.

GENERAL SCHOLIUM

§12. Before we may deduce rather special Corollaries from our theory about the motion of fluids from cylindrical vessels, it is fitting here to examine to what extent the assumed hypotheses agree with the nature of the matter and what other causes diminishing the fluid motion, of which we took no account in the computation, could intervene.

At first, as far as applies to the principle of *the conservation of live forces* or *of the perpetual equality between potential ascent and actual descent*, I see nothing here which can be a notable impediment to it, if only we disregard friction, viscosity, resistance of air, and other obstacles of this sort. But certainly it occurs often that the principle cannot be applied without limitation, which we shall show in the following; namely, when the individual particles of water are carried by a different motion, as a result of which it occurs that at every instant something from the motion, or, if preferred, from the *potential ascent*, is lost. But in the present case nothing similar happens, since indeed almost all the particles are moved altogether similarly, and, especially when the orifice is very small, the motion of the internal particles is almost nil, and therefore no detriment can develop from this. Moreover, the other principle, by which it is assumed that the velocity of any particle whatever is that which corresponds to the inverse ratio of the area, is indeed affected by a twofold disadvantage: *first*, namely, because the motion near the sides of a vessel is a little slower than in the middle, and therefore all particles corresponding to the same area of a vessel are not carried at an equal velocity; and *second*, because water not greatly distant from the base cannot have that motion which this principle postulates. However, neither carries a noticeable error with itself inasmuch as in this simple problem the internal shape of the vessel is of hardly any consequence to the motion of the water flowing out. By the same reasoning it is understood that the motion of water flowing out in some other direction cannot be very different because, to be sure, the internal motion of the water in only the lower part of the vessel becomes different, and this difference can hardly be of any importance. Therefore, it appears that the hypotheses by which the computation of this Problem of ours is supported thus agree with the nature of the question; hence no error perceptible to the senses can arise. But surely the hindrances mentioned above, attrition, viscosity of the fluid, and other similar ones, are of greater importance particularly when an orifice through which fluids spring forth is rather small, or when the height of the water above the orifice is very great, or, finally, when a pipe is very slender, concerning which many experiments are found in the writings of Mariotte in his *Traité du mouvement des eaux*. But now I progress to examining the motion of water flowing out of Cylindrical vessels through orifices of any size whatever. We shall also consider vessels placed vertically by reason of a short cut and a more elegant solution.

CONCERNING THOSE THINGS WHICH PERTAIN TO THE EFFLUX OF WATER FROM VERTICALLY POSITIONED CYLINDERS THROUGH ANY OPENING WHATEVER WHICH EXISTS IN A HORIZONTAL BASE

§13. Geometers who have discussed water flowing from a vessel are accustomed to consider principally cylinders positioned vertically. Therefore, it will not be out of place at all to deduce those conclusions which pertain here from our general theory. Let the area of the cylinder be to the area of the orifice as m is to n, the height of the water above the orifice when flow begins be a, the height of the residual water be x, and the height due to the velocity of the internal water be v; there will be, in the canonic equation of §8, $y = m$, and $N = mx$ (through §6), which therefore transforms into the following equation:

$$mx \, dv - \frac{m^3}{nn} v \, dx + mv \, dx = - mx \, dx$$

or

$$\left(1 - \frac{mm}{nn}\right) v \, dx + x \, dv = - x \, dx.$$

Let this latter equation be multiplied by $x^{-mm/nn}$, so that there results

$$\left(1 - \frac{mm}{nn}\right) x^{-mm/nn} v \, dx + x^{1-mm/nn} \, dv = - x^{1-mm/nn} \, dx.$$

Now this equation can be integrated; but the addition of a constant in the Integration is to be attended to such that at the beginning of flow, that is, when $x = a$, the velocity of the fluid is null, and hence v itself likewise is null. So indeed there arises:

$$x^{1-mm/nn} v = \frac{nn}{2nn - mm} \left(a^{2-mm/nn} - x^{2-mm/nn}\right)$$

or

$$v = \frac{nna}{2nn - mm} \left[\left(\frac{a}{x}\right)^{1-mm/nn} - \frac{x}{a}\right].$$

§14. From this equation, therefore, the height generating the velocity of the internal water is known; here it deserves to be noted that, if the vessel is very wide, it can be directly reckoned that $v = \dfrac{nn}{mm} x$, certainly after the water descends just a little, that is, as soon as x is at once a little less than a. This rule fails notably only at the very beginning of motion, and if that first element of motion is considered (in which certainly the height $a - x$ can be considered as infinitely small), the equation indicates that now $v = a - x$. From this it follows that in the entire cylinder, whatever the orifice might be, the internal water is accelerated from the beginning of motion just like freely falling bodies. But if the motion continues a little while, then this Rule will err the less, the greater is the orifice and the higher is the water in the pipe. If, further, that height is desired which corresponds to the velocity of the water flowing out, which in §9 we set equal to z, there will be

$$z = \frac{mm}{nn} v,$$

or

$$z = \frac{mma}{2nn - mm} \left[\left(\frac{a}{x}\right)^{1 - mm/nn} - \frac{x}{a} \right].$$

§15. When $n = m$, that is, when the base is null, it appears from the very nature of the matter that the water falls and is accelerated freely in the manner of heavy bodies, which very thing the equation also indicates; indeed, it occurs in this position that $z = a - x$. But if the orifice is considered as infinitely small in proportion to the area of the vessel, which case we have already considered above, it is to be assumed that $n = 0$, and then it occurs that $z = x$, which indicates that the water flows out constantly at that velocity by which it can ascend to the total height of the water. Finally, with $mm = 2nn$, there develops $z = \dfrac{mm}{0}(x - x)$; since nothing can be learned from this value, one must go back to the differential equation of §13, which now is this:

$$- v\, dx + x\, dv = - x\, dx, \quad \text{or} \quad \frac{x\, dv - v\, dx}{xx} = \frac{- dx}{x},$$

which integrated, with the addition of the required constant, gives

$$\frac{v}{x} = \ln \frac{a}{x}, \quad \text{or} \quad v = x \ln \frac{a}{x}, \quad \text{and so} \quad z = 2v = 2x \ln \frac{a}{x}.$$

§16. The velocity of the water flowing out increases at the beginning and afterwards decreases, and is a maximum somewhere,

VELOCITIES OF FLUIDS FLOWING OUT OF A VESSEL

namely at that place at which the water descends to the height $a \Big/ \left(\dfrac{mm - nn}{nn}\right)^{nn/(mm - 2nn)}$; Mariotte, also having learned this through experience, indicated it in his *Traité du mouvement des eaux*, Part III, disc. 3, exp. 5, and the maximum velocity itself is that which is due to the height

$$\frac{mma}{mm - 2nn}\left[\left(\frac{nn}{mm - nn}\right)^{nn/(mm - 2nn)} - \left(\frac{nn}{mm - nn}\right)^{(mm - nn)/(mm - 2nn)}\right]$$

which quantity reduced becomes

$$\frac{mma}{mm - nn}\left(\frac{nn}{mm - nn}\right)^{nn/(mm - 2nn)}.$$

It is seen from these formulas that the time during which the velocity is changed from nil to a maximum is clearly imperceptible when the orifice is moderately small and the pipe is not very long, but that it becomes noticeable when the situation is otherwise, which we see in leaping fountains to which water is carried through long conduits; but these things which pertain to time intervals will be explained further in the following section, and at the same time it will be shown how little water is ejected from very large vessels before it flows at maximum velocity.

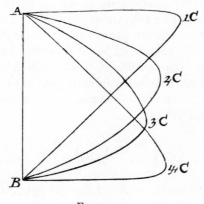

FIGURE 17

The nature of the velocities is better understood from the attached Fig. 17, in which, if *AB* represent the total height of the fluid above the orifice at the beginning of flow, the curves *A1CB*, *A2CB*, *A3CB*

$A4CB$ express the scales of the corresponding heights to which the fluid flowing out can ascend by its own velocity with different sizes of orifices. For example, the scale approaches the shape $A1CB$ if the orifice has a small ratio to the area of the vessel, and [approaches] the shape $A2CB$ when the base is assumed perforated by a greater opening; and if now the ratio of the orifice is to the area of the vessel as 1 is to $\sqrt{2}$, that scale will be as $A3CB$ (in which case the maximum velocity becomes less than in any other, and is expressly that which is due to the height $\dfrac{2a}{e}$, by understanding that e is the number the logarithm of which is unity, that is, to a height a little less than $\tfrac{3}{4}a$) and finally the scale will be as $A4CB$ when there is almost no base remaining.

§17. But now we will illustrate by a certain example that which was indicated above in §10: namely, that unless the orifice is very large, it can be considered in the calculation as infinitely small without very noticeable error, and therefore it can be assumed that $z = x$, as was mentioned in §§10 and 15. It seems that in the works of many Authors it prevailed that they reckoned only that no proportion of size of the orifice is ever to be taken, however great the orifice may be assumed, which matter is certainly ridiculous; at least up to this time no one whom I know has considered the size of the orifice correctly with regard to this matter. Therefore, let us consider a cylinder the diameter of which is only quadruple the diameter of the orifice, large orifices of which sort customarily occur rarely in hydraulic equipment; and let us consider that the surface of the water has descended through only one hundredth part of the entire initial height (indeed I assume that it has descended some little bit, because at the very beginning no motion can exist in the water, much less enough that the water flowing out can ascend by its own motion to the entire height); these assumptions make $m = 16n$ and $mm = 256nn$, and $x = \tfrac{99}{100}a$, from which there develops

$$z = \frac{128}{127}\left[\frac{99}{100} - \left(\frac{99}{100}\right)^{255}\right]a = \frac{92}{100}a,$$

which certainly differs somewhat from the quantity x or $\tfrac{99}{100}a$, but nevertheless not altogether greatly, and the difference becomes much less when the orifice is less and when the surface of the water descends a little further. Therefore, this Theory differs from the common one very greatly at the beginning of flow, at which time the motion is less than was stated; on the contrary, towards the end of flow the water is

VELOCITIES OF FLUIDS FLOWING OUT OF A VESSEL

thrust out at a greater velocity than it should be according to the usual principles.

§18. So far we have considered the motion of water having arisen from its own gravity; let us now consider: that the water has been ejected by some outside force apart from the force of gravity; that such a velocity has been communicated to the water flowing out that it can ascend to a much greater height than if the gravity of the water alone had produced the motion; that then that other force suddenly vanishes; and that the water is left by itself. But if this happens, experience shows that the velocity of the water decreases very quickly and soon is such that it is not notably greater than that velocity which would arise from the gravity alone of the water. Thus we see that it happens sometimes in leaping fountains (the true cause and the measurement of which I will discuss elsewhere) that the water leaps up to a triple or quadruple or even greater height than is customary; when this happens, that leap stops suddenly, and it does not exceed the customary height, as far as this can be perceived by observation; but I am speaking about pipes perforated by not very large orifices, for when an orifice is somewhat larger, the leap of the water does not decrease so suddenly. And so, we will examine now to what extent the theory agrees with these phenomena, and we will include the accurate measures of them which follow hence. In order that we may indeed follow the matter generally, we will again consider that the area of the cylinder is to the area of the orifice as m is to n; that the water is driven forth at that velocity by which it can surge to the height α; and that at that same instant the height of the water above the orifice is a, the gravity of which alone now expels the water; that then the surface of the water descends in the Cylinder through the vertical height $a - x$, so that the residual height is x, and then that the velocity of the ejected water is that which is due to the height z. With these having thus been set forth, we will make use of the general differential equation of §9, which is this:

$$nnN\,dt - mmzy\,dx + \frac{mmnnz\,dx}{y} = -mmyx\,dx$$

(where again, as was indicated in §13, y is $= m$ and $N = mx$), which in our particular case becomes

$$\left(1 - \frac{mm}{nn}\right) z\,dx + x\,dz = -\frac{mm}{nn} x\,dx,$$

which, multiplied by $x^{-mm/nn}$ and afterwards integrated accordingly, so that with $x = a$ one has $z = \alpha$, will give the final desired equation

$$z = \left(\frac{mm}{2nn - mm} + \frac{\alpha}{a}\right) a^{(2nn - mm)/nn} x^{(mm - nn)/nn} - \frac{mm}{2nn - mm} x$$

or

$$z = \frac{mma}{2nn - mm}\left[\left(\frac{a}{x}\right)^{1 - mm/nn} - \frac{x}{a}\right] + \left(\frac{x}{a}\right)^{(mm - nn)/nn} \alpha.$$

If this height is compared with that which was indicated in §14, the excess of one over the other is found to be $\left(\frac{x}{a}\right)^{(mm - nn)/nn} \alpha$; whence all these Phenomena are now confirmed which were just indicated; indeed, when the number m is much greater than n, that excess immediately becomes unnoticeable after the water descends just a little bit, that is, after a very short time, but nevertheless all of it never vanishes as long as the flow endures; and, finally, it is continuously more notable, the more the ratio of the number m to n approaches unity. For instance, let the diameter of the pipe be ten times greater than the diameter of the orifice, and let the water be expelled by such a force that by its own velocity it can spring up to a height which is quadruple the height a, or of the water above the orifice; it is sought to what height the water flowing can ascend by its own velocity after the aqueous surface has descended in the pipe through a thousandth part of a itself, if at the same time the water is stimulated to efflux through its own gravity alone, thence what the similar height would have been if the water had had no motion at the beginning. Therefore, $m = 100\, n$, $mm = 10{,}000\, nn$, $x = \frac{999}{1000} a$, $\alpha = 4\, a$, from which in the former case one obtains

$$z = \left[\frac{10{,}000}{9998}\left(\frac{999}{1000} - \left(\frac{999}{1000}\right)^{9999}\right) + 4\left(\frac{999}{1000}\right)^{9999}\right] a$$

or

$$z = \frac{99{,}915}{100{,}000} a + \frac{18}{100{,}000} a\, ;$$

but in the latter case it becomes

$$z = \frac{99{,}915}{100{,}000} a,$$

from which example it is evident how small and clearly unnoticeable is the excess of the former height above the other, and how suddenly that aqueous thrust is diminished since, indeed, the entire change occurs while the surface of the water descends through a thousandth part of the height a, which time in customary hydraulic machines cannot be other than very short. Thus what was stated above in §17 is confirmed as well, that certainly $z = x$ approximately when the orifice is rather small, since in the present case, wherein the motion begins from rest, the difference between z and x is only fifteen hundred-thousandths of the height a itself; since in the meanwhile the height z is a little greater than x, it is evident that the water flowing out can ascend to a greater height even after the water has flowed out for some time, than the height of the water above the orifice.

§19. Thus, inasmuch as we have deduced from our general Theory those things which attend the motion of fluids from cylinders placed vertically, we will now also consider pipes placed obliquely, which are customarily very long in leaping fountains. In these, certainly, it is singular that the acceleration of the motion does not occur as suddenly as when the Cylinders are vertical, and thus by observation one may perceive the accord of our Theory with the actual motion of water.

§20. Let us consider a conduit curved any way whatever, but nevertheless Cylindrical, the area of which again is in proportion to the area of the orifice as m is to n. Let the motion begin from rest, and let the vertical height of the water above the orifice at the beginning of motion be a. Let a certain quantity of water have flowed out, and let the vertical height of the residual water above the orifice be taken as x; let the length of the conduit, which at that very moment is full, be ξ, and then let the internal water (the individual particles of which I assume here to be carried in a motion parallel to the axis of the conduit) have a velocity which corresponds to the height v. Thus, with these things having been set forth, if we make use of a reasoning similar to the above for seeking indeed the increment of *potential ascent* while the volume element flows out, as we did in §6, and by assuming the same equal to the *actual descent*, the following equation is obtained:

$$\xi\, dv - \frac{mm}{nn} v\, d\xi + v\, d\xi = -x\, d\xi,$$

or

$$\left(1 - \frac{mm}{nn}\right) v\, d\xi + \xi\, dv = -x\, d\xi;$$

the integral of this, which is evident after the terms have been multiplied by $\xi^{-mm/nn}$, is this:

$$v = \xi^{mm/nn - 1} \int -x\, \xi^{-mm/nn}\, d\xi.$$

If there exists, for example, a straight conduit so inclined to the horizontal that the sine of the intercepted angle between the two is to the total sine as 1 is to g, there will be $\xi = gx$, from which

$$v = \frac{nna}{2nn - mm}\left[\left(\frac{a}{x}\right)^{(nn-mm)/nn} - \frac{x}{a}\right].$$

Since this equation does not differ from the equation given in §13 for vertical Cylinders, it follows that in each case the velocities of the water are the same when the vertical descents of the surface of the water are the same. Therefore, similar accelerations in homologous places on either hand are in proportion to the vertical heights, and only this distinction occurs, that in an inclined conduit it happens more slowly, and in proportion as 1 to g; therefore, these accelerations can be perceived easily by observation in greatly inclined conduits which cannot be [perceived] in vertical ones on account of the excessive speed of the changes. On the other hand, it is intrinsically evident from the fact that the frictions are increased by the length of the pipe that it cannot be that the velocities are not diminished, to which those should attend in whom there will be a desire to undertake experiments on this subject.

CONCERNING THE EFFLUX OF WATER FROM VERTICALLY POSITIONED CYLINDERS WHICH TERMINATE IN OTHER NARROWER AND SIMILARLY VERTICAL PIPES

§21. Experience shows that between two Cylinders wholly equal and similarly positioned, for one of which a rather narrow pipe corresponds to the orifice of the other, that the one is depleted more quickly which has the pipe attached, and indeed, the quicker it does so, the more the pipe increases in size from the place of insertion toward the extremity, which Mr. s'Gravesande showed to many in *Physices Elementa Mathematica*, lib. 2, cap. 8. Let us consider the entire matter in the following Problem.

PROBLEM

§22. Let there be a cylindrical vessel *AEHB* (Fig. 18) positioned vertically, perforated at *FG*, by which opening it connects with the conic tube *FMNG*, through the orifice *MN* of which water finally flows out. The velocity of the aqueous surface *CD* is sought after it descends from rest through *AC* or *BD*.

VELOCITIES OF FLUIDS FLOWING OUT OF A VESSEL 51

SOLUTION. Let the initial height of the water above MN, namely $NG + HB$, be a, the height of the aqueous surface at the position CD above MN, that is, $NG + HD$, be x, the length of the annexed pipe, or NG, be b, the area of the orifice MN be n, the area of the orifice

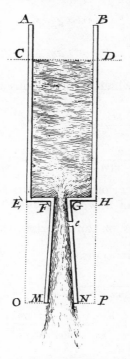

FIGURE 18

FG be g, the area of the upper Cylinder be m; let the velocity of the aqueous surface at CD be that which is due to the height v. In the general equation of §8 there will be $y = m$ and $N = m(x - b) + \dfrac{bmm}{\sqrt{gn}}$, which substitutions, it will be evident, are conformal to the derived calculation, since in §6, moreover, the remaining conditions are the same as before. Therefore, the equation of §8 is resolved to

$$m(x - b)\, dv + \frac{bmm}{\sqrt{gn}}\, dv - \frac{m^3 v\, dx}{nn} + mv\, dx = -mx\, dx$$

which, further, divided by m, and with one's having established

$$x - b + \frac{mb}{\sqrt{gn}} = z,$$

gives
$$\left(1 - \frac{mm}{nn}\right) v\, dz + z\, dv = -z\, dz - b\, dz + \frac{mb\, dz}{\sqrt{gn}}$$
which, multiplied by $z^{-mm/nn}$, yields
$$\left(1 - \frac{mm}{nn}\right) z^{-mm/nn} v\, dz + z^{1-mm/nn}\, dv$$
$$= -z^{1-mm/nn}\, dz - bz^{-mm/nn}\, dz + \frac{mb z^{-mm/nn}\, dz}{\sqrt{gn}};$$
after the integration of which, with the constant C having been added, there arises
$$z^{(nn-mm)/nn} v = C - \frac{nn}{2nn - mm} z^{(2nn-mm)/nn}$$
$$- \frac{nnb}{nn - mm} z^{(nn-mm)/nn} + \frac{mnnb}{(nn - mm)\sqrt{gn}} z^{(nn-mm)/nn}$$
in which the value of the constant quantity C is defined from the fact that at the beginning of flow, when indeed $x = a$ or $z = a - b + \frac{mb}{\sqrt{gn}}$, $v = 0$, because motion cannot arise in an instantaneous point of time; hence, therefore, it occurs that
$$C = \left[\left(a - b + \frac{mb}{\sqrt{gn}}\right) \frac{nn}{2nn - mm}\right.$$
$$\left. + \frac{nnb \sqrt{gn} - mnnb}{(nn - mm)\sqrt{gn}}\right] \left(a - b + \frac{mb}{\sqrt{gn}}\right)^{(nn-mm)/nn}.$$

From these equations, indeed, all things are defined; but because the calculation is more or less involved unless the area of the upper vessel, indicated by m, be so great that it can be reckoned as being infinite in proportion to the areas g and n, we will consider this case alone, and this the more so because a notable error does not arise from it, even if the number $\frac{m}{n}$ or $\frac{m}{g}$ be of moderate size.

§23. Thus, if hence we set $m = \infty$ and at the same time make use of the first differential equation of the last paragraph, and if in this it is assumed that $v = \frac{nn}{mm} s$, so that thus from the value of the letter s the height may be found to which water flowing out through the orifice MN can ascend by its own velocity, first there will be
$$\frac{nn}{m}(x - b)\, ds + \frac{bnn}{\sqrt{gn}}\, ds - ms\, dx + \frac{nn}{m} s\, dx = -mx\, dx,$$

and because $m = \infty$ it is easily foreseen that the ratio between s and x will be finite, and also between ds and dx, this same equation will be changed, after one's having rejected the appropriate terms into this again: $-ms\,dx = -mx\,dx$, or $s = x$, which was already proven as well in §10. But after that I decided to prove it again here because the present case could be seen as different from the former, about which there is a discussion in the aforesaid paragraph. These things having been understood, it is no task to explain to many the Phenomena in §21 concerning this matter indicated by the Author s'Gravesande; for it is evident that the water does not flow out through the composite vessel $AEFMNGHB$ otherwise than it does through the simple vessel $AOMNPB$ when, indeed, the orifice MN is very small; and that hence the velocity of the aqueous surface CD is greater than if the water were flowing through the vessel $AEFGHB$, after one has set the orifice $MN = FG$, and much more so if MN is greater than FG, which happens when the pipe increases in area towards the lower end. But, nevertheless, it must be observed that at the beginning of motion the water descends more slowly than has been thus defined, and that that rule does not hold until the surface CD has descended through some little space, which, however, occurs in a short time. We will examine the changes which occur in this case at the beginning of motion in the following section.

§**24.** The computation would be undertaken in the same way if in the vessel, which now we always consider to be of infinite area, were implanted a small pipe, not vertical but horizontal, just as in Fig. 19, or in any other direction whatever; moreover, it is always found that the water, after the surface of the water in the principal vessel descends some little bit, soon flows out at approximately that velocity which corresponds to the height of that surface above the orifice. Therefore, it is clear that, with the height of the water above the pipe GN as well as the orifice FG itself being maintained, the quantity of water flowing out in a given time is increased by the increased area of the orifice MN. Accordingly, therefore, we have given a description here of what was mentioned at the end of §5, Chapter I: that Frontinus had learned from experience, certainly, that *more water than is due is appropriated through a calix of both legitimate size and position to which pipes of larger size are directly attached.* And certainly quantities of water, all other things being equal, would be expended, approximately proportional to the orifices MN themselves unless there should be many hindrances; the latter may diminish this quantity greatly, about which I will speak soon. These hindrances can act so that the flow of water is increased very little on account of the increased final orifice; nevertheless it will always be increased some little bit.

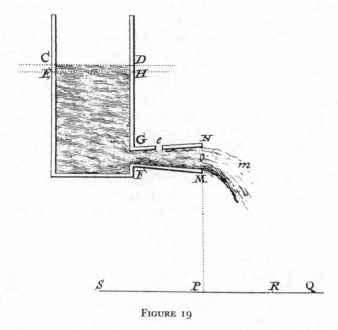

FIGURE 19

§**25.** From the previous discussion it is evident that the velocity by which the surface *CD* of the water in either case about which we spoke descends, the remaining things being equal, depends on the area of the orifices *MN*. Moreover, these things are supported by the hypothesis that the water adheres everywhere to the walls of the pipes *GN*, and it flows out from the full orifice *MN*, which hypothesis cannot hold further if that orifice should be increased too much. Hence also it is evident when water flows out through the vertical pipe in Fig. 18 that its flow is accelerated by an increased length of this pipe; nevertheless the latter can also be increased so that finally the water ceases to be continuous in the pipe, in fact, so that it is rather divided into columns, which may occur if the pipe has a length of more than thirty-two feet or even less if at the same time it increases in area toward *MN*. Thus if the orifice *MN* is double the other orifice *FG*, the length cannot be greater than eight feet without danger of separation of the water following in the uppermost part of the pipe, which matter I will show elsewhere. But there is an additional cause besides the excessive length of the pipe which can produce separation of the water, namely, that the height of the water *CEHD* be less than that which can enter the pipe quickly enough, by which it occurs that air together with water flows in above at the same time, while the surface of the water assumes the form of a cataract or of a bottomless hollow, such that not all of the orifice *FG* is covered by water. Indeed this causes the water to flow out in a lesser amount, but not at a lesser velocity, which a certain Italian Author, Carlo Fontana by name, considered later; he wrote the following about this matter in his own vernacular Language: "*But if here there were not,*" he says, "*as much water as would be sufficient to maintain the said pipe full, the water will attract air within itself in as great a quantity as water will be lacking to it for intermixing within the water on all sides; but the velocity of the water will be lacking as much as will be the height of all the air collected together that will be in that pipe.*" Anyone discerns the reasoning of this, because I stated that the velocity of the water can hence not be diminished, from the fact that otherwise the *potential ascent* could not be equal to the *actual descent*, and the matter will be confirmed easily by experiment, with the extremity *MN* of the pipe being bent so that the water flows out horizontally, and from the area of the jet the velocity of the water can be determined. Moreover, it may occur in any manner it pleases that, with none of the other conditions changed, the air is mixed with the water around the top of the pipe; thus remember, indeed, if there is a tiny opening in the pipe not at all far from the orifice *FG* (Figs. 18 and 19), and if, further, during the flow of water one has closed that little opening with a finger, pure water will flow through, and if one removes his finger, soon air will enter through the same little opening and will mix itself with the water flowing through. These things having been understood, it will be easy to present the reasoning of the Phenomena which are observed in chimneys, or smoke ducts; indeed smoke seeks height, because it is lighter than air, which is consistent with experiments performed on smoke in a vacuum, where it was seen to have descended. Therefore, it is the same for smoke ascending as for water descending; but in Fig. 18 the latter flows through the orifice *MN* more quickly, the larger it is, and the lower it is positioned; therefore, also, the smoke will travel through the chimney more quickly, the more the fire is kindled in the furnace, the higher the chimney is carried, and the more it diverges facing upward, if only it does not diverge too much; experience confirms each of these. I myself then learned in addition that if the chimney be perforated somewhere, it is not at all so that the smoke attempts an exit through that opening, but rather that air rushes in with a great impetus, and, mixing itself with the smoke, it rises through the chimney, and not otherwise than as we indicated that the air rushes into the pipe *FGNM* (Figs. 18 and 19). So indeed the smoke ascends certainly in a lesser amount, or at least with more difficulty, and the fire slackens.

Still, there are two causes in particular, the one extraneous, the other intrinsic in the nature of the matter, which can greatly retard the motion of the water in Figs. 18 and 19. The first is the adhesion of the water to the walls of the pipe, and the other is that when the pipe increases in area, the velocity of the water, nowhere constant to itself, is changing in every location in the pipe; if this change is considered to arise from infinitely small impulses of water moving more quickly into water moving less quickly, it appears that at every instant by these impulses of flexible bodies some of the *potential ascent* is lost, whence necessarily the efflux of the water is noticeably diminished.

§**26.** Finally, now I will say something about curved vessels from which not all the water flows: for the sake of brevity we will consider a cylindrical conduit, a certain part of which, that the aqueous surface does not cross, is straight.

Problem

For example, consider the cylindrical conduit *CEDB* (Fig. 20), the sufficient portion *CE* of which is straight, the remaining *EDB* being

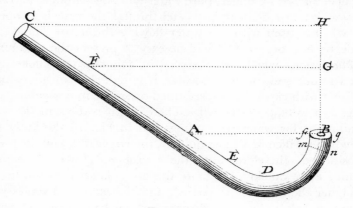

Figure 20

curved in any way whatever; the whole conduit is full of water that will flow out through the orifice *B*; after the surface of the water has fallen from *C* to *F*, the height corresponding to the velocity of the water at *F* is sought.

VELOCITIES OF FLUIDS FLOWING OUT OF A VESSEL

SOLUTION. Let the vertical BH and the horizontals CH, FG, AB be drawn, and let the sine of the angle HCE be to the total sine as 1 is to g. But now, if we consider the matter properly, we will see that the present problem is contained in the other more general one which we treated above in §20, where we had this equation:

$$v = \xi^{mm/nn-1} \int - x\xi^{-mm/nn} \, d\xi$$

where, for our present case, v represents the desired height corresponding to the velocity of the aqueous surface at the position F, ξ the length $BDEF$, x the height BG, and also $\frac{m}{n}$ the index of the ratio between the areas of the pipe and the orifice B. But if the length BDA is set equal to α, one will have $x = \frac{\xi - \alpha}{g}$, from which now

$$v = \xi^{mm/nn-1} \int - \left(\frac{\xi - \alpha}{g}\right) \xi^{-mm/nn} \, d\xi.$$

Let the length of the entire conduit $BDEC$ be indicated by β, and one will have

$$\int - \left(\frac{\xi - \alpha}{g}\right) \xi^{-mm/nn} \, d\xi = \frac{nn\alpha}{g(nn - mm)} \left(\xi^{(nn-mm)/nn} - \beta^{(nn-mm)/nn}\right)$$
$$- \frac{nn}{g(2nn - mm)} \left(\xi^{(2nn-mm)/nn} - \beta^{(2nn-mm)/nn}\right)$$

and therefore

$$v = \frac{nn\alpha}{g(nn - mm)} \left(1 - \left(\frac{\beta}{\xi}\right)^{(nn-mm)/nn}\right)$$
$$- \frac{nn\xi}{g(2nn - mm)} \left(1 - \left(\frac{\beta}{\xi}\right)^{(2nn-mm)/nn}\right). \quad \text{Q.E.I.}$$

SCHOLIUM

§27. Since these equations are somewhat involved, we will not tarry in the general contemplation of them, considering, rather, those particular cases which shorten the calculation and which cannot be defined by that last equation.

If we assume that the cover at B is wholly absent, it occurs that $m = n$, and (which must be determined separately for this and equally for the other case soon to be discussed)

$$v = \frac{b - \xi + \alpha \ln \xi - \alpha \ln \beta}{g}$$

and then the velocity is a maximum at A, and expressly that which corresponds to the height

$$\frac{\beta - \alpha + \alpha\ln\alpha - \alpha\ln\beta}{g}.$$

Finally, the point E corresponding to the maximum descent is obtained with the aid of this equation:

$$\xi - \alpha\ln\xi = \beta - \alpha\ln\beta.$$

The other case is to be calculated separately, when $mm = 2nn$, where there arises

$$v = \frac{\alpha\xi - \alpha\beta - \xi\beta\ln\xi + \xi\beta\ln\beta}{g\beta}$$

and also, if it is considered, e having been assumed as the number of which the logarithm is unity, that $\xi = e^{(\alpha-\beta)/\beta}\beta$, then the point of maximum velocity will be determined, of which the generating height is $e^{(\alpha-\beta)/\beta}\beta - \alpha$, while the maximum descent, which is proportional to the total water flowing out, is defined by making

$$\alpha\xi - \alpha\beta - \xi\beta\ln\xi + \xi\beta\ln\beta = 0.$$

I do not doubt but that these would correspond to practice exactly, if only the adhesion of the water to the walls of the pipe would not retard the motion; nevertheless, I consider that the results of the experiments can be such that they show the truth of these propositions well enough to the intelligent person, who has an understanding of these impediments.

§28. Finally, I will show the correct solution of a certain phenomenon which at first glance seems to be very much a paradox. Indeed, after it appears from all these things freely discussed up to now that it cannot happen that the water flows out at a much greater velocity than that which is due to the height of the water above the orifice (nevertheless they can be somewhat greater, especially if the orifices are large; *refer to what I said in §16 concerning maximum velocities*), it will seem to many perhaps a wonder *that it occurs sometimes in leaping fountains that for an instant water makes a far higher thrust than seems possible according to our rules.* It is far from true that these [rules] therefore lose some of their power; in fact, they are rather exceedingly strengthened. Moreover, the solution of the paradox consists in this: so far we have considered the water as continuous and not separated by any air void; and Mr. de la Hire rightly observed that irregular spurts of this sort do not occur unless air together with water has entered the pipe somewhat like bubbling water, which, as I indicated in §25, occurs frequently. Indeed, that air is carried together with the water right up to the orifice of efflux, through which it then erupts. While this occurs, the aqueous mass acquires an impetus which it employs exclusively in expelling the water, and in this way it produces an enormous thrust. Soon I will explain this cause of the phenomenon more clearly together with the required measurements, after I have presented some statements which appear concerning this matter in the *History of the Royal Academy of Science of Paris* for the Year 1702. It states in the place cited: "*It is seen sometimes that water discharging through an orifice springs forth three or four times as high as the height of the reservoir would permit, and that it also comes back quite quickly to the height which the laws of hydrostatics prescribe to it. But how could it deviate from it for an instant? Mr. de la Hire attributes this to the air enclosed in the conduit, which, being compressed by the continuously descending water and thus gaining a spring force, is released against the rising part of the water and imparts to it this instantaneous velocity.*"

And so Mr. de la Hire noticed correctly that the spurt is due to the air, and there is no doubt but that he would have extracted the correct reason by which air can produce this if he had considered more carefully the phenomenon to which he referred incidentally, and he would have easily perceived, certainly, that the air within the water sustains no pressure except that of the water lying above (on the contrary, not even this much in flowing water, as I will show below in Chapter XII), and that therefore the compressed air cannot expel the water preceding it more strongly than if water had been in its place. I certainly saw in advance (which I found often afterwards by very simple experiment) that it is not the water located in front of the air but that which follows the air that rises unusually high, which I will now show more clearly.

Therefore, in Fig. 20 let the aqueduct $CADB$ be cylindrical, as is customary, and let the whole of it be filled with water, except the small part mnB filled with air. Let the horizontal and vertical lines CH and HB be drawn; let us assume for the sake of brevity that the gravity of air can be considered as null with respect to the gravity of water, so that the transition of the air through the orifice B offers no resistance to the flow of the water, although concerning the remaining it would be easy to take into account the inertia of the air, unless we wish to avoid an abundance of calculation in a matter in which we require no precision. Let the length of the conduit $CADf$ or $CADm$ (indeed we assume the differential mf filled with air to be very small) be β; mf or ng be δ, HB be a, the area of the pipe be m, the area of the orifice B be n. Finally, let us state that the water has no motion when the surface is at mn; that the height due to the velocity which the surface mn has when it arrives at the position fg is to be sought; let that height be v, and the *potential ascent* of all the water at that very instant will also be v; moreover, the *actual descent* is, through §7, equal to the fourth proportional with respect to the total mass of water, the volume element of water $mngf$, and the vertical height HB, that is, $\frac{\delta}{\beta} \cdot a$; therefore $v = \frac{\delta}{\beta} a$. Indeed, this height is at once diminished faster than stated, and the water is forced to flow through the orifice B, which I showed in §18; but, nevertheless, at the first instant the water will retain the motion which it acquired, and thus the volume element closest to the orifice will be ejected at a velocity which is due to the height $\frac{mm\delta}{nn\beta} a$. However, this height can be not only triple or quadruple a itself, but howsoever great; indeed, with glass tubes I created thrusts ten or twenty times as high as a itself at will. For instance, if $\delta = 100$ feet, $\beta = $ one inch, but the diameter of the tube is tenfold the diameter of the orifice, then one will have $\frac{mm\delta}{nn\beta} = \frac{10,000}{1200} a$, so that in these circumstances the first volume element must spring forth, with the resistance of the air removed, to a height more than eight times as great as the customary height a. In addition, there are many hindrances, and these are of greatest importance, which restrain huge thrusts; indeed, something from the motion is lost by the impulse of the aqueous surface mn against the wall fg, then also by the enormous friction which the water experiences, having been carried so quickly through the little orifice, which has to be very small; much is also lost from the fact that the water $CADm$ is not moved with all its velocity on account of the adhesion of the water to the walls of the pipe, which adhesion is clearly noticeable in so long a reach.

Meanwhile, there can be no doubt that this is the correct solution of the phenomenon, and the experiments which I performed satisfy that solution in every extent. Then, as well, by this theory the other aspect of the phenomenon is solved correctly, namely, that that thrust is quasi-instantaneous, and after the shortest little time interval it is no greater than usual, according to observation. Thus in the present case that we just considered, if, with the rule of §18 changed a little (for there the only case discussed is that concerning vessels placed vertically), we investigate how much water must flow out in order that the thrust does not exceed the customary thrust by more than a thousandth part (which in any case can by no means be observed in experiments of this sort), if it was eight times as great as the same at the start, we find that that quantity is so small that the time in which the whole of it is ejected can in no way be perceived.

EXPERIMENTS WHICH PERTAIN TO CHAPTER III

Foreword

Indeed there are many things in this Chapter, and these quite extraordinary, which can hardly be subjected to experiments *immediately*. And indeed, since Authors up to now have not considered any motion in the efflux of fluids other than that which occurs through very small orifices, and accordingly, since our theory which we gave for arbitrary areas of orifices is new, this is the very thing the confirmation of which should be most gratifying. But I do not see in what way in vertical Cylinders, which we treated the most, the velocity of the water flowing out can be observed, especially when the orifice is very large (indeed, on the contrary, some judgment of the velocities can be made from the time of depletion). Thus, considering this, I reasoned at last that §§16 and 20 could be useful to our objective: in the former the maximum velocity of water flowing out of cylinders placed vertically had been determined; in the other, moreover, it was shown that the motion is the same from obliquely placed and vertical cylinders if both vertical heights are assumed alike. Therefore, we will make suitable use of cylinders placed obliquely in order that from the maximum area of the aqueous thrust the maximum velocity of the water or the height due to the same can be obtained by experiment; and indeed, by this reasoning that maximum velocity, whatever it really is, can be investigated, even if the orifices are as large as one wishes. Accordingly, if this is observed to agree with our rules, no doubt can remain regarding the entire theory. But before I attack the matter itself, the mechanics theorem that follows is to be set forth in advance.

Lemma

Let the line AB (Fig. 21) be vertical, BD horizontal; in addition let the line AD have any direction whatever, in which direction a body at A is understood to be projected, describing the arc AC of a parabola, the tangent of which at A is certainly the straight line AD; the height due to the velocity at which the body at A was projected

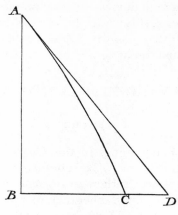

FIGURE 21

will be $\dfrac{(BC)^2 \cdot (AD)^2}{4AB \cdot BD \cdot CD}$; and if AD is horizontal, or BAD is a right angle, that same height will be $\dfrac{(BC)^2}{4AB}$. But now I will show those things observed by me.

CONCERNING THE MAXIMUM VELOCITIES OF FLUIDS FLOWING OUT THROUGH VERY LARGE ORIFICES

Pertaining to §§16 and 20

First experiment. I placed the Cylindrical Pipe *FA* (Fig. 22), of a length of four inches, obliquely to the horizon, and I secured it in that position; moreover, the area of the pipe was to the area of the opening at *A* as 2 is to 1, and the diameter of the pipe was equal to seven lines, more or less. Then, after measurements of the lines *FE*, *AB* and *BD* had been taken in equal units (the law of which is evident intrinsically from the figure itself), I found them to be 81, 619, and 740.

With these things so prepared, I filled the pipe with water, having closed the orifice *A* meanwhile with a finger, and with this suddenly removed, all the water flowed out in the shortest bit of time; however, I was able to observe that the first and the last [portions] had fallen nearer to the vertical *AB* than the intermediate [ones]; moreover, that the drops projected the furthest fell at the point *C*. And I found after rather frequent repetition of the experiment that *BC* was 235 of the units which I had used before.

But now, if through the previous lemma the height EG is desired to which the drops ejected at the maximum velocity can ascend, it is found that $EG = 56$ units; however, by dint of §§16 and 20 it should be 62, unless the friction of the water and its adhesion to the sides of the pipe contribute an impediment to the motion; I did not expect a greater agreement.

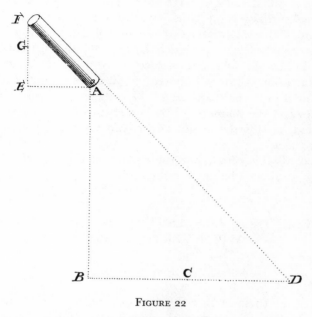

FIGURE 22

II. With things arranged as previously, only with the orifice A diminished to half so that the area of the pipe was quadruple the area pertaining to the opening, I observed that $BC = 252$. Hence it is deduced through experiment that $EG = 68$, but by theory it should be $= 70$; these numbers differ less than the preceding because here friction was a far lesser impediment on account of the diminished velocity of the internal water. However, each experiment actually confirms the theory excellently.

CONCERNING THE VELOCITY OF WATER FLOWING FORTH FROM A VERY LARGE VESSEL

Pertaining to §17

In that paragraph we say that if a vessel is very large, soon after the internal surface descends some little bit, water flows forth at a velocity which corresponds constantly to the height of the water above the orifice. However, one may allow that the water flows in any direction whatever (for indeed, in very large vessels any direction of the stream cannot change the velocity), and one may observe at any arbitrary instant at what distance from the vertical the stream impinges on the horizontal, and from there one may seek through the previous rule the height corresponding to the velocity of the water flowing out at that instant; thus one always finds that height equal to the height of the water above the center of the orifice, if only one overlooks the first few drops which, by dint of §16, must flow out and actually do flow out at a lesser velocity. And the hindrances, which we have mentioned rather frequently, will cause no noticeable delay to the flow if only the diameter of the orifice equals at least two or three lines and the diameter of the vessel itself is not less than a few inches, and, finally, the height of the water is not excessive, such as very many feet.

I tested all this often, but the nature of the experiment is too trivial for it to merit being described fully.

CONCERNING VESSELS WHICH ARE PROVIDED WITH VERTICAL PIPES

Pertaining to §§22 and 23

Concerning these things the honored s'Gravesande, in *Physices Elementa Mathematica*, undertook experiments which I repeated; in fact, those which apply to the present matter are brought out especially at this point.

Namely, in Figs. 23, 24, 25 and 26, the individual apertures denoted by the letter A are equal to each other, with B alone being little greater, in the proportion of 16 to 25, and also the areas as well as the heights of the cylinders are equal, except for the last, the height of which is quadruple; however, the pipes annexed to the two intermediate cylinders have triple the length of the cylinders. Therefore, after these vessels were filled with water, it was observed concerning its efflux:

I. That the surface of the water from the beginning does not descend more quickly in Fig. 23 than in Fig. 24; but that, after some water has flowed out from each, the motion becomes much quicker in the composite vessel than in the simple; I predicted both at the end of §23. But the matter is understood better and more accurately

VELOCITIES OF FLUIDS FLOWING OUT OF A VESSEL

from the differential equations which we gave in §§22 and 23 if we make use of them for finding the first increments of motion in the simple cylinder of Fig. 23 as well as in the composite one of Fig. 24, and if to this end we assume the areas of the cylinder and the pipe to

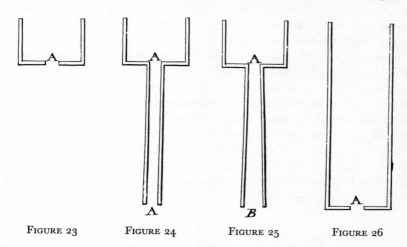

FIGURE 23   FIGURE 24   FIGURE 25   FIGURE 26

be as m to n, the increment, which we called dv in the simple vessel, will be to the increment in the composite vessel as $1 + \frac{3m}{n}$ is to 4, and therefore greater by far in the former case than in the latter. Therefore, if it should be granted to perceive that first motion correctly, we would observe at once that that is quicker which occurs in the simple Cylinder. Since, in fact, in §§15 and 23 it was demonstrated further that the water surfaces, after they have descended a little in each vessel, are approximately such that they correspond to the heights $\frac{nn}{mm} x$, by understanding by x the heights of the water above the orifices through which it flows; it follows directly that the water descends at a much greater velocity in Fig. 24 than Fig. 26. Thus, therefore, the Theory clearly agrees with the observations.

II. That the aqueous surface descends considerably more quickly in Fig. 26 than 24, so that the velocity in the case of Fig. 24 is somewhat halfway between the cases of Figs. 23 and 26. Here, indeed, it is evident again that the first accelerations occur much more slowly in the cylinder of Fig. 24 than that of Fig. 26. Therefore, in this respect the theory itself indicates what was observed; certainly something is missing that as great a difference as I found can hence develop, and it should no longer be noticeable after either surface descends a little, according to §23. However, that difference must be attributed to a hindrance which arises from the friction of the water in Fig. 24; indeed, the water is carried at a great velocity through the pipe AA, and thus as much on account of the increased velocity as on account of the diminished area of the vessel a very effective impediment is offered to the motion of the water.

III. Finally, that the aqueous surface descends very quickly in the Cylinder of Fig. 25, if one excludes the first instant, and notably more quickly than in Fig. 26.

Indeed, this conforms to those things which have been shown in §23; however, soon after the common beginning of motion, namely, after the heights of water above the orifices of efflux have been set almost equal, the velocities in Figs. 25 and 26 should be approximately as the areas of the orifices B and A, that is, as 25 to 16; and since a smaller difference of velocities is observed, this is again to be attributed to the impediment of friction more than to the other cause indicated at the end of §25.

CONCERNING THE SAME VESSELS, IN WHICH HORIZONTAL PIPES ARE INSERTED

Pertaining to §24

When water flows from a rather large vessel such as CDG (Fig. 19) through the horizontal pipe GM larger at the extremity NM than at the origin GF, the former is carried through the orifice GF at a greater velocity (if again one overlooks the first drops) than if the pipe were either absent or Cylindrical. Even Frontinus, taught by experience without doubt, affirmed this, but several modern men have denied it.

Therefore, as something worth the effort, I undertook to investigate the matter by experiment. Now the height of the vessel which I used was $5\frac{1}{3}$ English inches above the axis of the pipe, the length of the pipe GN was 2 inches 5 lines, the diameter of the orifice GF was 3.36 lines, the diameter of the aperture MN was 5.48 lines; thus the areas of the orifices were approximately as 3 to 8; the area of the vessel was large enough that it could be considered infinite with respect to the area of the pipe. I have wished to dispatch all measurements so that anyone can repeat the experiment. Now, after this vessel had been filled with water, I observed the area of the jet, and from this, after I had acquired all the measurements needed, I made a calculation of the height which should be due to the velocity of the water flowing through at GF as well as at NM; I found the latter to be approximately eleven lines, and thence the other to be 1 inch $6\frac{2}{9}$ lines, which same heights I found as well in another type of experiment. But since the height 6 inches and $6\frac{2}{9}$ lines is greater than $5\frac{1}{3}$ inches, our theory is confirmed about the acceleration of internal water by amplification of the pipe towards the extremity, although, as I predicted chiefly because of the two reasons dispatched in §25, it may be far from actually being accelerated as much as it should according to §24, after the obstacles have been subtracted which have not been considered in the calculation.

Pertaining to §25

In this paragraph I mentioned in passing that it can occur in many ways that air is mixed with water flowing through pipes. But from this it will happen that water flows out in a lesser amount certainly, but not at a lesser velocity; in order that I might prove one as well as the other, I first made a very small orifice in both the pipes AA and AB (Figs. 24 and 25) not far from their Origin; it is a fact that the water was carried through the pipes with some noise and flowed out in a turbulent state; moreover, the surface descended much more slowly than is customary. Then I perforated the pipe of Fig. 19 somewhat similarly, not far from G, and again I observed that the internal surface descended a little more slowly, of which matter I was certain, since I counted the oscillations of a certain pendulum while the surface descended through a given length; but with an understanding of the flowing out of the water, I saw that sometimes the water flows out from a full orifice and then the water is less clear than usual, but it makes an ordinary jet or one greater by a little bit than ordinary; however, most often the water and air are carried side by side, the former in the lower part of the pipe along the wall FM, the latter in the upper, along GN, and then the water is clear and is ejected at a velocity not only by no means smaller than the usual but even much greater; I had foreseen that this could occur by no means obscurely. Concerning this matter, in the following Chapter I will discuss another experiment undertaken with greater precision.

However, a place will probably be given elsewhere for showing that water mixed with a sufficient quantity of air flows out almost at that amount at which it would flow from a pipe cut off at that place where it is perforated, to which matter I noticed that my own experience corresponds also.

CONCERNING CURVED CONDUITS

Pertaining to §27

The horizontal MN (Fig. 27) having been drawn on a wall, I placed the cylindrical pipe CDB, having both legs parallel to each

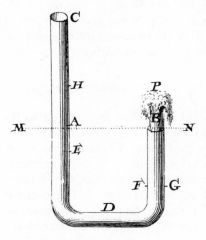

FIGURE 27

other, wholly filled with water, so that the one extremity B would just touch the horizontal MN, and also that the legs would be vertical, while at the same time I blocked the orifice C with my finger, thus restraining the flow of water.

Then I observed, after the finger had been removed, the maximum height BP to which the water flowing out ascended, and at other times I noticed the point E to which the surface of the water descended; however, I performed the experiment under two different conditions; indeed, in the first instance I had not placed a cover at B, then I used a cover perforated by such an opening that it had an area in ratio to the area of the pipe as 1 to $\sqrt{2}$. Meanwhile, the measurements were such: $CA = 345$; $ADB = 530$; $BP = 33$; and $AE = 88$ units, 375 of which were equal to the length of a London Foot. Things were so in the prior case, but in the other, with the rest of the things being unchanged, I observed $BP = 64$ and $AE = 54$. I will note here in passing that, desiring to determine the maximum descent AE in another way, I inclined the pipe after the experiment was ended until the water seemed just precisely at efflux through B, at which instant

VELOCITIES OF FLUIDS FLOWING OUT OF A VESSEL 69

I measured the distance of the surface from the point A noted previously; that distance which I considered to be the same as the maximum descent AE was far less than expected; whence I learned that a part of the water which had already flowed out through B in the experiment had entered the pipe again.

Thus, after those things had been observed, I sought the magnitudes of BP and AE by calculation, according to §27, by setting first $m = n$, and then $mm = 2nn$; but I found in the former case that $BP = 79$, which in the experiment did not exceed 33, and I discovered the maximum descent AE approximately equal to 250, which the experiment gave as 88. Next, for the case of $mm = 2nn$, BP appears more or less double that which had been observed, and $AE = 186$, which had been observed as 54 units.

I attribute these enormous differences for the most part to the adhesion of the water to the walls of the pipe, which adhesion in cases of this sort can certainly exert an incredible effect. In fact, I used a pipe of hardly more than two lines in diameter, and certainly I will experience a greater agreement with a larger pipe. Meanwhile, it is likely that the curvature of the pipe in the lower region also takes something away from the motion.

Pertaining to §28

I made use of the same curved pipe which I just described; but I placed a cap at B perforated by a very small orifice. I filled the whole thing with water except the small region FGB, in which location I detained the water with the help of a finger placed on the orifice C. After the finger was removed, the water descended, and when it had arrived at the position HDB, a number of drops were as if exploded at so great an impetus through the little orifice at B that they ascended to a height of more than ten feet, although the height HA hardly exceeded a height of half a foot. However, on account of the smallness of the little orifice the water encountered so much resistance while it went through the orifice that, after the impetus had been weakened, the water not only did not ascend to the height AH (above which nevertheless, with all hindrances removed, it should have continuously sprung a little) but hardly a drop or so was pressed out in a noticeable passage of time, so that I am convinced that if, apart from impetus, so great a thrust were to be produced from the natural pressure of the water alone, this would not occur except under a height of at least one hundred feet.

Further, I observed as well that the thrust of the water is diminished more, the smaller is the space GB before the experiment; all these things conform to the theory. It would have been superfluous to take measurements, because on account of the excessive hindrances certainly the thrust of the water cannot be as great as it would be with the [hindrances] removed. But nevertheless, in order that I might confirm that these things in the experiment also agree with the formulas, I took a larger pipe *CDB* in order to eliminate the hindrances of adhesion for the most part; the region *DFB* was very small, and the region *GB*, which I left free of water in the experiment, even smaller; and finally the cap was perforated by an orifice not altogether small. And then I saw that the leap was not very much less than the height $\frac{mm\delta}{nn\beta} a$, which I gave in §28 for this situation, and I even remember that I had predicted the height of the leap correctly to a Friend who was present after I had considered approximately how much should be given to the hindrances in the calculation.

One will obtain a similar instantaneous explosion of water very easily, and this arises from a similar cause with fountains which eject water through a pipe with a full orifice. If, for instance, one places a finger suddenly over the orifice of the pipe, so that part of the orifice remains open, one will see directly that the water is expelled with a great impetus, and soon the thin thread of water is reduced to within the original limits of the velocity. One will observe also that the water is projected further and with a greater impetus the less one leaves the orifice open with the finger, and, for the same orifice having been left open, that the unusual thrust is drawn forth more (but always very quickly) and is made more noticeable to the eye, the longer is the pipe, so that in leaping fountains to which water is carried from a reservoir through very long conduits, if the conduits are not very large and if the water flows out of the full orifice, I do not doubt but that through a noticeable period of time a vigorous thrust of water can thus be produced, returning gradually to the usual velocity. All this conforms to what has been shown in §§28 and 18.

I remember that I performed this experiment at some time or other, and for the first time, indeed, in the presence of the most honorable gentlemen, Messrs. De Maupertuis and Clairaut, with whom I had previously gotten into a violent discussion on those hydraulic matters. But although on this occasion there is no air which can be blamed, in truth, nevertheless, this phenomenon does not differ from that which Mr. de la Hire has observed, and each develops from the fact that the motion of the water contained in the conduit, which the enormous thrust of the water itself constitutes, or at least part of that motion, cannot be lost without any effect arising thence.

FOURTH CHAPTER

Concerning the Various Times which can be Expected in the Efflux of Water

§1. It will seem to many to be a completely Geometrical matter, which certainly has no concern with any physical consideration, that, when water flows from a given vessel through a known aperture at velocities determined in every position, the time be defined in which a given quantity of water flows out. Nevertheless, experience shows the contrary; for water flows out through orifices which exist in a thin place at a much lesser quantity than should follow from the simple consideration of the velocities, and this for the most part (for the matter was not self-consistent in different circumstances) in the ratio of 1 to $\sqrt{2}$; this moved Newton to affirm in the first edition of *Principia Mathematica Philosophiae Naturalis* that water flows from a vessel at that velocity which is generated by half the height of the water above the orifice, which opinion all experiments undertaken on velocities contradict immediately. Exploring the origin of this contradiction a little later, this great Man himself observed that it was located in the contraction of the aqueous stream, which contraction customarily occurs immediately in front of the orifice. Also, another change in the stream, now similar, now contrary to the former, was observed by me. Indeed, when water flows out not through a simple orifice but through a pipe, the stream is again contracted if the pipe converges toward the exterior, but is dilated if the same diverges. Concerning the contraction of the aqueous stream flowing out through convergent pipes, Giovanni Poleni performed very accurate experiments in the book *de castellis* [reservoirs] p. 15ff. The contraction of the stream was observed by this Most Celebrated Man to be greater, the greater was the internal orifice of the conic pipe, the external orifice and the length of the pipe being maintained, which is the reason that a similar quantity of water will flow out more slowly, the remaining things being equal, the greater the internal orifice will be, although the impediments from the adhesion of the water to the sides of the pipe continually will have a lesser effect; however, those diminutions of the impediments would cause the water to flow at a greater velocity at the place where the stream was contracted the most, and [the water] would be expended no less sparingly; truly, that is understood to occur from the observed times of efflux and the areas of the streams where they are contracted the most. Therefore, since the crux of the matter turns on these changes of the stream, one will be able from this to examine and explain the phenomena more fully.

§2. Let us assume, for instance, a vertical cylinder which has an orifice in the middle of its horizontally situated base, but let the internal water be considered as divided into horizontal strata. With these things thus assumed, we consider that the motion of every stratum whatever is the same, and certainly such that a horizontal position is preserved in them; however, I have warned that this hypothesis cannot be extended to the strata near the orifice, but that, since thence no noticeable error can arise by reason of the velocity of the flowing water, it is not worth the effort to take this matter into account. But now, since other phenomena depend upon the oblique motion of the internal water, especially such as when it is in the previously mentioned strata near the orifice, we will illustrate this in a few words.

§3. Moreover, it seems to me that the motion of the internal water is to be considered such as it would be if the water were carried through infinitely small pipes placed next to each other, of which the intermediate descend nearly directly from the surface towards the orifice, the remaining being curved gradually near the orifice, as Fig. 28a shows, from which it appears that the individual particles descend in this way with a motion very nearly vertical until they approach the base closely, and they then turn their course gradually toward the orifice, so that the particles near the base flow with an almost horizontal motion, the others more vertically, toward the orifice. I was able to observe this sort of motion often by eye when wax particles, which they call Spanish, were immersed in the water. Thence it is also known that the individual particles existing at the orifice cannot preserve their entire direction, and nevertheless they do not turn it so that they assume a motion clearly parallel to the axis, but rather the stream of water flowing out will be contracted right to *de*, where accordingly [the stream] will be noticeably more slender than at the origin near the orifice *ac*. But this contraction of the vertically flowing stream is not to be confused with the other contraction which occurs from the acceleration of the water. Next, it is also evident that when the direction of the individual particles near the orifice is different, unavoidably from the impetus which the same particles create mutually between themselves, the stream is compressed,

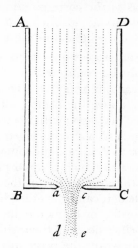

FIGURE 28a

and thus it thins. And from that compression it occurs, which otherwise would involve a contradiction, that the water already gone out is accelerated even in front of the orifice, and thus the *potential ascent* increases, even if we pay no attention to the other acceleration common to all falling bodies, as if not pertinent here, and we will not make mention of it from now on. But unless I am mistaken, this matter ought to be treated further in the following way.

I. At the outset the stream of water is to be considered while the velocities of the particles are not being changed further, which, although it never happens in all rigor, nevertheless is to be understood to occur not far from the orifice, such as at *de*. But if this were to be so and the water were assumed to flow out of the vessel *ABCD* through the orifice *ac*, in place of the simple vessel *ABCD* there is to be understood some other composite one *ABadecCD*.

Therefore, anything that was set forth in the preceding section for determining velocities everywhere will apply fully if in place of the proposed vessel a vessel is considered, as I said, provided with a small contracted pipe. Nevertheless, by reason of our previously mentioned method of determining velocities of water flowing out, this correction cannot produce a noticeable change on account of the shortness of the small pipe *adce*, but it can produce an exceedingly noticeable one with respect to the quantity [of water], because the water is to be considered as flowing out through *de* rather than through the orifice *ac*.

II. Thus the velocities in different places in the stream itself will be reciprocally as the areas of the corresponding sections, and since in very wide vessels the velocity at *de* is such that it compares to the total height of the water, and at the same time it is known from experiments that the areas *ac* and *de* are approximately as $\sqrt{2}$ is to 1, thus Newton thought his theory could be confirmed in which he stated that the water truly flows out from an orifice at a velocity that is due to half the height of the water above the orifice, although, in progressing, the velocity of the water increases; with regard to this, it seems to me that he adhered too much to a preconceived opinion: for neither is the ratio of the orifice *ac* to *de* always the same, nor can the motion of the water from the vessel to which a small pipe is attached thus be explained; in a word, the attenuation of the stream is by all means accidental, for the whole of it can be avoided by applying a very small cylindrical pipe to the orifice or by merely increasing the thickness of the plate in which the orifice lies, and then without that correction the theorems which were shown in the preceding section apply as much with respect to velocities as to quantities [of water].

III. It is also evident from the very explanation given above on the *contraction of the stream* that it cannot remain unchanged by diverse circumstances; thus, experiments show that the same is diminished by an increased thickness of the walls of the orifice; I am not sure whether the height of the water above the orifice contributes anything. I should almost believe that the contraction increases some little bit on account of the increased height of the internal water, although I readily foresee that it will be slight. Also, it is probable that the lesser will be the contraction of the stream, especially a vertical one, with all remaining things equal, the greater a ratio the area of the orifice will have to the area of the cylinder, because the motion of the internal water near the base becomes less oblique there, so that if the orifice occupies the entire area of the cylinder, certainly no attenuation of the aqueous stream can develop. To this I wish those would pay attention who will think perhaps that this contraction should be taken into account in the very determination of velocities. For when the orifice is not much less than the area of the vessel, no noticeable contraction can arise, and when the orifice is small, again hardly any difference arises concerning velocities whether the orifice is increased a little or is diminished.

§4. The reasoning is almost the same for water flowing out horizontally, so I am saying nothing about other directions: for water flows toward the orifice in a similar way from every region; finally, it ascends even from a lower region right up to the orifice so that it can flow out, which I myself often observed. Therefore, for a similar cause the attenuation will become similar in a stream flowing out, which is rather easy to observe by eye, because here the other attenuation developing from the acceleration of the water already having gone out does not apply. And on account of this reason, if anyone begins to make observations on the contraction of a stream, he will do better in my judgment by using streams flowing out horizontally rather than in any other direction.

§5. Moreover, it is possible to determine how great the contraction is, that is, what ratio exists between the area of the orifice and the minimum section of the stream flowing out horizontally, either by actually taking measurements of the diameters corresponding to those areas, or also indirectly from the quantity of water flowing out in a given time and at given velocities, where, nevertheless, the velocities are to be deduced not so much from the height of the water above the orifice as from the area of the jet, since certainly the hindrances, now greater, now lesser, never permit the full velocity of the water which it should acquire by dint of the theory, in which no account is taken of these hindrances.

§6. Now I think it is evident enough from the previous statements that there will be a perfect agreement between the quantity of water flowing out and its velocity, if only there is substituted for the orifice which is in the vessel some other orifice diminished just to that degree that it would not exceed the section of the maximum contracted stream; and it will be equally so at whatever place in the stream or at whatever depth from the surface of the water this orifice is understood to be, whether at *ac* or at *de*, since indeed the velocities will always correspond approximately to the total height of the water above that place at which the orifice is assumed; henceforth I shall call the area of this orifice, to be conceived in the mind, the *Section of the contracted aqueous stream*.

§7. Thus, if that *Section* about which we just spoke would have a constant ratio to the orifice, the orifice of efflux would have to be considered diminished in the same way, and afterwards a calculation of the quantity of water flowing out in a given time would have to be undertaken. Thus, indeed, with that ratio taken as $\frac{1}{\alpha}$ and with the area of the orifice called *n*, the *Section of the solid stream* would have to be considered as $\frac{n}{\alpha}$.

But since it is variable under different circumstances, rules cannot be given a priori in this matter; moreover, the section is changed greatly by an increase or decrease of the thickness of the plate in which the orifice lies; something, although only a little, can also be contributed by the size of the orifice, the areas of the vessel (and these are absolute as well as relative), and perhaps by the height of the water above the orifice. Meanwhile, there having been assumed that the wall is thin, that the vessel is very wide, and that the orifice develops to 4 or 6 lines in diameter, the ratio between the orifice and the *Section of the contracted stream* does not customarily depart much from that which Newton stated, namely as $\sqrt{2}$ to 1. But often it has been observed to be more by some and also less by others.

§8. But whatever it may be, in any case we will indicate it, as before, by $\frac{\alpha}{1}$. And now, for this situation we will develop the computation for the times; but for the sake of brevity we will consider only cylindrical vessels, and in these we will examine two kinds of time especially: the first which defines the point of maximum velocity, the other which corresponds to depletion. But in each case we will assume that the motion begins from rest.

§9. Therefore let a cylindrical vessel be placed vertically, full of water, and let the height of the water at the beginning of flow be a, the area of the cylinder be m, the area of the orifice be n, the *Section of the solid stream* be $\frac{n}{\alpha}$; let the water already have flowed out for the time t; and then let the residual height of the water above the orifice be x, and at the same instant of time let the surface of the internal water have a velocity which corresponds to the height v; the velocity itself will be $\sqrt{v}$, but the element of time dt is proportional to the element of space $-dx$ divided by the velocity $\sqrt{v}$, from which $dt = \frac{-dx}{\sqrt{v}}$.

Indeed, the value of v itself was determined in Chapter III where we used the same notation which we now use. But, since for a correct measure of the expended water it is required that the *section of the contracted stream* $\frac{n}{\alpha}$ be substituted for the orifice n, it follows that in the value of v itself the same substitution is made, and thus it is stated that

$$v = \frac{nna}{2nn - mm\alpha\alpha}\left[\left(\frac{a}{x}\right)^{1-mm\alpha\alpha/nn} - \frac{x}{a}\right].$$

TIMES TO BE EXPECTED IN THE EFFLUX OF WATER 77

But if this value is substituted in the equation $dt = -\dfrac{dx}{\sqrt{v}}$, there appears:

$$dt = -dx \Big/ \sqrt{\dfrac{nna}{2nn - mm\alpha\alpha}\left[\left(\dfrac{a}{x}\right)^{1-mm\alpha\alpha/nn} - \dfrac{x}{a}\right]}$$

with the help of which equation all desired times can be defined through approximations or series, if only at individual points the value of α itself be known. But let us assume it to be of constant value, since indeed in the present case there is nothing by which it could be changed except the different heights and velocities of the fluid, which contribute little or not at all to this aspect of the problem, as far as it can be perceived by observation.

§10. Now, in order that the desired equation can be expressed through a series, we will consider the quantity

$$1 \Big/ \sqrt{\dfrac{nna}{2nn - mm\alpha\alpha}\left[\left(\dfrac{a}{x}\right)^{1-mm\alpha\alpha/nn} - \dfrac{x}{a}\right]}$$

in the following form:

$$\left(\dfrac{nnx}{mm\alpha\alpha - 2nn}\right)^{-1/2} \cdot \left[1 - \left(\dfrac{x}{a}\right)^{mm\alpha\alpha/nn - 2}\right]^{-1/2}$$

and the latter factor we will resolve through the customary rules into this series:

$$1 + \dfrac{1}{2}\left(\dfrac{x}{a}\right)^{mm\alpha\alpha/nn - 2} + \dfrac{1\cdot 3}{1\cdot 2 \cdot 4}\left(\dfrac{x}{a}\right)^{2mm\alpha\alpha/nn - 4}$$
$$+ \dfrac{1\cdot 3 \cdot 5}{1\cdot 2 \cdot 3 \cdot 8}\left(\dfrac{x}{a}\right)^{3mm\alpha\alpha/nn - 6} + \text{etc.}$$

from which now is obtained a very slightly changed form of the equation:

$$dt = \dfrac{dx\sqrt{mm\alpha\alpha - 2nn}}{n\sqrt{a}} \cdot \left[\left(\dfrac{x}{a}\right)^{-1/2} + \dfrac{1}{2}\left(\dfrac{x}{a}\right)^{mm\alpha\alpha/nn - 5/2}\right.$$
$$\left. + \dfrac{1\cdot 3}{1\cdot 2 \cdot 4}\left(\dfrac{x}{a}\right)^{2mm\alpha\alpha/nn - 9/2} + \dfrac{1\cdot 3 \cdot 5}{1\cdot 2 \cdot 3 \cdot 8}\left(\dfrac{x}{a}\right)^{3mm\alpha\alpha/nn - 13/2} + \text{etc.}\right].$$

This equation is to be integrated so that for $x = a$ there will be $t = 0$; there arises, accordingly,

$$t = \left[2 + \frac{nn}{2mm\alpha\alpha - 3nn} + \frac{3nn}{16mm\alpha\alpha - 28nn} + \text{etc.}\right]$$
$$\times \frac{\sqrt{(mm\alpha\alpha - 2nn)a}}{n} - \left[2\left(\frac{x}{a}\right)^{1/2} + \frac{nn}{2mm\alpha\alpha - 3nn}\left(\frac{x}{a}\right)^{mm\alpha\alpha/nn - 3/2}\right.$$
$$\left. + \frac{3nn}{16mm\alpha\alpha - 28nn}\left(\frac{x}{a}\right)^{2mm\alpha\alpha/nn - 7/2} + \text{etc.}\right] \cdot \frac{\sqrt{(mm\alpha\alpha - 2nn)a}}{n},$$

where $2\sqrt{a}$ expresses the time which a body uses while it falls freely through the height a. But if in that equation it is assumed that

$$x = a\bigg/\left(\frac{mm\alpha\alpha - nn}{nn}\right)^{nn/(mm\alpha\alpha - 2nn)},$$

which is the height of the water when the velocity is a maximum (through §16, Chapter III and §8, Chapter IV), then the time is obtained which elapses from the beginning of flow right up to the point of maximum velocity; and when it is assumed that $x = 0$, the time appears in which the whole vessel is depleted; and, finally, if x is assumed equal to any quantity c whatever, t will express the time which the surface takes in descent through the height $a - c$. Moreover, we will see for these cases what should happen when the vessel is very wide, and the number m thus contains the other one, n, several times.

§11. At first let the number $\dfrac{m}{n}$ be infinite, then the height of the water corresponding to the point of maximum velocity will be

$$a\bigg/\left(\frac{mm\alpha\alpha - nn}{nn}\right)^{nn/(mm\alpha\alpha - 2nn)} = a\bigg/\left(\frac{mm\alpha\alpha}{nn}\right)^{nn/mm\alpha\alpha};$$

but since $\dfrac{mm\alpha\alpha}{nn}$ is an infinite number, this might be considered as

$$\left(\frac{mm\alpha\alpha}{nn}\right)^{nn/mm\alpha\alpha} = 1 + \left(\ln \frac{mm\alpha\alpha}{nn}\right)\bigg/\frac{mm\alpha\alpha}{nn},$$

of which matter the proof is this: let an infinite quantity A be proposed, and, as in our example, let $A^{1/A}$ be considered, and anyone readily sees that the latter quantity is little greater than unity, and indeed the excess, which we will call z, is infinitely small; and so one has $A^{1/A} = 1 + z$; let the logarithms be taken on both sides, and there will be $\frac{\ln A}{A} = \ln(1+z) = $ (on account of the infinitely small value of z itself) z. Therefore, $A^{1/A} = 1 + \frac{\ln A}{A}$; and from there, similarly, as we said,

$$\left(\frac{mm\alpha\alpha}{nn}\right)^{nn/mm\alpha\alpha} = 1 + \left(\ln\frac{mm\alpha\alpha}{nn}\right)\bigg/\frac{mm\alpha\alpha}{nn}.$$

Furthermore, because this quantity added to unity is infinitely small,

$$a\bigg/\left(\frac{mm\alpha\alpha}{nn}\right)^{nn/mm\alpha\alpha},$$

or

$$a\bigg/\left[1 + \left(\ln\frac{mm\alpha\alpha}{nn}\right)\bigg/\frac{mm\alpha\alpha}{nn}\right] = a - a\left(\ln\frac{mm\alpha\alpha}{nn}\right)\bigg/\frac{mm\alpha\alpha}{nn},$$

therefore the distance through which the surface of the water descends, while the maximum velocity develops from rest, is

$$a\left(\ln\frac{mm\alpha\alpha}{nn}\right)\bigg/\frac{mm\alpha\alpha}{nn} \quad \text{or} \quad \frac{2nna}{mm\alpha\alpha}\ln\frac{m\alpha}{n}.$$

This equation indicates that the descent of water in an infinitely wide vessel is infinitely small when the water has already attained its maximum level of velocity. Moreover, this notwithstanding, it could have been questioned whether or not in the meantime a finite quantity of water flows out, since indeed a cylinder erected above an infinite base of infinitely small height might have an infinite volume; but it follows from our equation that this quantity is also infinitely small, and nominally equal to $\frac{2nna}{m\alpha\alpha}\ln\frac{m\alpha}{n}$.

And this indeed agrees splendidly with the phenomena which we discover in the efflux of water from reservoirs through a simple orifice for a whole day. For when we cover the orifice with a finger, as soon as the finger has been removed we determine that the water flows horizontally, and we observe that halfway between the longest thrust and the point which lies on a plumb line from the orifice not a drop has fallen to the ground.

§12. Just as in the last paragraph we determined quantities howsoever small, such as the descent of internal water and of water flowing out while the water reaches the maximum level of velocity, so now we will show the same thing by reasoning of the time. But I say in the equation of §10, expressing the time, that it suffices that only the first term be accepted in each series, which will be evident if one extends the calculation to two terms; therefore the desired time interval is

$$t = \left(2 - 2\sqrt{\frac{x}{a}}\right) \cdot \frac{\sqrt{(mm\alpha\alpha - 2nn)a}}{n},$$

whence, after the value pertaining here, which we defined in the preceding paragraph, has been entered for x, there occurs

$$t = \left[2 - 2\sqrt{1 - \left(\ln\frac{mm\alpha\alpha}{nn}\right)\bigg/\frac{mm\alpha\alpha}{nn}}\right] \cdot \sqrt{\left(\frac{mm\alpha\alpha - 2nn}{nn}\right)a}$$

or, after $1 - \left(\ln\frac{mmaa}{nn}\right)\bigg/\frac{2mm\alpha\alpha}{nn}$ has been entered for the corresponding quantity under the radical sign, it yields

$$t = \left[\left(\ln\frac{mm\alpha\alpha}{nn}\right)\bigg/\frac{mm\alpha\alpha}{nn}\right] \cdot \sqrt{\left(\frac{mm\alpha\alpha - 2nn}{nn}\right)a};$$

but finally, after the quantity $2nn$ under the radical sign has been rejected, there appears $t = \frac{2n\sqrt{a}}{m\alpha} \ln \frac{m\alpha}{n}$.

But this interval of time is infinitely short, because, as is known, the logarithm of an infinite quantity is infinitely less than the quantity itself. But if, indeed, from the beginning of flow, water is expelled at once at its maximum velocity, it will seem remarkable at first glance to some, perhaps, that a finite motion is generated in an instant; nevertheless, no one will consider it absurd that an infinite mass, of which sort the quantity of water contained in an infinite vessel is, in an infinitely short time can produce finite motion, and this by the action of gravity alone.

§13. If, furthermore, in the case of the infinitely wide vessel we wish to express the time of depletion, which will of course be infinite, it will have to be assumed, as it was indicated above, that $x = 0$ in the equation of §10, and at the same time only the first term of the series is to be applied, and again $m\alpha$ is to be used for $\sqrt{mm\alpha\alpha - 2nn}$; and thus it occurs that $t = \frac{2m\alpha}{n}\sqrt{a}$.

Then at last the time which is spent in the descent of the surface through the height $a - c$ is expressed in a similar hypothesis by the following equation:

$$t = \frac{2m\alpha}{n}(\sqrt{a} - \sqrt{c}).$$

§14. The previously mentioned equations are satisfied, certainly not accurately, but closely, when the vessel is not of an infinite but nevertheless of a very great area; finally, they are not very greatly deficient when the number m moderately exceeds the number n. It may be possible here to add some words about the experiment which I indicated at the end of §11, and let this indulgence be given to our purpose which is most strongly directed toward illustrating and examining the phenomena of motions discovered by experience. But I said in the cited paragraph that, when the water flows out horizontally, the first volume element at once obtains the entire length of the thrust, and indeed theory certainly indicates that same thing for very wide vessels; but, truly, in moderately wide vessels a few drops must flow out at a lesser impetus before the point of maximum velocity appears, and these drops should strike at some point half way between the maximum thrust and the point which corresponds vertically to the orifice; and I observed that this even occurred from vessels of area about ten times as large as the orifice. Indeed, when at one time I undertook an experiment concerning a vessel half a foot high which had an area more or less one hundred times the area of the orifice, not even the least particle of water, as much as I was able to observe, withdrew noticeably from the full thrust of the water. And so we may see what quantity of water should flow out in this case before the instant of maximum velocity [is reached]; indeed it will be as great as that which a cylinder of the same area contains in the height

$$a - a \Big/ \left(\frac{mm\alpha\alpha - nn}{nn}\right)^{nn/(mm\alpha\alpha - 2nn)}$$

(*see* §10 *at the end*); nor does this very small height differ greatly from the following much shorter one, namely $\dfrac{2nna}{mm\alpha\alpha} \ln \dfrac{m\alpha}{n}$ (*see* §11), where now by $\dfrac{n}{m}$ is understood $\dfrac{1}{100}$ and by a half a foot, while for α one may substitute $\sqrt{2}$ (for we do not desire the greatest accuracy here) and through ln is indicated the hyperbolic logarithm; but thus there occurs

$$\frac{2nna}{mm\alpha\alpha} \ln \frac{m\alpha}{n} = \frac{1}{20{,}000}\left(\ln 100 + \frac{1}{2}\ln 2\right) = 0.0002475 \text{ foot}$$

or 0.00297 inch, and since I found the area of the vessel equal to $6\frac{1}{5}$ square inches, I knew that the desired quantity of water which indeed should have flowed out before the maximum thrust developed was equal to about a fifty-second part of one cubic inch, or, after it has been assumed that an average drop contains six cubic lines, more than five drops. But in the experiment I observed none, the reason of which matter I suspect to be that the first drops, although already ejected, are nevertheless still propelled by the water that is following; for the others follow too quickly for it to be possible that the first ones are removed from them. But it happens here that the interval of time from the beginning of flow right up to the maximum expulsion (which indeed through §13 is approximately $\dfrac{2n\sqrt{a}}{m\alpha} \ln \dfrac{m\alpha}{n}$, where through $2\sqrt{a}$ is understood here the time in which a body falls through a height of half a foot, which is about $\tfrac{2}{11}$ second), I say, that interval of time does not extend beyond the one-hundred-fifty-eighth part of one second.

Perhaps the fact that one's finger cannot be removed from the orifice quickly enough contributes something. But it pertains here particularly that the greatest part of that water which flows out before the maximum velocity is attained so approaches the maximum thrust that no difference can be observed, and thus hardly a single drop would have defected by a noticeable interval from the former if it could have separated itself freely from the water following.

§15. So much for water flowing out through orifices; let us progress now to the efflux of water from vessels through either converging or diverging cones. Moreover, if water flows thus through a converging pipe, the same ratio sought in §3 from the converging motion of particles explained for simple orifices dictates that the stream of water will be contracted in front of the orifice and its particles will still be accelerated, and thus the quantity of water flowing out in a given time is less than the measures of the orifice of efflux and the velocities indicate if the contraction of the stream has not been taken into consideration. But that contraction is customarily small in rather long pipes. In diverging pipes all things occur in a reverse way: for the stream is dilated in front of the orifice, the motion of the water is retarded, and a greater quantity of water flows out in a given time than would follow from the observed area of the orifice and the velocities of the water flowing out through the former without that dilation. Finally, the aqueous stream flowing out from cylindrical pipes is neither contracted nor dilated.

And thus one must properly attend to these contractions or dilations in estimating the quantities of water flowing out in a given time, which question we will treat in passing at the end of the section.

TIMES TO BE EXPECTED IN THE EFFLUX OF WATER

But now it is pleasing to study the changes which occur in the efflux of water from the beginning of motion. But in these things, for the sake of brevity, we will not consider the changes of the stream; for neither is the matter so well established that it can be confirmed accurately enough by experiments, nor are the previously mentioned changes of great moment here; but the matter itself is worth being sought after persistently so that the nature of it can be understood correctly in one's mind.

Concerning vessels which have pipes attached, we just discussed them in the above Chapter, §§31, 32 and 33; and indeed in §31 we gave rather general equations, whatever the ratio might be between the areas of the vessel and pipe; but they are overly involved, and they require highly troublesome calculation. In the paragraph which follows that one, I treated the hypothesis which makes the vessel everywhere of infinite area in proportion to the pipe, in which hypothesis I said that the water flows out at a velocity by which it can ascend to the full height of the water above the orifice of efflux; nevertheless, at the end of the paragraph I expressly warned that *at the beginning of motion the water descends more slowly than was thus defined, and that that rule does not apply before the surface has descended through some little space*, which matter is intrinsically evident, since indeed the maximum velocity cannot be produced in an instant from a state of rest in a pipe, although it may occur in a vessel perforated by a simple orifice.

So, holding these things in my mind, I began to investigate the initial changes and to reduce them to certain measurements. But the previously mentioned rule, in which those initial changes are not taken into consideration, does not suffice for this at all, although otherwise it is exactly true in an infinitely wide vessel; for all changes which precede the state of maximum velocity occur while the surface descends through an infinitely small space: nevertheless, if only the vessel is infinite in a Geometric sense, then that descent not only does not occur in an infinitely small time, as in the case of a simple orifice, but occurs in an infinitely great time, and meanwhile an infinite quantity of water flows out as well, while through an orifice an infinitely small quantity flows out, the other remaining things being equal. But in order to show these things, I made the effort of bringing forth another equation from the general equation of §23, Chapter III, which is this very simple one: $s = x$, with s taken for the height corresponding to the velocity of the water flowing out and x for the height of the water above the orifice of efflux; but anyone understands that for our purpose the matter is to be brought about thus so that the ratio of the increments of velocity is obtained, which was not required previously.

§16. Therefore, let there be the cylinder *AEHB* (Fig. 18) as in §22, Chapter III, and let this be considered as infinitely wide and full of water, and let it have the pipe *FMNG* attached, of finite area and in the form of a truncated cone, either increasing or decreasing in area toward the orifice *MN* through which water flows. Let the initial height of the water above the orifice *MN*, namely *NG* + *HB*, be a; the height of the aqueous surface at the position *CD* above *MN*, that is *NG* + *HD*, be x; the length of the attached pipe, or *NG*, be b; the area of the orifice *MN* be n; the area of the orifice *FG* be g; the area of the cylinder, which is infinite, be m; and finally let the velocity of the aqueous surface at the position *CD* be such that it conforms to the height v, which height indeed will be infinitely small. After these things had been established, we saw in the place cited that the following equation generally obtains:

$$m(x - b)\,dv + \frac{bmm}{\sqrt{gn}}\,dv - \frac{m^3}{nn}v\,dx + mv\,dx = -mx\,dx$$

in which it is evident that now the first term $m(x - b)\,dv$ can be neglected with respect to the second $\dfrac{bmm}{\sqrt{gn}}\,dv$, just as the fourth $mv\,dx$ with respect to the third $-\dfrac{m^3}{nn}v\,dx$, and thus there can be assumed

$$\frac{bmm}{\sqrt{gn}}\,dv - \frac{m^3 v}{nn}\,dx = -mx\,dx$$

in which equation if again the first term is neglected, which can be done unless the changes are desired as well which occur during the first descent, even if it is infinitely small, the common rule will arise of the *potential ascent* of the water flowing out to the full height of the water; but now for our purpose, in which we desire those first changes, that [first] term will have to be retained, and thus the last equation will have to be treated in its entire extension. However, for separating the unknowns from one another, let $\dfrac{mm}{nn}v - x = s$, or $v = \dfrac{nn}{mm}(s + x)$, and $dv = \dfrac{nn}{mm}(ds + dx)$, and so it will occur that $dx = \dfrac{-nnb\,ds}{nnb - ms\sqrt{gn}}$, which is to be integrated so that, with $x = a$

TIMES TO BE EXPECTED IN THE EFFLUX OF WATER 85

having been established, it produces $v = 0$, and hence $s = -a$; but thus it occurs that

$$x - a = \frac{nnb}{m\sqrt{gn}} \ln \frac{nnb - ms\sqrt{gn}}{nnb + ma\sqrt{gn}}$$

and the assumed value $\frac{mm}{nn}v - x$ having been taken for s, there results

$$x - a = \frac{nnb}{m\sqrt{gn}} \cdot \ln \frac{n^4b - m^3v\sqrt{gn} + mnnx\sqrt{gn}}{n^4b + mnna\sqrt{gn}}.$$

Here again in the quantity under the logarithmic sign the term n^4b, certainly infinitely less than the term $mnnx\sqrt{gn}$, can be eliminated from the numerator, and indeed from the denominator the term n^4b, likewise infinitely less than the other, $mnna\sqrt{gn}$, [can be eliminated]. And so there occurs

$$x - a = \frac{nnb}{m\sqrt{gn}} \ln \frac{nnx - mma}{nna}.$$

From this there is obtained, after e has been employed for the number of which the logarithm is unity,

$$v = \frac{nnx}{mm} - \frac{nna}{mm} e^{m(x-a) \cdot \sqrt{gn}/nnb};$$

or, with $a - x = z$ having been assumed, so that z denotes the distance through which the surface of the water has already descended, this form can be obtained for the equation:

$$v = \frac{nn(a - z)}{mm} - \frac{nna}{mm} \bigg/ e^{(mz/nb)\sqrt{g/n}}$$

from which again it is clear that, when z has a minimum ratio to b, the denominator of the other term may become infinite, and

$$v = \frac{nn(a - z)}{mm} = \frac{nnx}{mm};$$

but truly the matter is otherwise as long as the descent z is infinitely small, which case we now consider.

§17. With these things having been set forth, it is now easy to define through what little distance the fluid descends while it acquires the maximum velocity, namely by making $dv = 0$, or,

$$\frac{-nn\,dz}{mm} + \frac{na}{mb}\sqrt{g/n}\Big/e^{(mz/nb)\sqrt{g/n}} = 0$$

that is,

$$z = \frac{nb}{m}\sqrt{n/g}\cdot\ln\left(\frac{ma}{nb}\sqrt{g/n}\right).$$

But this height multiplied by the height m of the cylinder gives the quantity of water flowing out in the meantime, namely $nb\sqrt{n/g}\cdot\ln\left(\frac{ma}{nb}\sqrt{g/n}\right)$, which quantity, as I hinted above in §15, is infinite, although only logarithmically, an infinity of which sort is less than the root of any sort of dimension given from the same infinity: that is to say, $\ln\infty$ is less than $\infty^{1/n}$, however great a number may be assignable to n. And it should hence be understood that, if we reason from a true infinity to very great quantities, this quantity of water becomes small enough. Finally, the corollaries to the formula are these:

I. If the attached pipe is cylindrical, it occurs that $z = \frac{nb}{m}\ln\frac{ma}{nb}$. Therefore, with the remaining things being equal, this quantity appears as the length of the attached pipe, which is generally true also: for from a changed value of b itself the quantity $\ln\frac{ma}{nb}\sqrt{\frac{g}{n}}$ is to be considered as not changed on account of the infinite value of the number $\frac{m}{n}$.

II. For the same orifice g, and with the remaining things also equal, the quantity z follows a three-halves-power ratio of the final orifice: and if the same pipe with first a narrower and then a wider orifice is applied to a vessel, the quantity of water in the former case will be to the similar quantity in the latter as the square of the wider orifice is to the square of the narrower orifice.

III. Finally, it is to be observed that the whole reasoning is valid for all directions of the pipe, which anyone will perceive who will properly examine §22, Chapter III. Therefore, the pipe can be furnished either horizontally or in any other direction whatever, and however curved, to which one will have to pay attention especially in undertaking experiments. However, let it always be understood that

b is the length of the pipe, and a is the vertical height of the water above the extreme orifice.

§**18.** Now I come to the time in which those changes from rest to the maximum velocity occur. But I say that in a calculation of times of this sort one can simply set $v = \frac{nn}{mm} a$. For the remaining quantities in the last equation of §16 vanish, however small the height z is assumed, if only it has the minimum assignable ratio to that infinitely small height which corresponds to the maximum velocity, namely to $\frac{nb}{m} \sqrt{n/g} \ln\left(\frac{ma}{nb} \sqrt{g/n}\right)$. Thence it follows that the time is predicted, which I will call

$$t = \frac{b\sqrt{n}}{\sqrt{ga}} \cdot \ln\left(\frac{ma}{nb} \sqrt{g/n}\right)$$

and that in a like manner it is infinite, although the same time is extremely small when the area of the vessel is not infinite but very large, which again is to be deduced from the nature of the infinite logarithm.

§**19.** Because the height of the velocity, as we saw in the last paragraph, can at once be reckoned as $\frac{nn}{mm} a$ (that is, equal to the maximum when the surface descends through the minimum assignable portion of the infinitely small descent, after which the full maximum velocity is present), it follows that most changes from rest to the state of maximum velocity are not noticeable (that is, infinitely small), by all means not only the majority, but also all except an infinitely small portion; to be sure, the matter occurs thus: the velocity is clearly null at the very beginning, and, after the water descends through an infinitely small distance, it is already very nearly maximum; then, while it descends again through some little distance, infinitely small but nevertheless infinitely greater than the former [distance], it continues to be moved at its own velocity, taking on infinitely small increments, and then at last it truly attains a maximum velocity. But since these latter or infinitely small changes cannot be perceived by observation, we will treat differently those theorems which we gave in §17 by considering, in place of the changes from rest right up to the point of maximum velocity, the same changes right up to a given rate of speed.

§**20.** And so we will investigate through how great a distance z the surface of the water descends from the state of rest, and how much water flows out, and finally how much time must pass in order that the internal water be moved at a velocity which is generated by free fall through the given height, which we will call $\frac{nn}{mm}c$, so that c itself denotes the similar height for the velocity of the water flowing out. For this it is required that in the last equation of §16 $\frac{nnc}{mm}$ be entered for v, so that there will be

$$\frac{nnc}{mm} = \frac{nn(a-z)}{mm} - \frac{nna}{mm}\bigg/ e^{(mz/nb)\cdot\sqrt{g/n}}$$

and hence it is deduced that $\frac{mz}{nb}\sqrt{g/n} = \ln\frac{a}{a-c-z}$; but when c is assumed here to be noticeably less than a, the letter z under the logarithmic sign can be rejected, whence there is obtained

$$z = \frac{nb}{m}\cdot\sqrt{n/g}\ln\frac{a}{a-c}.$$

But this equation now indicates a space which is infinitely small and through which the surface of the water descends while the velocity of the water flowing out from rest is that which is due to the height c; and this little distance is to that indicated in §17, by which indeed the velocity becomes maximum, as $\ln\frac{a}{a-c}$ is to $\ln\left(\frac{ma}{nb}\sqrt{g/n}\right)$, so that the first is infinitely less than the other, although both are infinitely small. If, further, the defined quantity z is multiplied by m, one obtains the quantity of water flowing out while that velocity due to the height c is produced, which quantity, accordingly, is equal to

$$nb\sqrt{n/g}\ln\frac{a}{a-c}$$

and thus is of finite magnitude, and indeed is greater the longer the pipe is assumed and the greater a thrust is expected.

And, finally, the time in which the same occurs, if the terms to be rejected are selected correctly, is discovered to be equal to

$$2\sqrt{\frac{nbb}{ag}\ln\frac{a}{a-c}}$$

and thus is finite but very small, and in no case is it to be extended easily beyond one second.

§21. I wished to examine and pursue all these things accurately, first because the solution of many phenomena which are customarily

TIMES TO BE EXPECTED IN THE EFFLUX OF WATER   89

observed in the efflux of water depend thereon, then as well in order that we might understand correctly in our mind those changes which are clearly imperceptible through observation. There have been many who, not having followed correctly the transition from the infinite to the finite or, in turn, from the finite to the infinite in flowing water, were not able to extricate themselves from the many difficulties which elsewhere easily permit a solution; but if in place of the almost infinite vessel, of which there is none, a very wide vessel is assumed, or even, since it suffices in many cases, a moderately wide vessel, the formulas will be approximately true, and they will approach the truth more or less in accordance with the nature of the question; concerning these matters I will point out certain things in the following experiments. Thus, meanwhile, it is already apparent enough from theory, because I had prepared to explain especially why water flows from a simple very wide vessel at once at the entire velocity, and why it is different for water ejected from a vessel through a pipe. But the precise measures concerning these questions are to be deduced from the equations themselves.

§22. Finally it is evident, since it pertains to the time of depletion, that, when the area of the vessel moderately exceeds the area of the attached pipe, the former can be considered without noticeable error to be $\frac{m\alpha}{n}\theta$, by understanding by θ the time in which a body by falling freely from rest falls through the height which the water had at the beginning of flow above the final orifice of the pipe, and by assuming for $\frac{m\alpha}{n}$ the ratio which exists between the area of the vessel and the *section of the stream*, whether *contracted* or *dilated*. But the hindrances which accidentally occur in addition in these cases increase that time somewhat. But if the time is desired in which the surface of the water descends through a given height, that [time] is to be taken as $\frac{m\alpha}{n} \cdot (\theta - T)$, after there has been assumed for T the time which a body takes in falling freely through the height which the water has above the orifice at the end of the flow.

EXPERIMENTS WHICH PERTAIN TO CHAPTER IV

Since a large part of this chapter was employed in the contraction of an aqueous stream flowing through an orifice in a thin plate, I undertook to begin accurate experiments concerning that contraction, certainly not by undertaking measurements of the diameters, which method I found cannot be used with sufficient accuracy, but by observing the *actual* velocities from the extent of the thrust and the quantities flowing out in given times. In the experiments I used an automaton which pulsed 144 times in the period of the first one minute, and I assumed it to be thus in the following [minutes].

PERTAINING TO THE THEORY OF THE CONTRACTION OF AQUEOUS STREAMS

EXPERIMENT 1. I furnished a cylindrical pipe, the diameter of which was 4 inches 3 lines, English measure, made of a thin plate and which had an orifice in the side, that is, in the cylindrical surface; the diameter of the orifice was $4\frac{52}{125}$ lines. The water flowed out horizontally from the vertically positioned cylinder, and the height of the water above the center of the orifice was 4 inches 8 lines at the beginning of flow, and the similar height at the end of flow was 3 inches. However, the whole flow took place in the interval of eleven pulses of the automaton, which constitutes a time of approximately $4\frac{1}{2}$ seconds.

Further, after the experiment had been repeated often, and after the height of the orifice above the horizontally placed table and the extent of the thrust, and this both in the beginning and at the end of flow, had been observed, I saw from the *Lemma indicated in the beginning of the Experiments of the preceding Chapter* that the velocity of the water flowing out at the place of the maximum contracted stream had constantly been that, indeed as much as one could judge by observation, which should be due to the height of the water above the very place which is at the same height as the orifice.

Therefore, if we assume that the contraction of the aqueous stream was the same everywhere, we apply to this case the last equation of §13, namely $t = \dfrac{2m\alpha}{n}(\sqrt{a} - \sqrt{c})$, there will have to be established $t = 4\frac{1}{2}$ seconds; $\dfrac{m}{n} = 133$; $2\sqrt{a}$ (the time which a body takes in falling freely through the initial height of the water) $= 0.1483$, and $2\sqrt{c}$ (the similar time for the final height of the water) $= 0.1246$; it occurs that $4\frac{1}{2} = 3.15\alpha$, from which $\alpha = 1.43$. Hence it is a consequence that the area of the orifice was to the section of the contracted stream as 143 to 100; this ratio is a little greater than that which exists between $\sqrt{2}$ and 1, that is between 141 and 100; but if the velocities could have been observed very accurately, there is no doubt but that they would have been a little less than those which are due to the entire height of the water; and when this matter is taken into account, it is discovered that the value of α itself is thus to be diminished a little bit; therefore, from the entire experiment it can be concluded very safely that the previously mentioned ratio was as $\sqrt{2}$ to 1.

EXPERIMENT 2. Next I wished to find out by experiment whether in all jets flowing in any direction whatever the contraction is the same, and to this end I reckoned that the matter was to be attacked in such a way that, except for the change of that direction, all the remaining circumstances should be the same thereafter. Indeed, I obtained it in this manner.

Obviously, I used the same cylinder as previously, but I attached it to a prismatic box, positioned vertically, so that the axis of the cylinder would be horizontal, and thus I revolved the attached [cylinder] so that the center of the orifice selected for the efflux of water would occupy first the highest place, then the middle, then the lowest; in the first case the water flowed out vertically upward, in the second horizontally, in the third vertically downward; but in the individual [cases] I made the heights of the water in the box above the center of the orifice perfectly equal; the result was this:

I observed that in equal times the surface of the water in the individual cases descends through equal distances in the box. Therefore, in streams projecting upward, the water above does not offer noticeable resistance to the water following below, which same thing I understood in a different way, because certainly if I intercepted the aqueous stream in any direction whatever at a small distance from the orifice, such as 3 lines, say by a coin, so that the stream would strike against the coin perpendicularly, the efflux of the water would not be retarded. Further, neither does the water below in streams descending vertically draw the sequent water after itself: and the very contraction of the stream is everywhere the same, since the retardation and acceleration of the water ejected upward or downward, which cause the stream either to swell or to become slender at some distance from the orifice, were not considered. For here certainly the discussion is only about that contraction which arises from the oblique motion of the particles in the region of the orifice.

EXPERIMENT 3. I used the same device, prepared in the previously mentioned manner, for finding out whether the contraction of the stream, all remaining things being equal, would be changed by an increased height of the water above the orifice. To this end I fixed two needles to the internal walls of the box along a plumb line; the first projected above the center of the orifice 13 inches 10 lines, the other 12 inches $1\frac{3}{5}$ lines, English measure; the area of the box was to the area of the orifice as 404 is to 1. Moreover, I saw that the surface of the water had descended from the upper needle to the lower after an interval of 24 pulses of the automaton, which gives a time of 10 seconds.

But if indeed the same time is sought according to the Hypothesis that the stream had not contracted and that at the same time the water had flowed out at the whole velocity which it should have had by dint of the theory with no alien hindrances being present, that [time] is ascertained to be $6\frac{7}{8}$ seconds.

Thus, therefore, it can be concluded that the area of the orifice was to the *section of the contracted stream* as 10 to $6\frac{7}{8}$; that is, $\alpha = 1.45$ while in the first experiment for the same orifice, all circumstances having been considered, one found $\alpha = 1.41$.

After I had tried this in that way, it remained to discover whether the water would flow out at the whole velocity, according to observation, about which matter I doubted all the more, because with the velocity of the water increasing, the hindrances increase at the same time, and accordingly they can be noticeable at the greater heights of water while they are not so at the lesser.

And so with all care applied (because it is required especially for the precision of the experiment), I made the water flow out in a perfectly horizontal direction, and after the measures of both the extent of the thrust and the height of the orifice above the horizontal table had been taken, I saw, after performing a calculation, that when the height of the water was 13 inches and 10 lines, or 166 lines, the water would flow out, or rather would flow across the *section of the contracted stream* at a velocity which corresponds to a height of $158\frac{1}{2}$ lines; therefore, the velocity is to be diminished in the calculation in proportion to the square roots of these heights, and the discovered value of the letter α, which thus is a little less than 1.42, or again 1.41, decreases in approximately the same ratio, and thus it is allowable to deduce that the changed height of the water does not alone change the contraction of the stream according to observation.

EXPERIMENT 4. I used a cylindrical pipe of height 4 inches, the section of which through the axis is represented (Fig. 28*b*) by *CABD*; the area of the cylinder was to the area of the orifice *ac* as 110 to 1. This entire cylinder full of water was evacuated in a time of $21\frac{1}{2}$ seconds. However, it must be noted that efflux is not to be granted to the water until no turbinate motion is observed in it; for otherwise the water is soon changed into a whorl, somewhat swift during efflux, and the efflux is greatly retarded, and all the more so the faster the internal water is driven in a circle; further, because all the water never flows out, I considered the time of efflux [to extend] until it began to flow out drop by drop.

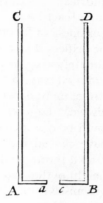

FIGURE 28b

This experiment indicates that here the contraction of the water was less than in the ratio 2 to 1. I had expected the time of evacuation to be only about 23 seconds, but the occurrence was a little different, as I said, of which matter I noticed a little later that the reason was that the elongated lips of the orifice more or less formed a short pipe, although very short, as the Figure shows, which hindered the contraction of the stream; however, the length of those lips did not attain two thirds of a line.

EXPERIMENT 5. I made water flow horizontally through a small pipe from a very wide vessel; but the pipe was very short, indeed not exceeding a length of 3 lines, and it was almost 5 lines in diameter.

A given quantity of water flowed out in a time of $11\frac{1}{4}$ seconds which should have flowed out in $10\frac{2}{3}$ seconds if it is assumed that neither was the stream contracted nor were any hindrances present.

I did not consider the [determination of the] true velocities of the water to be a task that I should undertake, not doubting that they were such as they must be in order that a given quantity of water might flow in an observed time through an observed orifice, with no attention having been given to the contraction of the stream.

In addition, I furnished other small pipes of different diameter and length, and I saw that the quantities of water flowing out in a given time and at given velocities correspond directly to the orifices of efflux, but that the velocities were more deficient from the velocity due to the entire height of the water, the narrower and the longer was the pipe, and also the higher was the water.

Pertaining to the Theory of Water Flowing out through Pipes

Experiment 6. The cylindrical vessels, of which Figs. 24 and 25 represent the sections through [each] axis, had a height of 4 English inches and pipes attached of one-foot length; the areas of the cylinders were to the areas of the orifices A as 110 to 1. But the orifice B was to the orifice A approximately as 25 to 16; the time of evacuation before the cylinders were emptied was, in Fig. 24, $6\frac{1}{2}$ seconds, in the other, $4\frac{1}{3}$ of these units of time, more or less.

In these cases the vessels were large enough with respect to the attached pipes that they could be considered as infinite; and accordingly the water should have flowed out, according to the Rules indicated by us within the text, through the final orifices at velocities corresponding to the total height of the water, if only one excludes the first instants of flow, which themselves are so short here that they cannot be observed. And since in addition, as I advised in passing, the quantity of water flowing out in a given time through the pipes is to be estimated simply from the speeds and the size of the orifices, I found, through the rule shown in §22, the time of evacuation in the first case as $4\frac{1}{3}$ seconds, in the latter, almost 3 seconds.

The fact that these things were observed to be a little greater in the experiment in Fig. 24 is to be attributed for the most part to the adhesion of the water to the sides of the pipe, but in Fig. 25 to a different reason in addition, indicated in §34, Chapter III.

Other Phenomena are to be noted in these vessels: namely, when the vessels are not quite evacuated, a certain sound is perceived from the air which then mixes with the water in the upper [part of the] orifice; in fact, I heard this sound to the last instant of flow; further, it occurs easily that the efflux of water may be permitted before the latter has been reduced to perfect rest (for the water is agitated by filling and moved in a whorl); but then the efflux is retarded very much, and a kind of cataract is formed internally, and air is continually intermixed with the water flowing out. Thus the efflux can be retarded at will if the water is agitated into a vortex before it flows out.

Experiment 7. I used a Prismatic vessel, to which a small pipe was attached horizontally as in Fig. 19. The orifice GF was precisely 5 lines in Diameter, the other, NM, $6\frac{1}{2}$ lines. Accordingly, the very areas of the orifices GF and NM were as 100 to 169, but the area of the vessel contained the area of the orifice NM 201 times. The length of the small pipe GN was 4 inches.

Then I filled the vessel with water right up to CD, the height of which above the axis of the pipe was 13 inches 10 lines. After the orifice NM had been opened, the water flowed out, and the surface descended right to EH in a time of $8\frac{1}{3}$ seconds, and the difference of the heights, CE or DH, was 2 inches 8 lines.

After the calculation has been reduced according to the pattern of §22, where attention was paid neither to the hindrances nor to the change of the stream, we see that the predicted time of descent should have been approximately 5 seconds, almost $5\frac{1}{2}$. Therefore, it is to be stated in this way: that the mean total velocity was to the entire velocity which the theory indicates as $5\frac{1}{2}$ is to $8\frac{1}{3}$, or approximately as 2 to 3; and hence it can be concluded that the water flowed out through the orifice MN at a velocity which compared with $(\frac{2}{3})^2$, or four ninths of the height of the water above the orifice MN, but through the other orifice GF it flowed at a velocity due more or less to five fourths of that same height.

And so it appears again that the efflux of water is increased by the increased area of the orifice of the pipe toward the exterior, although neither the opening at which the pipe is implanted in the vessel nor the position of the pipe is changed.

Further, on the horizontally placed table PQ I observed the amplitude of the thrust PQ for the height oP, which was 4 inches 8 lines. Moreover, I found $PQ = 9$ inches 6 lines.

It follows from this observation that if consideration of the dilation of the stream is set aside, the water would be required to have a velocity at NM which is due to a height of 4 inches 10 lines, while, nevertheless, by dint of the aforementioned experiment it certainly had a velocity due to a height of almost 6 inches 2 lines. This observation confirms what I said in §15, namely that in divergent pipes the aqueous stream is dilated as at m, and the motion of the same is retarded. But in the present case, in order that both observations might agree, it will have to be said that the stream was so dilated that it had an area in proportion to the orifice NM reciprocally as the aforementioned velocities or reciprocally as the roots of the heights due to these velocities, namely as $\sqrt{74}$ to $\sqrt{58}$, and thus that the diameters of the dilated stream and of the orifice were as $\sqrt[4]{74}$ to $\sqrt[4]{58}$, or as 100 to 94.1.

EXPERIMENT 8. I performed another experiment which, although it does not yet pertain to this, nevertheless I will recount: namely, at the origin near the orifice GF [Fig. 19] I perforated the pipe with an opening e of almost two lines, and again I observed the descent of the surface from CD to EH, the water flowing through NM, and at the same time I examined the amplitude of the thrust.

I saw these two things which at first glance seem almost to contradict one another: the descent from *CD* to *EH* was made more slowly than it was in the preceding experiment, and now it lasted 10 seconds; and nevertheless the thrust *PQ* was greater for the same height *oP*, for now *PQ* was 10 inches 10 lines.

I explain both Phenomena thus: on account of the orifice *e*, which was made near *GF*, because it allows the free transit of air, the pressure is removed which at other times the water exerts within itself in the pipe, and the water accordingly does not flow differently where the small opening *e* is than if the pipe were cut off at that very place; but the water would flow more slowly, which I showed in passing, if the pipe *GNMF*, as if diverging, were made shorter. Further, since the water can flow, although at a lesser quantity, nevertheless with a greater impetus, through the unchanged orifice *NM* without implicit contradiction, the reason is the mixing together of air with water; for air perpetually rushes into the pipe through the small opening *e*, and together with the water it flows out through *NM*. And, finally, it seems to me that the phenomenon according to which the water flows actually through *MN* more quickly with the orifice *e* opened rather than closed cannot be explained otherwise than that the hindrances from outside act less on water rarefied by air than on natural [water].

Pertaining to the Theory of Water which Flows out of Very Large Vessels from the Point of Rest Right up to a Given Degree of Speed

Experiment 9. When water flows from a very large vessel through an orifice made in a thin plate, the first drop bursts forth immediately at the entire velocity which is due to the height of the water above the orifice.

This conforms to the theory indicated in §11 if the vessel is truly infinite; and even though it may not be infinite in a Geometric sense, as long as it is very large, in a like manner no drop can be observed at the beginning of flow which will not flow out at the maximum velocity; I explained this Phenomenon in §14, since indeed, by dint of the theory in some particular case reviewed there, scarcely one or two drops should have defected noticeably from the maximum thrust. I said that such a little quantity of water could not separate itself from the subsequent water on account of the mutual attraction or cohesion of the aqueous particles.

EXPERIMENT 10. When in fact the water flowed out from a very large vessel through a pipe inserted horizontally in the vessel, I observed before the stream flowing out formed the maximum thrust omQ (see Fig. 19) that a noticeable enough quantity of water fell on the horizontal table situated below, halfway between P and Q; that the greater is this quantity, the longer is the pipe GN and the more it diverges toward N; and finally that that water is distributed unequally, and that it obviously falls away more abundantly in the place which is more remote from the point P than in that which is nearer; however, in the understanding of the time in which all these changes take place, I saw that that was very short and such that the measure of it cannot be obtained.

All these phenomena follow as a unit from propositions which we gave from §11 right up to the end of the section. But the measures shown in that place cannot be confirmed directly by experiments, especially those which are indicated in §§15, 16, and 17, where, as you know, the formulas are communicated which express the quantity of water flowing out while the maximum thrust is attained from rest; the reason is: first, because the first drops which should have fallen near the point P on the table did not separate themselves freely from the water following; second, because the quantity of water following the stream oQ (which certainly constitutes the maximum portion according to the theory itself) cannot be intercepted; and finally, because the motion of water through pipes is customarily retarded very much by outside hindrances, particularly if the pipes diverge, and thus the real motion is very different from the motion which the water should have if all the hindrances were removed. The remaining measures indicated by us are subjected to fewer difficulties, and these of lesser moment; but they are contained in §20, and they express especially the quantity of water which flows out at the first instant of motion while the water attains a given degree of speed.

Although on account of the reasons just mentioned, especially in the case of divergent pipes, a perfect agreement of theory with experiments cannot be expected at all, nevertheless I found such success that I would understand that complete agreement would have occurred easily if all hindrances together with the mutual adhesion of the aqueous particles could have been prevented. Indeed, I performed experiments on a divergent pipe as well as on a cylindrical one; let me now explain them individually.

EXPERIMENT 11. In Fig. 19 a pipe in the form of a truncated cone was inserted horizontally in a vessel; I filled the vessel itself with water right up to *CD*, so that the height of it above the axis of the pipe was equal to 433 parts identical to those which I used in the entire experiment. For that height I sought by experiment the point *Q* corresponding to the maximum thrust, and *PQ* was 287 parts while the height *oP* was 146 parts. Thus I saw that the motion of the water, both on account of the adhesion of the water and on account of the shape of the pipe, had been greatly retarded, which must occur in these cases, as I warned several times. However, it should have been, if nothing had opposed the motion, that *PQ* was 503 parts.

Next I placed a Pan on the horizontal table, the edges of which were at *S* and *R*; but first I moistened the Pan, and I allowed all the water to rain down from the former [height] again; and after the measure of *PR* was taken, I found that to be 206 parts.

And finally the diameter *GF* was 13 parts, and *MN* 17 parts, but the length of the pipe was 125 parts.

After all these things had been thus prepared, during which time I covered the orifice *MN* with a finger, after the finger had been suddenly removed, the water was ejected, and some part of this fell on the pan; I collected this anxiously in a cylindrical glass pipe of which the diameter was $8\frac{1}{2}$ parts; that pipe was filled to a height of 210 parts, therefore the quantity of water having fallen on the pan was 11,922 cubic parts.

But now that quantity, through §20, should be $nb\sqrt{n/g} \ln \dfrac{a}{a-c}$, where by *n* the area of the orifice *NM* is understood, or 227 square parts, by *g* the area of the orifice *GF*, or 133 square parts; further, *b* denotes the length of the pipe, which was 125 parts; by *a* is properly understood the height of the surface *CD* above the axis of the pipe, but here rather the height complying with the velocity of the water striking at the point *Q* is to be understood, or 141 parts, and similarly for *c* is to be assumed the height complying with the velocity of a particle striking at the point *R*, namely 73 parts. Finally, the abbreviated expression ln signifies the Hyperbolic logarithm. After these numerical substitutions have been made, there occurs

$$nb\sqrt{n/g} \ln \frac{a}{a-c} = 227 \cdot 125 \cdot \frac{17}{13} \cdot \ln \frac{141}{68} = 26{,}830.$$

Therefore, the quantity of water found in the experiment was to the quantity which the theory indicates after the consideration of hindrances has been set aside as 11,922 is to 26,830, which numbers, although they differ by not a little, nevertheless confirm the theory splendidly, which very thing I now place clearly before one's eyes.

In the formula $nb\sqrt{n/g} \ln \frac{a}{a-c}$, we took for a the height due to the maximum velocity of the water flowing out, as in fact it was in the experiment, not as it would have been if the obstacles had been removed; certainly we set $a = 141$, but in theory $a = 433$. However, if that latter value is assumed, by retaining the value of the height $c = 73$, one has $nb\sqrt{n/g} \ln \frac{a}{a-c}$ equal to approximately 6700, which number is now much less than the number found through the experiment, since before it was so much greater. But so it occurs when the height c is assumed to preserve its value: just as in fact the height a was increased from 141 to 433, so also the height c is certainly to be increased, and each height should be increased in the same ratio if the hindrances resist equally to the first drops and those following; but, the remaining things being equal, the particles experience less resistance the slower they are moved, and accordingly also the drops which fall on the near side of the limit R are retarded less than those crossing that limit. From this it is easy to conclude that the height c is to be increased in a lesser proportion than the height a, but the proportion itself we cannot give, unless a posteriori by obviously making the theory agree with the experiment; thus it is ascertained that c is to be set equal to 120, which number clearly suffices, after all circumstances have been well considered.

Thus, therefore, it seems manifest to me that the success of the experiment was such that it clearly agrees with the theory. But examples of this sort show wholly that we have transmitted the true laws of motions into fluids, and these [examples] I selected from among infinite others, because they have no relationship nor any affinity to the common rule which states that fluids flow everywhere at a velocity due to the entire height of the water above the orifice, and they cannot be solved by the usual principles. And in remainder, since the motion of the water in this experiment was retarded, I wished to undertake another one in which all hindrances would be altogether diminished, so that thus it would appear that the numbers of the experiment and of the rules would agree the more with each other, the less would be the hindrances.

EXPERIMENT 12. And so I now used a cylindrical pipe through which the flow might pass more easily, and for that very reason it was wider: in addition, the box to which the pipe was attached was much wider, and finally, the height of the water contained in the box above the axis of the pipe was much less, so that the water would flow out at a lesser velocity, and thus it would encounter obstacles of lesser moment. The remaining things were as before.

Therefore the height of the water above the axis of the pipe was 130 parts, $oP = 553$ parts, $PQ = 453$ parts, $PR = 297$, the diameter GF or $MN = 19$ parts, and the length of the pipe was 130 parts.

I saw that the water, having fallen into the pan, had filled up the cylinder, which had a diameter of $8\frac{1}{2}$ parts, to a height of 281 parts, the capacity of which was then 15,950 cubic parts. In this case a is to be set equal to $\dfrac{453 \cdot 453}{4 \cdot 553} = 93$ parts, $c = 40$ parts, $n = g = 284$ square parts, and $b = 130$. After these substitutions have been made, it occurs that

$$nb\sqrt{n/g}\ln\frac{a}{a-c} = 284{,}130\ln\frac{93}{53} = 20{,}760,$$

to which the number in the experiment, 15,950, as we saw, corresponds. But the latter number is almost four fifths of the other, and thus it agrees approximately to the same, since in the preceding example on account of the applied reasons a similar number defected from a similar one by more than half.

Therefore, now it is amply evident that it is to be attributed to outside obstacles only that the experiments do not correspond to the formulas correctly; meanwhile, nevertheless, the experiments are such that they cannot demonstrate the strength of these formulas any better.

FIFTH CHAPTER

Concerning the Motion of Water from Constantly Full Vessels

§1. Vessels are maintained full when as much water is continuously poured in as flows out: but the pouring in can be in the same direction as the motion of the aqueous surface and at the same velocity at every instant, as if certainly a new surface were created continuously which already possesses the velocity of the adjacent water, or [it can be] lateral and without impetus, just as if the surface which is assumed continuously to be created anew were provided with no motion and finally is to be stimulated into motion by the water below. I will pass over the remaining methods of supplying new water, which are infinite.

Meanwhile, the rule about this motion, especially in the latter phase, is accepted: that the water flows out at a velocity complying with the height of the surface above the opening; nevertheless, it is easy to see in advance that this cannot be valid unless for a vessel infinitely wide everywhere, but in the remaining [vessels] it happens that the motion beginning from rest will be increased very gradually through some interval of time, and finally after an infinite time it will acquire the entire velocity. Nevertheless, if one is to say what the reason is, those accelerations for the most part occur so quickly that the entire velocity is not present for only the shortest time. But the situation is otherwise in very long aqueducts, in which the increases in velocities do not escape notice, and they can be observed with separate measurements.

But whatever the matter may be, since mathematical accuracy is never objectionable, I undertook to consider and follow the motion of water from the beginning to some given limit.

§2. All properties of this motion may be reduced to essentially three equations: first, between the quantity of water ejected and the corresponding velocity; second, between time and velocity; and third, between the quantity of water and the time. If one of these equations is obtained, the remaining follow from it spontaneously.

Therefore, we will scrutinize only the first one rather closely. But here let us be mindful of those things which were advised in the preceding section about the contraction of a stream flowing out through simple orifices or converging pipes, and the dilation of the same when it is ejected through diverging pipes. However, we indicated in §3, Art. I, Chapter IV, that the stream is to be considered until that time when the velocities of the particles (diverting one's mind from the changes which gravity produces on the particles beyond the vessel) are not changed any further, and all that portion of the stream is to be considered as moving inside the vessel, just as if the surface of the stream became uniformly hardened there. Therefore, from now on when the discussion will be about a vessel through which water flows, that ideal vessel is to be considered, the orifice of efflux of which is the section of the stream subjected to no further change except that which is due to the descent or ascent of the stream.

Problem

§3. To find the velocity of water flowing out of a constantly full vessel after a given quantity of water has already flowed out.

SOLUTION. There are two methods of supplying water especially worthy of considering, either of which postulates another solution to the problem: for either the water is assumed to rain down vertically into the vessel, and such indeed that it flows in at precisely the same velocity which the surface of the water has, or the water flows in laterally and thus lacks the impetus by which the water could follow the surface on its own and finally is to be excited into motion.

Case I

In order that for the first case we may find the equation between the quantity of water ejected and the corresponding velocity, this is to be pursued, with a single circumstance changed, along the same paths which we followed in the first paragraphs of Chapter III.

Therefore, as in §6, Chapter III, let the vessel *aimb* (Figs. 15 and 16) be proposed, which is kept constantly full right up to *cd* by the inflow of water; but let water flow out through the orifice *pl*; and let it be established that that quantity of water has already flowed out which can be contained in a cylinder erected above the orifice *pl* to a height x, but that the last drop has flowed out at the velocity by which it can

MOTION OF WATER FROM CONSTANTLY FULL VESSELS 103

ascend to the height qs or v; thus, the equation between x and v will now have to be shown.

Let the curve CGI be the scale of the areas, such indeed that, for HL denoting the height above the orifice, HG expresses the area of the vessel at that place. Next let a third curve tru be drawn, the ordinate Hr of which is everywhere continuously equal to the third proportional with respect to GH and PL, or, the ordinate Hr of which is $(PL)^2/GH$.

Let the space $DCIL = M$, the space $DtuL = N$, and the *potential ascent* of the water contained in the vessel, after the previously mentioned quantity has already flowed out (through §2, Chapter III), will be $\frac{N}{M}v$. Further, let the particle $plon$ be understood to flow out, and the surface cd to descend to ef; now the height of the velocity for the particle $plon$ will be $v + dv$; and if now the parallelogram $LxyO$ is constructed, the side LO of which is lo, and the other, Lx, is PL, the *potential ascent* of the same water in the position $efmlonpie$ will be equal to the fourth proportional with respect to the space $EFLONPIE$ (which again is M, because $PLON$ expresses the magnitude of the volume element $plon$, while $CDFE$ expresses the minimum quantity $cdfe$ equal to that volume element), the space $wuxyOLF$ (which is equal to the space $N - DtwF + LxyO$, from which, if PL or Lx is set equal to n, $CD = m$, and $LO = lo = dx$, there will be $Dt = \frac{nn}{m}$, $DF = \frac{n}{m}dx$, hence the small space $DtwF = \frac{n^3}{mm}dx$, and the space $LxyO = ndx$ and finally the space $wuxyOLF = N - \frac{n^3}{mm}dx + n\,dx$), and the height $v + dv$. Therefore the *potential ascent* just mentioned is $\left(N - \frac{n^3}{mm}dx + n\,dx\right) \cdot (v + dv)/M$, equal to the rejected differentials of the second order $\frac{N}{M}v + \frac{N}{M}dv - \frac{n^3}{mmM}v\,dx + \frac{n}{M}v\,dx$, such that the increment of *potential ascent*, which was added to the water while the volume element $plon$ was flowing out, is $\frac{N}{M}dv - \frac{n^3}{mmM}v\,dx + \frac{n}{M}v\,dx$, where the spaces N and M are of constant size on account of the continuous pouring in of water. In this first case we do not consider the *potential ascent* of the volume element $cdfe$, which is filled while the other equal [volume] $plon$ flows out, because that ascent is not generated by an internal force, nor indeed is the lower water considered to draw the particle *cdfe* after itself, but rather we consider this to be poured in continuously by a certain outside force, and this at neither a greater nor a lesser velocity than that which pertains to the surface *ef*. Therefore the entire increment to be considered here is, as we said,

$$\frac{N}{M} dv - \frac{n^3}{mmM} v\, dx + \frac{n}{M} v\, dx.$$

But that increment must be equated to the *actual descent* of the center of gravity. And that descent, after DL has been set equal to a, is, from §7, Chapter III, $\frac{na\, dx}{M}$; therefore, the following equation results:

$$\frac{N}{M} dv - \frac{n^3}{mmM} v\, dx + \frac{n}{M} v\, dx = \frac{na\, dx}{M},$$

or

$$dx = N\, dv \Big/ \Big(na - nv + \frac{n^3}{mm} v\Big).$$

But if this is so integrated that v and x vanish together, it gives

$$x = \frac{mmN}{n^3 - nmm} \ln \frac{mma - mmv + nnv}{mma}$$

which equation, after e has been established as the number the logarithm of which is unity, is equivalent to

$$v = \frac{mma}{mm - nn} [1 - e^{(n^3 - nmm)x/mmN}].$$

But this solution is suitable for the first case, where the water is poured in from above with a motion which is common with the descent of the nearest surface.

Case II

Thus if now the particle *cdfe* is considered continuously to be poured in laterally, then on account of its own inertia it stays behind the motion of the lower water, and accordingly the *potential ascent* of the same enters differently into the computation. Moreover, first the *potential ascent* of the aqueous mass *cdmlpic* increased by the volume element soon to be poured in is to be considered; then the *potential ascent* of the same water at the position *cdmlonpic* is to be investigated, indeed, after the volume element has already flowed out, and their difference is to be equated to the *actual descent* $\frac{na\, dx}{M}$. Truly the

MOTION OF WATER FROM CONSTANTLY FULL VESSELS

potential ascent of all the previously mentioned water before the pouring in of the particle and after the pouring in of the same is thus determined: certainly the *potential ascent* of the water *cdmlpic* is $\frac{Nv}{M}$, and the *potential ascent* of the particle ready to be poured in is nil, because, poured in laterally, it does not yet have a common motion with the lower mass. Therefore, the *potential ascent* of each [volume of] water (which one obviously determines by multiplying the respective mass by its own *potential ascent* and dividing the sum of the products by the sum of the masses) is $\left(M \cdot \frac{Nv}{M} + n\,dx \cdot 0\right) \Big/ (M + n\,dx) = \frac{Nv}{M + n\,dx}$. But at that very time when the particle $n\,dx$ was poured in from above, it acquired a common motion with the water just below, and thus the *potential ascent* of the same water in the position *cdmlonpic* becomes equal to the fourth proportional with respect to the space $CDLONPIC \cdot (M + n\,dx)$, the space $DtuxyOLD$ $(N + n\,dx)$, and the height $v + dv$, that is, $\frac{(N + n\,dx)(v + dv)}{M + n\,dx}$, the excess of which over the prior *potential ascent* is $\frac{N\,dv + nv\,dx + n\,dx\,dv}{M + dx} = \frac{N\,dv + nv\,dx}{M}$ (after the differentials of the second order have been rejected). Therefore, the following equation is obtained

$$\frac{N\,dv + nv\,dx}{M} = \frac{na\,dx}{M}$$

which, handled as previously and carried out to the end, gives

$$x = \frac{N}{n} \ln \frac{a}{a - v}$$

or

$$v = a(1 - e^{-nx/N})$$

which solution is valid for lateral pouring in.

Scholium I

§4. These equations are altogether different from each other; moreover, the greater the difference, the less the area of the vessel; and if indeed the uppermost area of the vessel at *cd* is more or less infinite with respect to the area of the orifice, n vanishes with respect to m, and it occurs in the former case, just as in the latter, that

$$v = a(1 - e^{-nx/N}).$$

Therefore, in this hypothesis the motion is the same on either hand, which everyone could have seen in advance with no difficulty at all. But the motion in the former pouring in is always swifter than in the latter, the remaining things being equal.

Here it is convenient to explain the matter physically as well, in order that we can perceive it more distinctly in all phenomena.

In place of any vessel whatever having any direction whatever, for the sake of a shorter delineation, let there be a vertical cylinder with an orifice at the base, namely *GHND* (Fig. 29), and then let the vessel

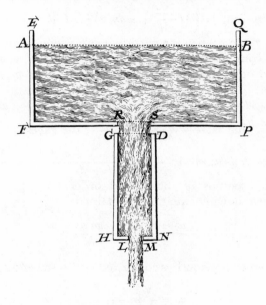

FIGURE 29

EFPQ be perforated at *RS*; let the orifices *RS* and *GD* be assumed perfectly equal and corresponding to each other perfectly at a minimum distance apart, so that all the water flowing out from the upper vessel flows into the cylinder placed below.

Let the water begin to flow from each vessel, but let it be assumed to flow constantly from the upper at that velocity which the surface of the water in the cylinder below has.

Thus it is evident that this is satisfactory for the first condition of filling. But now, to see whether they agree with the preceding, we will investigate the phenomena of this motion.

Therefore, let us consider the upper vessel as if it were infinite, so that the water flowing through *RS* at every instant has a velocity

MOTION OF WATER FROM CONSTANTLY FULL VESSELS 107

which complies with the height PB or FA; thus it will have to be considered that this height PB is infinitely small at the beginning, because then the water must flow at an infinitely small velocity, but that it then increases gradually, and this continuously more and more, until after an infinite time the motion remains uniform; however, it is asked whether the height PB of the water will finally become infinite or if indeed it will not pass beyond a certain limit. This is determined as follows.

Let the height GH or RH (for it is not to be considered that they differ from one another) be a, $AF = x$, the area of the orifice LM be n, the area of the orifice RS be m; because, indeed, as is manifest, the two vessels can be understood to cohere and thus become one, the velocity of the water at LM after an infinite time (from §23, Chapter III) will be $\sqrt{a+x}$, and at RS it will be $\sqrt{x}$ (which is evident afterwards if now the vessels are considered separated again, for either can be assumed without error), but the velocities must be in the inverse ratio of the areas of the orifices; and so $\sqrt{a+x} : \sqrt{x} :: m : n$, from which $(a + x) : x :: mm : nn$, or $a : x :: (mm - nn) : nn$; therefore, $x = \dfrac{nna}{mm-nn}$ and $a + x = \dfrac{mma}{mm-nn}$; therefore, we see that the height due to the velocity of the water at LM is, in this way, $\dfrac{mma}{mm-nn}$, certainly after an infinite quantity of water has already flowed out; but above we had the same height, or

$$v = \frac{mma}{mm-nn}\left[1 - e^{(n^3-nmm)x/mmN}\right],$$

where if one sets $x = \infty$ (for in an infinite time an infinite quantity flows through), the exponential term vanishes if only m is greater than n, and thus equally there appears $v = \dfrac{mma}{mm-nn}$. That agreement is remarkable, because the paths which we followed are greatly different. On the other hand, if m is not greater than n, the motion never becomes steady, not even after an infinite time, for the velocity then increases to infinity, while otherwise the height of the velocity never surpasses the height $\dfrac{mma}{mm-nn}$. Therefore, concerning the latter cases there is nothing we may say.

Scholium 2

§5. Now another question occurs here worthy of being noted: indeed, what can be the mechanical method of filling in order that the vessel above remain full to the required height during the entire flow? That Problem would be difficult on account of the inconstancy of the desired height unless a peculiar artifice were applied here, which I shall now treat.

However, it goes beyond the fact that the water in the minimum space *RSDG* undergoes no compression, either positive or negative, because from the hypothesis it is moving at a common velocity with the water just below, and thus no particle tends to propel or retain any other.

Therefore, let each vessel be made as I said, and let a pipe be attached to the vessel above (for no other purpose than demonstration did we previously consider it separated), but let the pipe have a small opening at the highest point *a* (Fig. 30), to which the short pipe

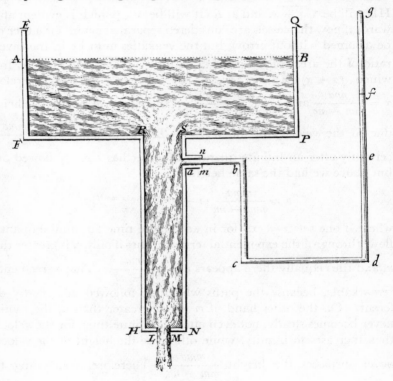

FIGURE 30

am corresponds; in this short pipe let the curved glass tube *abcdg*, sealed at the mouth *mn* by wax, be inserted; let the horizontal *ae* be constructed, and the point *e* be marked. With these things prepared

MOTION OF WATER FROM CONSTANTLY FULL VESSELS

thus, one will have to make the highest level of the water during the entire experiment remain constantly at the point e; and for this, one will see that it is required that at the beginning the surface of the water be near the base FP, further, that it be elevated continuously, and finally that after some time, albeit infinite, it never surmounts the height $\dfrac{nna}{mm-nn}$; but it will be easy to moderate the pouring in of water so that the surface does not diverge very much from the point e, if only the circumstances are not so matched that the water is to be added overly quickly at the beginning.

But thus, if one notices that the surface in the small tube is elevated above the point e, one must restrain the pouring in a little, which I will show is to be done elsewhere; if it should be the opposite, one must pour in the water more abundantly.

That kind of experiment, of which sort I performed often, involves no difficulty, but, lest any error creep into the experiment, the capillary effect of the glass tube is to be examined; one finds this effect if, after the orifice LM has been blocked, and before the tube has been moistened, the cylinder is filled with water right up to the summit, and thus one finds that the surface of the water in the tube extends right up to f, a point certainly higher than e; however, one substitutes this point f for that about which we just spoke in order to disregard the character of capillary tubes.

Therefore, in this way the filling will be properly done according to the rule of our hypothesis, and finally experiments can be performed concerning this motion. But thus, after we have explained the matter freely enough, I think it is unnecessary to warn that the vessel above does not pertain otherwise to the cylindrical vessel below, which we consider alone, than to the extent that the cylinder there is kept full in that manner in which it must be done, and thus by m is not to be understood the area of the upper vessel, but the area of the orifice RS, which, especially to us, is the surface of the water, since the water above RS serves only for the proper supplying to the cylinder below.

SCHOLIUM 3

§6. Here I must not overlook the fact that thus a case occurs which pertains to *hydraulico-statics*, concerning which science I advised certain things in Chapter I, §8: certainly we know now at what velocity the water must flow by at a in order that its pressure against the sides of the pipe be precisely null. But while I was writing these things, I had already formulated the general laws of *hydraulico-statics*, and I saw not without pleasure that that case deduced as a corollary from a clearly different theory acquires a similar solution from the general theory. Thus all things conform everywhere to a mutual relationship, and they show a legitimate application of the principles.

Scholium 4

§7. Now certain things follow concerning another method of adding water. Let the cylinder *RHNS* be assumed as any vessel whatever, and let it be kept constantly full by lateral pouring in; this can be done by injecting a sufficient quantity of water through the short pipe *ma*; but although this does not occur without motion, nevertheless, because it is horizontal here, soon all is removed, and on its own it neither advances the flow through the cylinder nor retards it; but there is, furthermore, another method which we find, after the calculation has been performed correctly, to reduce to the same: namely, if we consider the vessel *EFPQ* infinitely large, and we understand the base of it [to be] continuously covered with water, but such that the height of the water in the vessel above is to be taken as infinitely small, the vessel above will furnish water to the pipe attached to it, and no other motion will arise thereby than from lateral pouring in, if only the orifice *RS* always remains covered over; but it easily occurs that a certain cataract is formed there if the orifice *LM* is large and the pipe *RSNH* is long. Since here the other method must put forth the same effect as the former in the motion of water, everyone sees from this that in each method the inertia of all the water flowing into the pipe is to be exceeded by [that of] the water below. But the same could also be shown a priori by inquiring into the motion which must arise thence according to the equation of §8, Chapter III, which is this:

$$N\,dv - \frac{mmvy\,dx}{nn} + \frac{mmv\,dx}{y} = -yx\,dx.$$

But it will be accommodated to the present case if for m, x, and $-dx$ one substitutes, respectively, n, a, and $\dfrac{n\,dx}{y}$ (the reason for which act will be evident if one matches these with the others), and at the same time one sets y to be infinite; for then the third term of the equation vanishes, and it occurs indeed that for the present treatment above we find

$$N\,dv + nv\,dx = na\,dx.$$

MOTION OF WATER FROM CONSTANTLY FULL VESSELS 111

After we showed in these scholia the nature of each motion, as much as a simple physical consideration of the matter permits, and the difference between them, and at the same time we treated the mechanical method of advancing those hypotheses to law, it remains that the rest of the rather notable phenomena be indicated as well, which I will now do.

§8. COROLLARY 1. If in the vessel $RSNH$ the entire base is absent, the orifice LM will equal the orifice RS; the latter can even exceed the former if indeed the sides of the vessel diverge. But in these cases the height v has no limit in the equation

$$v = \frac{mma}{mm - nn} \left[1 - e^{(n^3 - nmm)x/mmN}\right]$$

and it becomes infinite if the quantity of ejected water, indicated by nx, is infinite.

This indeed is evident intrinsically from the equation when n is greater than m; but when the areas of the orifices are equal, one is to return to the differential equation of §3, from which the next equation was deduced, namely,

$$\frac{N}{M} dv - \frac{n^3}{mmM} v\,dx + \frac{n}{M} v\,dx = \frac{n}{M} a\,dx,$$

which, for $n = m$, gives $N\,dv = na\,dx$, that is, $v = \frac{nax}{N}$, where v is manifestly infinite if x is infinite.

§9. COROLLARY 2. But if the proposed vessel has a base and an orifice in it, the area indicated by n of which is less than the area of the orifice RS, expressed by m, v has a value which indeed it never attains, but nevertheless reaches approximately, and to which it converges so quickly after a minimum perceptible time of flow that it is not noticeably different unless by special effort vessels contrived contrary to this matter are furnished. But that term is as follows: $v = \frac{mma}{mm - nn}$; therefore, in the case of Scholium 2 of §5, the last term PB is $v - a = \frac{nna}{mm - nn}$. I will illustrate by example the very quick accession of the velocity to its ultimate limit, after I have applied the equation between v and the time corresponding to the height v.

§10. COROLLARY 3. In the case of the pouring which we call lateral, the ultimate height becomes $v = a$, whatever ratio may exist between both orifices of the vessel.

§**11.** COROLLARY 4. If the vessel is cylindrical and its length is set equal to b, there occurs (see §3) $\mathcal{N} = \dfrac{nnb}{m}$; but let it be noted that the values of the letters a and b are not to be confused, for the first expresses the height of the uppermost orifice above the lower, the other the length of the conduit. And so the values thus agree with each other at least in that case in which the axis of the vessel is a straight line and vertical; but if the axis is tortuous, or at least not vertical, they differ from each other. Therefore, I wished to advise this expressly, lest anyone be misled by the shapes of the vessels, the axes of which I made straight and vertical everywhere. If, therefore, for cylindrical vessels one sets $\mathcal{N} = \dfrac{nn}{m} b$, for vertical pouring in

$$v = \frac{mma}{mm - nn}\left(1 - e^{(nn-mm)x/mnb}\right)$$

and for lateral, $v = a(1 - e^{-mx/nb})$.

PROBLEM

§**12.** To find the velocity of the water flowing out from a constantly full vessel after the flow has taken place for a given time.

SOLUTION. With the hypotheses and all the notation retained which we applied in §**3** and, further, with the time elapsed from the beginning of flow having been taken equal to t, we will have to change the equations given in that paragraph into others which express the relation between t and v, after the quantities x and dx have been eliminated. But the element of time difference dt is proportional to the very small space dx which it passes through divided by the velocity $\sqrt{v}$; therefore, we will set $dt = \dfrac{\gamma\, dx}{\sqrt{v}}$, and thus the equation

$$dx = \mathcal{N}\, dv \bigg/ \left(na - nv + \frac{n^3}{mm} v\right)$$

which was given for determining the required velocity for vertical pouring in, will be changed into the following:

(I) $\qquad dt = \mathcal{N} \cdot \gamma \cdot dv \bigg/ \left(na\sqrt{v} - nv\sqrt{v} + \dfrac{n^3}{mm} v\sqrt{v}\right);$

but the other, serving for lateral pouring in, namely, $dx = \mathcal{N}\, dv/(na - nv)$, is changed into the following after the same substitution:

(II) $\qquad dt = \mathcal{N} \cdot \gamma \cdot dv / (na\sqrt{v} - nv\sqrt{v}).$

But these equations, integrated in the required manner, give for the first,

(α) $$t = \frac{mN\gamma}{n\sqrt{mma - nna}} \cdot \ln \frac{m\sqrt{a} + \sqrt{mmv - nnv}}{m\sqrt{a} + \sqrt{mmv - nnv}}$$

and for the other, which is deduced from the former, after one has set $m = \infty$,

(β) $$t = \frac{N\gamma}{n\sqrt{a}} \cdot \ln \frac{\sqrt{a} + \sqrt{v}}{\sqrt{a} - \sqrt{v}}.$$ Q.E.I.

Scholium

§13. If the vessel which is being discussed is cylindrical and twisted and inclined in any way whatever, and the length of it is set equal to b, the height of the aqueous surface above the orifice remaining equal to a, there will be again, as in §11, $N = \frac{nn}{m} b$.

But since, as is well known, $2\gamma\sqrt{A}$ expresses the time which a body takes in falling freely and from rest through the height A, it is evident that the quantity $\frac{2mN\gamma}{nn\sqrt{a}} \left(= 2\gamma\sqrt{\frac{bb}{a}} \right)$ expresses the time in which a body beginning to be moved from rest descends freely through the height $\frac{bb}{a}$; we will accept that time as a common measure and we will set the same equal to θ, and the equation (α) for vessels or cylindrical conduits will be changed into this:

$$t = \frac{n\theta}{2\sqrt{mm - nn}} \cdot \ln \frac{m\sqrt{a} + \sqrt{mmv - nnv}}{m\sqrt{a} - \sqrt{mmv - nnv}};$$

the other suitable one, designated by (β), becomes the following:

$$t = \frac{n\theta}{2m} \ln \frac{\sqrt{a} + \sqrt{v}}{\sqrt{a} - \sqrt{v}},$$

from each of which it appears that the water cannot but acquire almost the full velocity in a very short time, and this all the more quickly the larger is the pipe, the shorter, and the more nearly vertical; and that the accelerations are not perceptible in any way, unless the aqueducts are made very long; and then also almost all grades of accelerations are passed through in a short time, each of which I shall now illustrate by an example.

I. The time is sought in which fluid from a constantly full vertical cylinder, 16 English feet long, the diameter of which is five times the diameter of the orifice, acquires a velocity which is due to the height $\frac{99}{100} a$ according to the hypothesis to which the second equation pertains; thus $\frac{n}{m} = \frac{1}{25}$, $v = \frac{99}{100} a$, and $b = a$, whence the time which a body takes in falling freely through the space $\frac{bb}{a}$ or θ equals one second; hence t becomes $\frac{1}{50} \ln 399$, that is, approximately the ninth part of one second, which short time is indeed imperceptible. But when the time is considered notable, the changes of the heights v become unnoticeable. If the similar time (in which certainly the velocity due equally to $\frac{99}{100}$ of the height which results after an infinite time is generated) is sought in the first hypothesis, namely, the time in which there is obtained $v = \frac{99}{100} \cdot \frac{mma}{mm - nn}$, then that is found to be slightly greater than the preceding, but of unnoticeable excess; whence it is evident in vessels of this sort that the water cannot be poured quite quickly enough into the upper vessel to satisfy the hypothesis, and therefore, by reason of the same, that other experiments cannot be performed with regard to the hypothesis in order to show whether the height *BP* in Fig. 30 is truly as great as it should be by virtue of §5, in order that the point *e* or *f* retains the position during flow which it had before flow when the orifice *LM* was blocked off and no water existed in the upper vessel.

II. Now the same time is sought again for the second hypothesis, if the pipe is of the same area and has the same orifice attached to it and has an oblique position and a length *b* of 184 poles or 1104 Paris feet, while the height of the aqueous surface above the orifice of efflux is 16 Paris feet. Thus it will occur that $b = 1104$, $\frac{bb}{a} = 76{,}176$, and $\theta = 72$ seconds, more or less, from which the mean desired time is between eight and nine seconds, which certainly is noticeable enough. But if the time is desired in which the height v is equal to only a fourth part of the height a, that will be found equal to $\frac{72}{50} \ln 3$, or approximately one second and a half.

I do not know whether these things agree with those observed by Mariotte to which he refers in his *Traité du mouvement des eaux*, Part V, Disc. 1, where he mentions a certain leaping fountain, which is at

MOTION OF WATER FROM CONSTANTLY FULL VESSELS

Chantilly, to which the water descends through a conduit 184 poles long, if only I concluded correctly from the antecedent [discussion], and the greatest height of the aqueous surface above the orifice of efflux had been indicated by a as 16 feet; the diameter of the aqueduct was 5 inches, but the orifice had a diameter of one inch. Mariotte seems to me to speak thus, as if the accelerations had been much slower than is indicated by our formula, although I do not know whether this is to be attributed to the fact that the water perhaps may have had some other exit in addition to the orifice which is being discussed here, or that the aqueduct had not been full of water when flow was beginning, which many things occurring afterwards caused me to believe; if it has been neither, I believe that the phenomena, as they had been observed by Mariotte and can be observed again daily, clearly agreed with our calculation. Finally, these words are Mariotte's: "*Furthermore, this,*" he affirms, "*happens to the same individual thrust: when an orifice has been blocked by hand for a space of time of ten or twelve scruples of a second, and afterwards the same has been opened, water does not burst forth at once, but, surging slowly, the thrust ascends to 3 inches, afterwards to a height of a foot, and finally to two feet, at successively noticeable intervals. . . . But, nevertheless, finally the water springs forth with its full impetus.*"

Problem

§14. To find the quantity of water flowing through a given vessel, constantly full, in a given time.

SOLUTION. With the positions and notation of §§3 and 12 applied again, now the equation between x and t is to be found; but because, as we saw in §12, dt is $\dfrac{\gamma\, dx}{\sqrt{v}}$, $\sqrt{v}$ will be $\dfrac{\gamma\, dx}{dt}$, and this value has to be substituted in the integrated equations which we gave in §3; the former of these equations was this:

$$v = \frac{mma}{mm - nn}\left[1 - e^{(n^3 - nmm)x/mmN}\right],$$

which, according to the previously followed custom is changed into this:

(I) $$\frac{\gamma\gamma(dx)^2}{(dt)^2} = \frac{mma}{mm - nn}\left(1 - e^{(n^3 - nmm)x/mmN}\right).$$

The other of the mentioned equations from §3 was the following:
$$v = a(1 - e^{-nx/N})$$
which therefore supplies the following in the present case:

(II) $$\frac{\gamma\gamma(dx)^2}{(dt)^2} = a(1 - e^{-nx/N}).$$

Now the equations (I) and (II) have to be integrated, which is certainly easy, and because the former contains the latter (for each is the same if $m = \infty$), we will treat that one alone, and we will now consider it in the following form:

$$dt = \frac{\gamma\sqrt{mm - nn}}{m\sqrt{a}} \cdot dx \bigg/ \sqrt{1 - e^{(n^3 - nmm)x/mmN}}.$$

But let it be established, so that the method of integration is explained all the more, that $e^{(n^3 - nmm)x/mmN} = z$, and, accordingly, $dx = \frac{mmN\,dz}{(n^3 - nmm)z}$, from which, for the sake of brevity, the constant quantity $\frac{\gamma\sqrt{mm - nn}}{m\sqrt{a}} \frac{mmN}{n^3 - nmm}$, or $\frac{-\gamma mN}{n\sqrt{(mm - nn)a}}$, is indicated by α, and there will result $dt = \frac{\alpha\,dz}{z\sqrt{1 - z}}$, in which, if in addition it occurs that $1 - z = qq$, or $z = 1 - qq$, $dz = -2q\,dq$, there develops

$$dt = \frac{-2\alpha\,dq}{1 - qq} = \frac{-\alpha\,dq}{1 + q} - \frac{\alpha\,dq}{1 - q}$$

the integral of which is

$$t = \alpha \ln(1 + q) + \alpha \ln(1 - q) = \alpha \ln\frac{1 - q}{1 + q}.$$

And there is no need of a constant, since indeed from the nature of the matter t and x must vanish simultaneously; but for $x = 0$ it occurs that $z = 1$ and $q = 0$; therefore, equally t and q must begin simultaneously from zero, which condition the derived equation $t = \alpha \ln\frac{1 - q}{1 + q}$ satisfies. It remains that we reassume the original values in reverse order; in fact, it thus occurs that

$$t = \alpha \ln\frac{1 - \sqrt{1 - z}}{1 + \sqrt{1 - z}}$$

or

$$t = \frac{\gamma m N}{n\sqrt{(mm-nn)a}} \ln \frac{1 + \sqrt{1-z}}{1 - \sqrt{1-z}}$$

or, finally,

(I) $\quad t = \dfrac{\gamma m N}{n\sqrt{(mm-nn)a}} \ln\left[1 + \sqrt{1 - e^{(n^3-nmm)x/mmN}}\right]$
$\quad\quad - \ln\left[1 - \sqrt{1 - e^{(n^3-nmm)x/mmN}}\right].$

And that equation, after one has set $m = \infty$, gives the other desired equation,

(II) $\quad t = \dfrac{\gamma N}{n\sqrt{a}} \ln\left[1 + \sqrt{1 - e^{-nx/N}}\right] - \ln\left[1 - \sqrt{1 - e^{-nx/N}}\right]$

Q.E.I.

§15. COROLLARY 1. If one sets $x = \infty$, in order that the nature of the matter might appear when an infinite quantity of water has already flowed through, and m is assumed greater than n, just as is mostly customary, the exponential quantity is to be considered to vanish in both places if the logarithm has been assumed positive, and ln 2 will occur on either hand. But indeed, if the logarithm has been assumed negative, it is to be stated that

$$\sqrt{1 - e^{(n^3-nmm)x/mmN}} = 1 - \tfrac{1}{2}e^{(n^3-nmm)x/mmN}$$

and accordingly

$$\ln\left[1 - \sqrt{1 - e^{(n^3-nmm)x/mmN}}\right] = \ln \tfrac{1}{2}e^{(n^3-nmm)x/mmN}$$
$$= \frac{n^3 - nmm}{mmN}x - \ln 2.$$

If these substitutions are properly made for the first method of pouring in which we devised,

(I) $\quad t = \dfrac{\gamma m N}{n\sqrt{(mm-nn)a}} \left(2\ln 2 + \dfrac{mmn - n^3}{mmN}x\right)$

which, for $m = \infty$, gives for the other case

(II) $\quad t = \dfrac{\gamma N}{n\sqrt{a}}\left(2\ln 2 + \dfrac{n}{N}x\right).$

It follows from the above formulas: that certainly water flows at a lesser quantity than if it would flow from the beginning at the entire velocity which it acquires in either case after an infinite time; that the difference, however, never surpasses a certain limit; and that after an infinite time it is described in finite terms.

§16. COROLLARY 2. When we convert the derived equations, we obtain

(I) $$x = \frac{2mmN}{mmn - n^3}\left[\ln(1 + e^{-t/\alpha}) - \ln 2 + \frac{t}{2\alpha}\right]$$

(II) $$x = \frac{2N}{n}\left[\ln(1 + e^{-t/\beta}) - \ln 2 + \frac{t}{2\beta}\right]$$

where, as above, $\alpha = \dfrac{-\gamma mN}{n\sqrt{(mm-nn)a}}$, and $\beta = \dfrac{-\gamma N}{n\sqrt{a}}$.

If in addition, as in the last corollary, one sets $t = \infty$, unity vanishes with respect to the exponential quantities, which are infinite beyond all degree, and there results

$$\ln(1 + e^{-t/\alpha}) = -\frac{t}{\alpha}, \quad \text{and} \quad \ln(1 + e^{-t/\beta}) = -\frac{t}{\beta};$$

from this then, after the values of the letters α and β have been considered again,

(I) $$x = \frac{mt\sqrt{a}}{\gamma\sqrt{mm - nn}} - \frac{2mmN}{mmn - n^3}\ln 2$$

(II) $$x = \frac{t\sqrt{a}}{\gamma} - \frac{2N}{n}\ln 2.$$

Therefore, if suddenly from the beginning of flow the water would flow constantly on either hand at the entire velocity which it can acquire, its quantity should not exceed after an infinite time the quantity corresponding to the theory for the same time except by the very small quantity which in the first case is expressed by $\dfrac{2mmN}{mm-nn}\ln 2$ and in the second by $\dfrac{aN}{n}\ln 2$. But if in place of an infinite time one takes a time of only a few scruples of a second, the same *theorem* will hold approximately, so that, for instance, if after the first ten seconds the quantity Q has flowed out, $Q + \dfrac{2mmN}{mmn - n^3}\ln 2$ will thereafter flow out in approximately the same number of seconds, or in the other case, $Q + \dfrac{2N}{n}\ln 2$.

Scholium

§17. The motion of water through siphons pertains as well to the theory set forth thus far. However, the theory indicates that the axis of a siphon can be inflected in any way whatever and the motion of the water thence will not be altered, if only the height of the aqueous surface above the orifice of efflux remains the same; since in addition aqueducts, siphons or *diabetes*, and other vessels of this sort are usually cylindrical, it is to be established, as I advised in §13, whenever this applies, that $N = \frac{nn}{m} b$, by understanding that b is the length of the conduit or siphon; also in the formulas of §§14, 15, and 16, the quantities are to be so interpreted, where the question concerns time, that $2\gamma\sqrt{A}$ represents the time which a body having begun from rest uses in a descent through a vertical height A.

For the rest, as I said in passing, the theory of this chapter indicates nothing unique which falls under observation except in very long aqueducts, greatly oblique to the horizontal, and having not very small orifices; for these three things combine to the end of retarding and thus effecting noticeable accelerations, the measures of which very strongly support the theory.

Nevertheless, even in these circumstances some average is to be observed lest the hindrances developing from the adhesion of the water are excessive.

As far as the pouring in of water is concerned, I seemed to notice for myself that, if it occurs vertically and with impetus, the motion hence is so far from being accelerated that, rather, it is retarded, unless the pouring in of water occurs equally over the whole surface, in the matter which I showed in §4; for, if it is poured in otherwise, the motion of the water in the vessel is disturbed, and this disordered motion retards the efflux.

§18. Finally, to this point there pertain to some extent the experiments undertaken by the celebrated Giovanni Poleni, as he reports in the first book of *De Motu Aquae Mixto*, p. 12ff., which, therefore, I thought should be brought forth here, since they show splendidly that everywhere the ultimate speed in constantly full vessels is that which agrees with the entire height of the water, if the vessels are not submerged, or with the difference of the heights of internal and external water in submerged vessels, although as for the rest there is nothing in them thus far which is new, because there no accelerations are considered.

Take a cylinder of length as if infinite, the axis of which has a vertical position; let the base be whole; but in a wall let there be a fissure parallel to the axis, rendering an orifice [in the form] of a rectangular parallelogram which extends from the base right up to the summit of the cylinder. Further, consider water to be poured into the cylinder steadily, so that in equal times equal quantities are injected; the water flows from the cylinder through the fissure; nevertheless, from the beginning it does not flow at the same quantity at which it is poured in from above, but at a lesser one; therefore, the surface of the water in the cylinder surges right up to a certain height asymptotically; but if this limit is now known to be reached, the height of the water will remain unchanged, and the water flows out constantly at the same quantity at which it is poured in. It appears also that the greater will be the height of the water in the cylinder, the more [water] is poured in. And so there is sought, for increased quantities of water to be poured in in a given time, in what ratio the heights to which the water surges in the cylinder must increase.

The solution is this: Let the height of the water, when it is in a permanent state, be α, and let the part which is x be terminated at the surface, together with the differential dx; let the width of the crack be n, and we will have [something] just like an orifice of area $n\,dx$, through which water flows at a velocity $\sqrt{x}$, therefore, the quantity of water flowing out here in a given time is, for instance, $n\,dx\,\sqrt{x}$, the integral of which is $\frac{2}{3}nx\sqrt{x}$; this expresses the quantity of water flowing out in a given time through the partial length x of the crack; and thus the quantity of water pouring out in the same time through the entire crack will be expressed by $\frac{2}{3}n\alpha\sqrt{\alpha}$; but only as much flows out as is poured in; hence if the quantity of water poured in in that given time is called q, there will be $\frac{2}{3}n\alpha\sqrt{\alpha} = q$. This indicates that the quantities of water to be poured in in that given time follow a three-halves-power ratio to the heights to which the water ascends from the base of the cylinder, or on the other hand that the heights follow a cube-root ratio to the squares of the quantities at which the water is poured in in a given time.

§**19.** With this problem having been solved, I come to the other considered by the celebrated Poleni.

Let the cylinder be the same, but submerged in water standing in a trench, as if in an infinite vessel, and let the depth of submersion be designated by a; now, with the same things having been assumed as before, again the equation is sought between the height α of the internal aqueous surface above the external, and the quantity q to be poured in in a given time.

MOTION OF WATER FROM CONSTANTLY FULL VESSELS 121

With respect to that portion of the crack α which ejects water and rises above the external water, we already saw that it expends the quantity $\frac{2}{3}n\alpha\sqrt{\alpha}$ in a given time; but the remaining submerged portion of the crack transmits water everywhere at a common velocity, as will be evident from things to be discussed below, and certainly at the velocity $\sqrt{\alpha}$, so that, after this velocity has been multiplied by the magnitude of the submerged crack na, the quantity $na\sqrt{\alpha}$ is obtained which it ejects in a given time. If both quantities are gathered in a sum, it will result that $(\frac{2}{3}\alpha + a)n\sqrt{\alpha} = q$.

With the help of this equation, q is known from the given heights a and α; or, on the other hand, the height α [is known] from the known quantities a and q.

Moreover, the very celebrated author himself, whose solution does not differ from this of ours, shows that this equation agrees very accurately with the experiments. It follows from the equation that the elevations α are the greater for the same pouring in of water, the less is the depth of submersion a.

EXPERIMENTS WHICH PERTAIN TO CHAPTER V

Pertaining to §5

I used the vessel described in §5 with a glass tube (Fig. 30). But first I covered the orifice LM and filled the pipe RN with water until its surface touched the little opening at a; then I observed that the water having gone into the tube reached the point f at the extremity; after the orifice LM had been opened and the water was flowing, I poured new [water] into the vessel $EFPQ$ above, care having been employed so that the extremity of the water at f meanwhile neither ascended nor descended. While these things were occurring, the surface AB was elevated, but it never exceeded a certain limit; indeed, the maximum height PB or FA, as far as I could see, was $\dfrac{nn}{mm-nn}a$, with $\dfrac{n}{m}$ denoting the ratio between the lower orifice LM and the upper RS, and a the vertical height of the latter orifice above the other.

But this is the only experiment which I myself undertook; although there are many propositions contained in this section which would merit attention, and these sufficiently unexpected, nevertheless I was not able to perform experiments concerning them; for things in the shorter vessels are so composed that whatever unusual [property] they possess escapes observation; moreover, I was not able to prove the matter aptly in long aqueducts; when the opportunity will arise for others to examine this theory, they should turn their attention to the following:

I. In leaping fountains the whole height of the thrust may be observed; after the orifice has first been closed, and the same soon opened, the quantity of water which flows out may be seen, while the water reaches half the height of the entire thrust, or any part whatever, which certainly will happen in the shortest time; let the measure of that quantity be the length of the cylinder erected above the orifice through which the water springs forth, which length we have called x, but we have called the entire height of the thrust a, and the observed height of the thrust which has not yet reached the total height we have designated by v. Then at last, after the calculation has been performed, let it be examined whether these quantities correspond correctly to the equations for either method of pouring in shown in §3.

II. Let all things be done as before, with this difference only, that in place of the quantity flowing out let the time of efflux be noted, so that thus the formulas of §13 can be examined, and finally let the quantity be compared with the time of flow, in order that this may show whether it compares properly to the formula of §14.

III. Then expecially let that kind of experiment be performed which I indicated in §16, by observing certainly the quantities of water corresponding to the half times; but I said that, however great a time is taken, the difference of these quantities never equals $\dfrac{2mmN}{mmn - n^3} \ln 2$ in the former method of pouring in which we established, or $\dfrac{2N}{n} \ln 2$ in the latter. But these differences, although they will never develop perfectly, will nevertheless be reached closely in a very short time.

The remaining things in this section are Corollaries and Scholia; anyone will see easily how they can be subjected to experiments. But let me wish, before judgment is made, that attention be paid to all circumstances with regard to the hindrances, the contraction of the stream, and other things which I do not care to repeat everywhere. To §§17 and 18: For the confirmation of the problem of §17 pertaining to vessels not submerged, see the experiments on p. 26, *lib. cit.* of G. Poleni.

But since in the submerged vessel the height a should be 55 Paris lines (which height is called dead by him), he undertakes five experiments in which the height which he calls live, or α, was successively

$8\frac{3}{4}$, 25, 42, 58, and $73\frac{1}{2}$ lines; after these values have been substituted in the equation shown in §18, it follows that the quantities of water poured in in a given time were as 100, 199, 299, 396, and 495; actually they had been poured in in proportion as 100, 200, 300, 400, and 500; the difference is so little that it cannot be doubted but that the agreement would have been perfect if all measurements had been taken most correctly.

Also, the remaining experiments undertaken by that celebrated man agree perfectly with the theory; the calculation of them is seen among the works of that same Author. But I undertook in the interest of the matter to include them here because they pertain to the argument of this very section, although as for the rest, I may say freely that I long more for those experiments which by calculation depend on a change of instantaneous [conditions], considered to this time by no one that I know, rather than those which assume the permanent state.

SIXTH CHAPTER

Concerning Fluids not Flowing out but Moving within the Walls of Vessels

§1. Up to this point we have considered water flowing out; but now we will contemplate the motions of water which does not flow beyond the bounds of vessels. Let me reduce all these motions to two kinds, each to be treated separately:

First: When the fluid in an infinitely long pipe is moved continuously in the same direction.

Second: When it is driven in reciprocal or oscillatory motion.

CONCERNING THE MOTION OF WATER THROUGH INDEFINITELY LONG CONDUITS

CASE I

§2. First let there be a conduit placed horizontally but varying with respect to [cross-sectional] areas according to some given law; let a fluid be so placed in it, as it customarily occurs in rather narrow pipes, that both end surfaces obtain an alignment perpendicular to the axis of the conduit and thus begin to be moved at a certain given velocity. If these things are so, and clearly no impediments to the motion are assumed to be present, it is obvious that there will be no end to the motion, in the same way that a sphere progressing freely on a horizontal table continues its motion without end. Nevertheless, a significant difference arises between the two motions; namely, all portions of the sphere progress continuously at a uniform velocity; in water they perpetually change motion. And it will not be difficult to define that motion, if we will consider that the motion must be such that the same *potential ascent* of the entire water which existed at the beginning of motion is conserved. But we have determined the *potential ascent* of water moved at a certain velocity in any conduit

FLUIDS MOVING WITHIN THE WALLS OF VESSELS

whatever in §2 of Chapter III: therefore, nothing further remains with respect to the solution of the problem. Nevertheless, it will not be useless to have introduced one or another example of this matter.

EXAMPLE 1. Let there be, for example, the conduit *BgfC* (Fig. 31),

FIGURE 31

which has the shape of a truncated cone; let the portion *BGFC* of it be considered filled by fluid moved toward *gf*; and let the particles of fluid at *GF* have a velocity due to the height v; and, finally, let the fluid have arrived at the position *bgfc*. With these things established, the velocity of the fluid at *gf* is sought. Moreover, I will assume that V equals the height due to the velocity of the water at *gf*. Let the vertex of the cone be at H, the diameter at $BC = n$, the diameter at $GF = m$, the length $BG = a$; if $GB = b$, the diameter *gf* will be $\frac{ma - mb + nb}{a}$. Hence, because the solid *BGFC* is equal to the solid *bgfc*,

$$(BC)^2 \cdot BH - (GF)^2 \cdot GH = (bc)^2 \cdot bH - (gf)^2 \cdot gH,$$

from which

$$(bc)^2 \cdot bH = (BC)^2 \cdot BH - (GF)^2 \cdot GH + (gf)^2 \cdot gH;$$

but $bH = \frac{BH}{BC} \cdot bc$, therefore,

$$(bc)^3 = (BC)^3 - \frac{(GF)^2 \cdot GH \cdot BC}{BH} + \frac{(gf)^2 \cdot gH \cdot BC}{BH}$$
$$= (BC)^3 - (GF)^3 + (gf)^3$$

or

$$bc = \sqrt[3]{n^3 - m^3 + \left(\frac{ma - mb - nb}{a}\right)^3}.$$

But, from §3, Chapter III, the *potential ascent* of the water in the position *BGFC* is $\frac{3m^3 v}{n(mm + mn + nn)}$; and similarly the *potential ascent* of the same water in the position *bgfc* is found equal to

$$\frac{3\alpha^3 v}{\beta(\alpha\alpha + \alpha\beta + \beta\beta)},$$

where, for the sake of brevity, α and β have been inserted for the determined values of the diameters gf and bc. Therefore,

$$V = \frac{m^3 \cdot (\alpha\alpha + \alpha\beta + \beta\beta)\beta v}{\alpha^3 \cdot (mm + mn + nn)n}.$$

From this formula it is easily understood that the anterior particles are moved at a continuously greater velocity, the posterior at a lesser and such that, if the little orifice gf is considered infinitely small, the velocity of the water at gf becomes infinite and at bc infinitely small.

EXAMPLE 2. Let there be a conduit composed of two cylindrical pipes BN and OP (Fig. 32) of unequal area; in the wider branch let

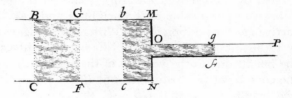

FIGURE 32

the fluid $BGFC$ be assumed to be moved toward P at a velocity which corresponds to the height v. Thus it is obvious that no change will occur in the motion before the surface GF will have reached MN; but from this point of time the motion will be varied continually until all the fluid will have entered the narrower pipe. Therefore, one seeks the velocity of the surface fg when the fluid occupies the position $bgfc$; moreover, we will designate the height required for this velocity by V.

Let the diameters GF and gf be as n and m; let the length $BG = a$; [if] $bM = b$, $Og = \frac{nn}{mm}(a - b)$; the *potential ascent* of the water at $BGFC$ is v; the *potential ascent* of the water at $bgfc$ is

$$\frac{n^4 a - n^4 b + m^4 b}{n^4 a} V;$$

therefore,

$$V = \frac{n^4 a}{n^4 a - n^4 b + m^4 b} v.$$

From these [relationships] it is understood that the velocity of the first drop bursting forth into the narrower pipe corresponds to the height $\frac{n^4}{m^4} v$, but that this velocity decreases very quickly, so that, after a very small portion of fluid has flowed through, it can then be considered that $V = \dfrac{a}{a-b} v$ and after all the fluid has flowed through, it assumes the former velocity. For example, let the diameter of the wide pipe be ten times the other, and the first drop will flow from the wider pipe into the narrower at a velocity due to the height $10,000\ v$; but if one assumes that $\tfrac{1}{10}$ of the fluid has already flowed through, one will find that the height which conforms to the velocity of the fluid moving in the narrower pipe is approximately equal to $\tfrac{10}{9} v$.

If one seeks the time in which the transflux of the fluid Of occurs, one finds that this is equal to

$$\frac{2(n^4 a - n^4 b + m^4 b)^{3/2} - 2m^6 a \sqrt{a}}{3mm(n^4 - m^4)\sqrt{av}}.$$

Therefore, all the fluid flows through in the time

$$\frac{2n^6 a \sqrt{a} - 2m^6 a \sqrt{a}}{3mm(n^4 - m^4)\sqrt{av}} = \frac{2(n^4 + mmnn + m^4)a}{3mm(nn + mm)\sqrt{v}},$$

where by $\dfrac{a}{\sqrt{v}}$ the time is defined in which the fluid, moving freely in the wider pipe, traverses the distance a. But, as I said, these things will behave this way if there are no impediments to the motion, and at the same time the velocities in the full extent of the composite conduit are assumed reciprocally proportional to the areas. Meanwhile I have already showed elsewhere that the water near the boundary MN cannot satisfy this law. And so, since the situation is such, the more nearly the actual motion will agree with theory, the longer the portion bm will be and the fewer obstacles will be present.

§3. Thus, if now the conduit is placed not horizontally but obliquely to the horizon, it is evident that all things occur similarly except that the *potential ascent* of the water in every position is to be equated to the initial *potential ascent* augmented by the *actual descent*, that is, by the vertical descent of the center of gravity. And so if the water begins to move on its own without any impulse, the *actual descent* will simply be equal to the *potential ascent*.

Therefore the water keeps on progressing continuously as long as the center of gravity is located at a lower point than it was at the beginning of motion. But when the pipe has been so formed and curved and filled with such a quantity of fluid that the center of gravity can assume its previous height again, then the fluid will develop a retrograde motion and will oscillate without end. We will soon discuss that motion comprising the principal portion of this section. Meanwhile it may be observed that it can occur that all water flows on its own accord from a lower place through a higher without previous suction if only all things occur in the required manner.

CONCERNING THE OSCILLATIONS OF FLUIDS IN CURVED TUBES

Case II

§4. In the *Commentaries of the Imperial Academy of Science of St. Petersburg*, Book II, my Father presented certain theorems which manifest the significant use which the theory of live forces possesses in mechanical matters. Indeed that which was given third is as follows:

Let there be a cylindrical pipe ABCH (Fig. 33) open at either end, bent into

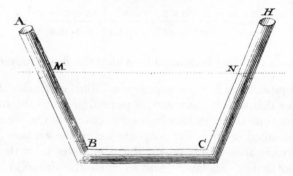

Figure 33

two legs BA and CH [attached] to the horizontal portion BC; let the sine of the angle ABC be p, and the sine of the angle HCB be q, with the total sine, certainly, being 1; further, let that pipe be filled with water right up to the horizontal MN, and let the length of the portion of the pipe MBCN filled with water be called L. All the oscillations of the fluid agitated within this tube, greater as well as lesser, will be tautochronous with and of the same duration as the very small oscillations of some simple pendulum the length of which is $\dfrac{L}{p+q}$.

The following is a corollary to this theorem, by the same author.

If ABC and HCB are right angles, which is the only case solved by Newton, the length of the simple pendulum which is isochronous to the oscillating water will be $\frac{1}{2}L$, *just as Newton found.*

§5. These are the things which have been communicated to the public up to this time on the oscillations of fluids, and certainly first

FLUIDS MOVING WITHIN THE WALLS OF VESSELS   129

by Newton, in order to show the nature of waves, and by my Father, in order to show the fruitfulness of the principle of *live forces*. But since it is our intention to give a more complete theory concerning the motions of water, it will be to the point to follow that type of argument to its full extent. Therefore, let me inquire by which ways the unequal oscillations of a fluid may become isochronous, and by which they do not. Then for the former I will give the length of the simple tautochronous pendulum, and for the other I will indicate the time of duration; moreover, I will consider pipes that are bent in any way whatever and unequally large.

LEMMA

§6. Let *cAd* (Fig. 34) be a leather bag or a conduit of any given shape whatever, full of water, terminating at either end in two

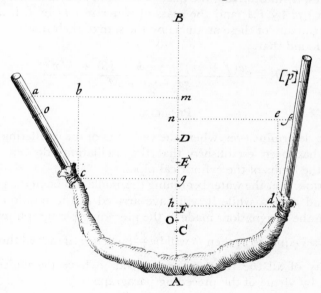

FIGURE 34

cylindrical conduits *ac* and *fd* inclined to the horizon in any way whatever and of any area whatever, one of which let me assume full of water up to *a*, the other up to *f*; let it be necessary to determine the height of the center of gravity of all the water from the given height of the center of gravity of the water contained in the bag *cAd*, with as many of the remaining things known in advance as is sufficient.

130   HYDRODYNAMICS, CHAPTER VI

SOLUTION. Let the center of gravity of the water contained in the vessel cAd be at C; and let it be understood that the vertical AB is drawn through that point C, and then let the horizontals am, cg, fn, and dh be drawn, together with the verticals cb and de. Let it be established that $ac = a$, $fd = \alpha$, $bc = b$, $ed = \beta$, the area of the pipe $ac = g$, and the area of the pipe $fd = \gamma$; further, let the aqueous mass or the volume of the conduit cAd be M, the line $Ag = f$, $Ah = \phi$, and $AC = m$. Let the lines mg and ng be divided in two at the points D and E, and thus the centers of gravity of the water contained in the cylindrical pipes will be at the heights of the points D and E.

After these things have been set forth, it occurs that $AD = f = \tfrac{1}{2}b$; $AE = \phi + \tfrac{1}{2}\beta$; the mass of water at ac is ga, and at fd it is $\gamma\alpha$. Therefore, if the sought center of gravity for all the water $acAdf$ is understood to be located at the height F, AF will be obtained, as is well known in mechanics, by multiplying the mass of water in ac by DA, the mass of water [in] fd by EA, and the mass of water in cAd by CA, and by dividing the sum of these products by the sum of their masses. From this it is found that

$$AF = \frac{ga(f + \tfrac{1}{2}b) + \gamma\alpha(\phi + \tfrac{1}{2}\beta) + Mm}{ga + \gamma\alpha + M}.$$

PROBLEM

§7. To determine everywhere the velocities of the oscillating water, after it has been established that the oscillations do not extend beyond the limits of the cylindrical pipes.

SOLUTION. Let the water beginning the oscillation be in the position $acAdf$, and after a while it will have arrived at the position $ocAdp$; and with the designations made in the preceding paragraph retained, let ao be set equal to x; then fp will be $\dfrac{gx}{\gamma}$; whence (if indeed the center of gravity of all the water is understood to have descended from F to O), by virtue of the preceding paragraph,

$$AO = \frac{g(a - x)\left(f + \tfrac{1}{2}b - \dfrac{bx}{2a}\right) + \gamma\left(\alpha + \dfrac{gx}{\gamma}\right)\left(\phi + \tfrac{1}{2}\beta + \dfrac{\beta gx}{2\alpha\gamma}\right) + Mm}{ga + \gamma\alpha + M}.$$

Hence the descent of the center of gravity or the *actual descent* is deduced

$$FO = \frac{(b - \beta + f - \phi)gx - \left(\dfrac{bg}{2a} + \dfrac{bgg}{2\alpha\gamma}\right)xx}{ga + \gamma\alpha + M}.$$

FLUIDS MOVING WITHIN THE WALLS OF VESSELS

Now let the velocity of the water in the pipe *ac* (namely, when the surface is at *O*) be such that it corresponds to the height v, and then the *potential ascent* of the water in the other pipe will be $\frac{gg}{\gamma\gamma}v$; similarly, the *potential ascent* of the water *cAd* will be proportional to the height v, and therefore we will set it equal to Nv (where N depends upon the shape of the bag *cAd* and can be determined through §2, Chapter III). But now if, after the *potential ascents* everywhere have been multiplied by their proper masses, the products are divided by the sum of the masses, the *potential ascent* of all the water *ocAdp* will be obtained as

$$\frac{\left(ga - gx + \frac{\alpha gg}{\gamma} + \frac{g^3 x}{\gamma\gamma} + MN\right)v}{ga + \gamma\alpha + M}.$$

And because this *potential ascent* is equal to the *actual descent FO* found a little earlier,

$$v = \frac{(b - \beta + f - \phi)gx - \left(\frac{bg}{2a} + \frac{bgg}{2a\gamma}\right)xx}{ga - gx + \frac{\alpha gg}{\gamma} + \frac{g^3 x}{\gamma\gamma} + MN}. \quad \text{Q.E.I.}$$

§8. COROLLARY 1. Because the line $mn = mg - nh + gh = b - \beta + f - m$, we will set $mn = c$, and at the same time we will multiply the denominator and the numerator by $2\gamma\gamma\alpha\alpha$. Thus indeed we will obtain

$$v = \frac{2g\gamma\gamma\alpha\alpha cx - (g\gamma\gamma\alpha b + gg\gamma\alpha\beta)xx}{2g\gamma\gamma\alpha a\alpha - 2g\gamma\gamma\alpha\alpha x + 2gg\gamma\alpha\alpha + 2g^3 a\alpha x + 2\gamma\gamma\alpha\alpha MN}.$$

§9. COROLLARY 2. If $v = 0$, it is evident that the value x then denotes the total displacement of the surface of the fluid in the pipe *ac*, which is thus found equal to

$$\frac{2\gamma a\alpha c}{\gamma\alpha b + ga\beta},$$

but in the other pipe it becomes

$$\frac{2ga\alpha c}{\gamma\alpha b + ga\beta}.$$

Therefore, the water in the narrower pipe can be elevated to any height whatever if only the ratio of the areas g and γ is taken large enough.

§10. COROLLARY 3. That portion of the vessel cAd which we assume is never reached by either of the surfaces contributes nothing to either increasing or diminishing those paths of the fluid; nevertheless, as is shown below, it can serve for accelerating and retarding the oscillations.

§11. COROLLARY 4. Let each pipe be assumed of a common size; there will be, namely, for $g = \gamma$,

$$v = \frac{2ga\alpha cx - (g\alpha b + ga\beta)xx}{2gaa\alpha + 2ga\alpha\alpha + 2a\alpha MN}.$$

In this case the maximum velocity of either surface occurs when they are located at the middle of the total displacement, but it occurs differently when the pipes are of unequal area.

It is to be noted also that the retardations and accelerations are similar to each other at similar distances of the surfaces from the points at the middles of the paths, that is, from the points of maximum velocities.

THEOREM

§12. When the areas of the cylindrical pipes are equal in the previously mentioned manner, greater as well as lesser oscillations will be Isochronous to each other if only the surfaces never descend below the orifices of these pipes.

PROOF. It is known from mechanics that if an oscillating mobile [object] has passed through a distance x, and if at individual locations it has the element of time $dt = \dfrac{m\,dx}{\sqrt{nx - xx}}$, with m and n understood to be constant quantities, this [object] makes its respective oscillations, whether greater or lesser, in the same time.

But because in our case

$$v = \frac{2ga\alpha cx - (g\alpha b + ga\beta)xx}{2gaa\alpha + 2ga\alpha\alpha + 2a\alpha MN},$$

and because the velocity itself is equal to $\sqrt{v}$, there will be

$$dt = dx \cdot \sqrt{\frac{2gaa\alpha + 2ga\alpha\alpha + 2a\alpha MN}{g\alpha b + ga\beta}} \Big/ \sqrt{\frac{2a\alpha cx}{g\alpha b + ga\beta} - xx},$$

where in a like manner all letters have constant values except x, which denotes the distance traveled through; it is evident that these oscillations of the fluid will also be isochronous. Q.E.D.

Problem

§13. To find the length of the simple pendulum which is tautochronous with the previously mentioned oscillations of a fluid.

SOLUTION. It is shown in mechanics that, with $dt = \dfrac{m\,dx}{\sqrt{nx - xx}}$, the length of a simple tautochronous pendulum is $\tfrac{1}{2}mm$. Therefore, in our case which is under discussion the length of the pendulum sought will be $\dfrac{ga a\alpha + g a\alpha\alpha + a\alpha MN}{g\alpha b + g a\beta}$. Q.E.I.

§14. COROLLARY 1. If the conduit cAd is assumed to be of the same area as the attached pipes, and if its length is called l, the mass of the water contained in it, which we have called M, will be gl; and the *potential ascent* of the water contained therein, which we have set equal to Nv, will be v, so that $N = 1$. Moreover, after those values have been substituted for the letters M and N, the length of the tautochronous pendulum becomes, for that particular case,

$$\frac{aa\alpha + a\alpha\alpha + a\alpha l}{\alpha b + a\beta} = \frac{a\alpha}{\alpha b + a\beta}(a + \alpha + l) = \frac{a + \alpha + l}{\dfrac{b}{a} + \dfrac{\beta}{\alpha}}.$$

But since $a + \alpha + l$ is the length of the entire system filled with water, and $\dfrac{b}{a}$ signifies the ratio of the sine of the angle bac to the total sine, and, equally, $\dfrac{\beta}{\alpha}$ denotes the ratio of the sine of the angle efd to the total sine, we see that our solution does not differ from that which my Father gave for that case, and which I recounted above in §4.

§15. COROLLARY 2. If the conduit cAd is assumed to be of infinite area everywhere, $MN = 0$ (through §2, Chapter III) and the length of the tautochronous pendulum will be $\dfrac{a + \alpha}{\dfrac{b}{a} + \dfrac{\beta}{\alpha}}$, certainly just as if the entire intermediate conduit cAd were absent and the cylindrical pipes were connected immediately to each other.

Nevertheless, something special is to be considered here, which I will show below.

Scholium

§16. This theorem includes all cases which cause tautochronous oscillations where the pipes ac and pd are straight; but when these pipes in which the surfaces of the fluid are traveling are curved, other cases of tautochronism are given in addition which would be easy to solve if we should wish to delay here for a longer time. Finally, when these pipes are of unequal area, the times corresponding to oscillations of different magnitudes become unequal also, and how such a time must be defined is apparent to everyone from §8, where we gave the velocity of the fluid at any arbitrary point.

But this concerns finite oscillations. If now we consider that the oscillations are very small, we will see that they all become mutually tautochronous for the same quantity of fluid and the same conduit being retained, whatever might be the shape of the conduit and the areas. Let me show this in the following paragraph.

Theorem

§17. Very small displacements of a fluid oscillating in any conduit whatever, although they are unequal to one another, are all Isochronous.

PROOF. When the oscillations are very small, those small portions of the conduit in which the surfaces of the fluid are agitated can be taken as cylinders; therefore, with all the designations kept the same, the value which we assigned to the letter v in §8 will remain, and from the same reasoning it follows that the letters a, b, α, β, and x can be neglected as being of infinitely small value with respect to $\dfrac{M}{g}$, so that in the present case it must be considered that

$$v = \frac{2g\gamma a\alpha cx - (g\gamma ab + ggab)xx}{2\gamma a\alpha MN}.$$

Therefore, by virtue of §12, all oscillations, as far as they are very small, are Isochronous to one another. Q.E.D.

Problem

§18. To determine the length of a simple pendulum tautochronous with the very small oscillations of a fluid agitated in any conduit whatever.

SOLUTION. Because in the entire motion the element of time is $dt = \dfrac{dx}{\sqrt{v}}$, now there will be

$$dt = dx \cdot \sqrt{\frac{2\gamma a\alpha MN}{g\gamma ab + ggab}} \bigg/ \sqrt{\frac{2\gamma a\alpha cx}{\gamma ab + ga\beta} - xx}.$$

Therefore, by virtue of §13, the desired length of the pendulum tautochronous with the previously mentioned oscillations will be $\frac{\gamma a\alpha MN}{g\gamma\alpha b + gga\beta}$. Q.E.I.

Scholium

§19. Although I may have already advised in passing what is to be understood by the quantities M and N, nevertheless let me set forth the entire construction here, so that the nature of the matter is all the more evident to everyone.

Let there be a conduit $ABCDE$ of any kind whatever (Fig. 35a

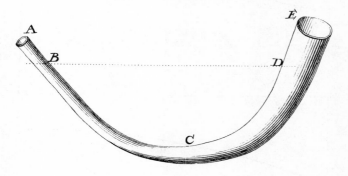

FIGURE 35a

and b), filled with water right up to B and D; let the total sine be assumed as 1, the sine of the angle DBC as $\frac{b}{a} = m$, the sine of the angle BDC as $\frac{\beta}{\alpha} = n$; the length of the tautochronous pendulum will be $\frac{\gamma MN}{mg\gamma + ngg}$, where g denotes the area of the conduit at B, and γ the area of it at D.

Now let the length BCD of the conduit filled with fluid be considered as extended in the straight line bcd, above which, as if it were an axis, is formed the curve FGH, which let be the scale of areas in homologous places, so that, after bc has been set equal to BC, cG is to bF as the area at C is to the area at B. Therefore, if bF represents the area at B, then the area $bdHF$ will represent the magnitude M. Then, from the same axis bd let another curve LMN be constructed, the ordinate cM of which is everywhere $\dfrac{(bF)^2}{cG}$, and $\mathcal{N}$ (from §2, Chapter III) will be the area $bd\mathcal{N}L$ divided by the area $bdHF$, so that $M \cdot \mathcal{N}$ is the area $bd\mathcal{N}L$, which, multiplied by $\dfrac{\gamma}{mg\gamma + ngg}$, will give the length of the tautochronous pendulum.

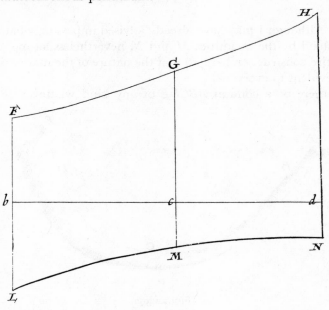

FIGURE 35b

§20. COROLLARY 1. If the pipe BCD is of the same area everywhere, and its length is called l, FH will be a straight line parallel to bd itself and $L\mathcal{N}$ equally; hence the distance $bd\mathcal{N}L = gl$ and the length of the tautochronous pendulum is $\dfrac{l}{m + n}$.

§21. COROLLARY 2. Let BCD be a conical conduit of length l; cG (after bc has been set equal to x) will be $\left[\dfrac{x}{l}(\sqrt{\gamma} - \sqrt{g}) + \sqrt{g}\right]^2$; from this $cM = gg \Big/ \left[\dfrac{x}{l}(\sqrt{\gamma} - \sqrt{g}) + \sqrt{g}\right]^2$; therefore, the area $bcML = \dfrac{ggl}{\sqrt{g\gamma} - g} - \dfrac{ggl}{\sqrt{\gamma} - \sqrt{g}} \Big/ \left[\dfrac{x}{l}(\sqrt{\gamma} - \sqrt{g}) + \sqrt{g}\right]$, and thence the entire area $bd\mathcal{N}L = \dfrac{ggl}{\sqrt{g\gamma} - g} + \dfrac{ggl}{\sqrt{g\gamma} - \gamma} = \dfrac{ggl}{\sqrt{g\gamma}}$. Therefore,

FLUIDS MOVING WITHIN THE WALLS OF VESSELS 137

the length of the pendulum tautochronous with the oscillating water is $\dfrac{l\sqrt{g\gamma}}{m\gamma + ng}$.

Hence it is understood, the remaining things being equal, that the water is oscillated very slowly when the areas at B and D are in reciprocal proportion to the sines of the corresponding angles DBC and BDC; hence, the longer the portion full of water and the less the angles just mentioned, the slower the oscillations become as well.

Further, after the cylindrical and conical pipes have been compared to one another, and the angles BDC and DBC have been set equal, it is clear that the water, the remaining things being equal, oscillates more quickly in the conical than in the cylindrical [pipe], namely, because $\dfrac{l\sqrt{g\gamma}}{\gamma + g}$ is always less than $\tfrac{1}{2}l$, whatever unequal ratio exists between g and γ. If, further, the previously mentioned angles are made unequal, it can happen that the water oscillation is slower as much as faster in one kind of pipe with respect to the other; so, in order that I may show this by an example, let me assume DBC to be a right angle, that is, $m = 1$, and the sine of the other angle BDC, or n is equal to $\tfrac{1}{4}$, then the length of the pendulum for cylindrical pipes will be $\tfrac{4}{5}l$. But if under the same remaining circumstances one substitutes for the cylindrical pipe a conical one which has an area at B four times as large as the area at D, one will have, after $\gamma = \tfrac{1}{4}g$ has been established, the length of the pendulum equal to l; therefore, the remaining things being equal, the tautochronous pendulum for the conical pipe is longer than for the cylindrical, and the oscillations in the former occur more slowly than in the latter; but if now, once again the remaining things constant, we assume the conical pipe narrower at B than at D, the situation will be the opposite; for example, let $\gamma = 4g$, and the length of the pendulum will be $\tfrac{8}{17}l$, and accordingly less than if the pipe were cylindrical; and again it will be less if one assumes the area at B altogether greater than it is at D; thus, if $\gamma = \tfrac{1}{64}g$, the length of the pendulum will be $\tfrac{8}{17}l$, as before. It is notable that we saw also in the preceding example that, with the area at B, the position of the conduit BCD and the length of the same being maintained, two distinct areas can always be defined at D for the same length of a tautochronous pendulum, except when the angles DBC and BDC are equal. A particular example of this matter is that, if either the area at D is equal to the area at B, or it has to the same a squared ratio of the sine of the angle BDC and the sine of the angle DBC, the oscillations of the fluid are completed in either pipe within the same time.

General Scholium

§22. I performed experiments on the oscillations of fluids in such a way that by trial I often found the length of the simple Isochronous pendulum, and I was able to observe in different cases that this length was more or less such as the theory in this section indicates; nevertheless, once I found that length to be a little greater than required; I saw with no difficulty that the reason for this situation is that the frictions of the fluid not only diminish the paths, but also retard them, and this because the pipes are customarily narrower in that place where they are bent. Subsequently, if this is avoided with all care, and if the very inflections are not made at an angle, but slowly, and if finally the purest mercury is used as the oscillating liquid, I have no doubt that the experiments will confirm the previously mentioned theory to the letter, so that I would not think it is worth the effort to inquire anxiously about them.

Nevertheless, I will add this as an explanation of the experiments undertaken by me, that before the experiment I accurately determined the areas of the pipes in their different places with the help of a column of mercury: while it slowly ran through the entire length of the pipe, it disclosed the variations of the areas everywhere by its different lengths, of which I continually took measurements. And certainly these areas are thus to be determined in the pipe after it has already been bent, for the areas are decreased somewhat by the bending. This was the reason that in the first experiment undertaken by me in this regard success failed my expectation: Indeed, I bent a glass tube of the sort that is customarily used for making barometers, wide enough and almost perfectly cylindrical, more or less as Fig. 27 shows, and then, after a very great portion had been filled with mercury, I saw that its oscillations occurred much more slowly than I had expected, because I did not pay attention to the fact that the tube was constricted significantly by the bending at D, especially where the angles are formed. Therefore, in order to take this matter into consideration, I made use from then on of gradually curved tubes, such as Fig. 35a shows, and in these I carefully determined the areas after the bending.

SEVENTH CHAPTER

Concerning the Motion of Water through Submerged Vessels, where it is Shown by Examples how Significantly Useful is the Principle of the Conservation of Live Forces, even in those Cases in which Continually some Part of Them is to be Considered Lost

FIRST PART: CONCERNING THE DESCENT OF WATER

§1. Assume there is a cylinder full of water, the base of which is perforated, submerged to a certain depth in standing water of infinite extent, and one easily realizes that the surface of the water contained in the cylinder will descend, and indeed below the surface of the exterior water, then it will ascend again, and so on. But these oscillations differ completely from the oscillations considered in the preceding chapter, in which certainly the motions are always reciprocal in inverse order to those motions which have preceded. But someone may presume here that the reflux of water, or the ascent, will be the same as the descent was. If one would state such things, he would certainly be in error, even if, for instance, the motion is diminished not at all by the adhesion of water to the sides of the vessel and other hindrances of this sort; but the rules of motions for percussion for elastic bodies are otherwise not very different from these which are valid for pliable bodies, in whatever manner in either case bodies are considered to be moved completely freely. I use the following similar [procedure] which illustrates our argument splendidly: For just as the rules of motions are determined correctly in pliable bodies, if after collision that part of the *live force* is considered lost which was expended in the compression of the bodies (for this is not restored to the progressive motion as in elastic bodies), so the ascent of the fluid will be defined no less correctly if one examines accurately how much of the *live force* in individual instants is communicated to the internal motion of the aqueous particles, never to return to the progressive motion, which is the subject of this discussion.

§2. Since, therefore, the matter is reduced to the fact that it should be investigated how much of the *live force* is lost continuously in those reciprocal motions, we will begin the investigation from this point.

But first it is evident that all the *live force* which the particles flowing out possess is transferred to the external water, and in no way does it promote the subsequent ascent or influx of external water into the pipe. This hypothesis is all too clear to warrant a greater explanation; but it pertains to the efflux of water and in this case it is the only one to be considered. Then we come to the other, which pertains to the influx of water.

Secondly, therefore, it is no less clear to me that with the water rushing in through the orifice at a greater velocity than that which is present in the internal rising water, that excess again produces a certain internal motion in the same internal water, adding little or nothing to the ascent. If this is so, and if the area of the orifice is set equal to 1 and the area of the cylinder equal to n, the *potential ascent* of a volume element flowing is equal to nnv, and its velocity is $n\sqrt{v}$, this particle will in its own motion, which it has in common with the remaining internal water, retain the velocity $\sqrt{v}$, and accordingly it will conserve the *potential ascent* v; but it must be considered that the remainder of the *potential ascent*, namely, $nnv - v$, was transferred to the internal motion of the particles. This hypothesis, although it is physical and only approximately true, nevertheless has great usefulness for determining the motions of fluids without noticeable error whenever in a vessel uniform continuity, which was assumed so far, is broken off, as when the water is forced to go through many orifices. Finally, I should believe that this [hypothesis] is unique, with the help of which amazing phenomena of this kind of motion can be explained correctly. On this account let me implore that it be pondered properly before the reader is diverted to other things.

§3. Therefore, we will now examine this very question by beginning with the descent of the water. Let the cylinder $AIMB$ (Fig. 36) be considered full of water up to XY and submerged in the infinite water $RTVS$, so that it is in a vertical position and its base has the opening PL through which water from the vessel can flow into the water surrounding. The velocity of the internal water is sought after its surface descends through the given space XC or YD, after one

MOTION OF WATER THROUGH SUBMERGED VESSELS   141

has set MY or $IX = a$, $MV = b$, $MD = x$, the area of the orifice equal to 1, and finally the area of the cylinder equal to n.

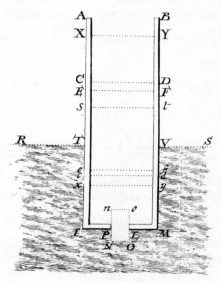

FIGURE 36

The solution will be the same as that which we gave for a similar but very general question in Chapter III; only let it be noted that, with the infinitely small particle of water $CDFE$ assumed equal to the volume element $PLON$ ejected in that same time, the *actual descent* must now be estimated from the height DV or CT, while in the other case it was to be defined from the total height DM.

Indeed, let the velocity of the aqueous surface CD be that which is due to the height v, and in the infinitely close position EF the same velocity will correspond to the height $v - dv$. And since the *potential ascent* of the water $CDMLPIC$ is v, the *potential ascent* of the same water in the next position $EFMLONPIE$ will be obtained if the mass $EFMLPIE$ $(nx - n\,dx)$ is multiplied by its proper *potential ascent* $(v - dv)$, as also the volume element $LONP$ $(n\,dx)$ by its proper *potential ascent* nnv in the same way, and the aggregate of the products is divided by the sum of the masses, (nx); and so *that potential ascent* is

$$\frac{(nx - n\,dx)(v - dv) + n\,dx \cdot nnv}{nx},$$

or

$$\frac{xv - v\,dx - x\,dv + nnv\,dx}{x}.$$

Accordingly the increment of *potential ascent* is $\dfrac{-v\,dx - x\,dv + nnv\,dx}{x}$ (see §6, Chapter III). But that increment must be considered equal to an infinitely small *actual descent*, which (accordingly to §7, Chapter III and the notation just given) is $\dfrac{(x-b)\,dx}{x}$. And so the following equation is obtained

$$-v\,dx - x\,dv + nnv\,dx = (x-b)\,dx$$

which, integrated in the proper way, is changed into this:

$$v = \frac{1}{nn-2}\left(x - \frac{x^{nn-1}}{a^{nn-2}}\right) - \frac{b}{nn-1}\left(1 - \frac{x^{nn-1}}{a^{nn-1}}\right).$$

But from that equation certain corollaries follow.

§4. Let the area of the cylinder be as if infinite in proportion to the orifice, and it must be considered that $v = \dfrac{x-b}{nn}$; and the very height corresponding to the velocity of the water, while it flows out, is $x - b$. Hence it is a consequence that the water flows out at the velocity which a heavy body acquires by falling from the height of the internal surface above the external, and it will flow out until both surfaces are placed at a level, and then all motion will cease; and therefore the water flows out by the same law as if the base changed the position *IM* with *TV*.

But when the orifice cannot be considered as infinitely small, the surface of the internal water descends below the external; and in order that it may become known to what depth xy the surface *CD* is to descend, v must become null, or

$$(nn-1)(a^{nn-1}x - x^{nn-1}a) = (nn-2)(a^{nn-1}b - x^{nn-1}b),$$

but the internal surface will never descend so far below the external surface as it had been elevated above the same; that defect arises from the *potential ascent* of the water ejected during descent, to which it must be proportional.

§5. It is noticeable that although the water descends more deeply in the cylinder the greater it has been elevated at the beginning of the descent and the greater the opening is by which the base is perforated, nevertheless all the water can never flow out of the cylinder, no matter how much it has been elevated before the descent and however small is the submerged portion of the cylinder, and at the same time the orifice itself or the entire base is assumed to be discharging.

§6. The velocity of the surface of the internal water is a maximum when it is assumed that

$$x = \left(\frac{a^{nn-1}}{nna - nnb - a + 2b}\right)^{1/(nn-2)}.$$

Accordingly, if $n = 1$, the orifice of the cylinder of course being fully opened, it occurs that $x = b$, and the velocity is a maximum when both surfaces are positioned at the same height.

But because there are many things which cannot be learned from these equations in the two cases, namely, when $nn = 1$ and $nn = 2$, and these have many particulars, let me now attack them separately.

§7. First, let $nn = 1$, whereupon $-x\,dv = (x - b)\,dx$ (by §3), or $-dv = dx - \frac{b\,dx}{x}$, which, integrated so that simultaneously one has $v = 0$ and $x = a$, gives $-v = x - a + b \ln \frac{a}{x}$, or $v = a - x - b \ln \frac{a}{x}$. From these the following can be deduced.

I. In order that the maximum descent be obtained, it must be established that $a - x - b \ln \frac{a}{x} = 0$; moreover, it is evident from that equation that the letter x never obtains a negative value; on the contrary, it certainly does not vanish completely without [causing] a contradiction, unless $\frac{a}{b}$ is set equal to ∞, which indicates that it cannot occur that all the water flows out during descent in that case, and much less in the others, which §5 confirms.

II. The maximum velocity is that which is due to the height $a - b - b \ln \frac{a}{b}$, and if the difference between a and b, which I set equal to c, is very small, the expansions of the fluid being indeed insignificant in proportion to the depth to which the cylinder is submerged, $\ln \frac{a}{b}$ could be considered as $\frac{c}{b} - \frac{cc}{2bb}$, and therefore the height itself due to the maximum velocity as $a - b - b \ln \frac{a}{b} = \frac{cc}{2b}$, which indicates that the motion will be very slow.

Moreover, I shall demonstrate in what follows that the entire motion remains the same, the rest of the things being equal, when the cylinders are considered infinitely submerged, by whatever orifice the base may be perforated, so that the motion of the internal water is not retarded by the diminished orifice; although this may seem at first glance to be altogether a paradox, nevertheless, the true physical reasoning of it cannot escape the mind that is attentive to these things. It is certainly concerned with the fact that the *live force* which is generated in the pipe is as if infinite with regard to the live force of the water flowing through the orifice, and therefore the consideration of this orifice does not make the computation different.

We will also show that the reciprocal motions are similar, and that both the greater and the lesser oscillations are Isochronous with each other, and for these we will determine the length of a simple tautochronous pendulum.

§8. Now let $nn = 2$ be assumed; but thus it occurs, by virtue of §3, that $v\,dx - x\,dv = (x - b)\,dx$, or $\dfrac{x\,dv - v\,dx}{xx} = \dfrac{(b - x)\,dx}{xx}$, which, correctly integrated, changes into this:

$$v = \frac{bx}{a} - b + x \ln \frac{a}{x}.$$

If it occurs that $\dfrac{bx}{a} - b + x \ln \dfrac{a}{x} = 0$, x will give the position of maximum descent; but the position of maximum velocity will be obtained by setting $x - ae^{(b-a)/a}$, where by e is understood the number of which the logarithm is unity.

After we have so glanced over various cases for different magnitudes of orifices, it remains that we also consider what can happen in different cases of the heights a and b.

§9. And first, certainly, if b is considered null with respect to a, which occurs when the base of the cylinder only touches the surface of the exterior water, then there results

$$v = \frac{1}{nn - 2}\left(x - \frac{x^{nn-1}}{a^{nn-2}}\right)$$

which equation, certainly, differs only in form from that which was given in §14, Chapter III, for that case in which the water is considered to be ejected from the cylinder into the air. And I also often found that the cylinder is evacuated in the same time whether the water is ejected into the air or the base is submerged a very little bit in standing water. This experience shows that the external air offers little or no hindrance to the efflux, since a resistance more than eight hundred times greater does not have a more noticeable effect. Therefore, because that case contains nothing in particular which has not been mentioned in the place cited, we will not dwell on it any further. Rather, we will inquire into what must occur when the elevation of the internal water above the external, such as it is at the beginning of the descent, is assumed very small and should be neglected with respect to the immersion of the cylinder; this hypothesis is satisfied when the excess of the height a over the height b (which excess again let us call c, as in §7) is very small.

§**10**. Therefore, if one sets $a - b = c$, one must also set $a - x = z$, and then both quantities, namely, c and z, are to be neglected with respect to the quantities a and b; but if $a - x = z$, then $x = a - z$ and

$$x^{nn-1} = (a - z)^{nn-1} = a^{nn-1} - (nn - 1) a^{nn-2} z$$
$$+ \left[\frac{(nn - 1)(nn - 2)}{2}\right] a^{nn-3} zz$$
$$- \left[\frac{(nn - 1)(nn - 2)(nn - 3)}{2 \cdot 3}\right] a^{nn-4} z^3 + \text{etc.}$$

This series is to be continued as much as it suffices for our purpose; however, up to three terms will be sufficient. Therefore, in the integrated equation which we gave in §3, we will assume $x = a - z$, and

$$x^{nn-1} = a^{nn-1} - (nn - 1) a^{nn-2} z + \left[\frac{(nn - 1)(nn - 2)}{2}\right] a^{nn-3} zz,$$

and thus there will be

$$v = \frac{1}{nn - 2} \left\{ a - z - a + (nn - 1)z - \left[\frac{(nn - 1)(nn - 2)}{2}\right] \frac{zz}{a} \right\}$$
$$- \frac{b}{nn - 1} \left\{ 1 - 1 + (nn - 1) \frac{z}{a} - \left[\frac{(nn - 1)(nn - 2)}{2}\right] \frac{zz}{aa} \right\}.$$

If in this equation the terms cancelling themselves are deleted, and $a - c$ is put in place of b, and the term which is affected by the quantity $\frac{czz}{aa}$ is rejected, there results simply

$$v = \frac{2cz - zz}{2a}$$

from which formula, since the letter n has vanished, we have evidence that the size of the orifice pertains not at all to the motion of the internal water, the origin of which matter I already indicated above (§7).

In the following [paragraphs] we will show further that this motion does not differ from the subsequent reflux motion, and hence that the oscillations become tautochronous. But before I proceed to other things, I consider that one must be advised that in this calculation the quantities $\frac{c}{a}$ and $\frac{z}{a}$ were taken as infinitely small, not only with respect to unity, but also with respect to $\frac{1}{nn}$, of which one has to take proper notice in undertaking experiments; certainly it is allowable to apply the theory of the infinitely small to experiments without noticeable error by greatly diminishing the quantities which have been considered as infinitely small in the theory, but it must be done so that all things in the experiment are subject to this law. Thus, for example, if in the cylinder the entire base is absent, $n = 1$ having been established, and it is considered submerged to a height of 35 inches, the experiment may be regarded as sufficiently accurate if the water has been elevated before the oscillations only to a height of one inch above the surface of the water surrounding; the error will not yet be noticeable even if the orifice below is half obstructed, with $\frac{c}{a}$ then being to $\frac{1}{nn}$ as 1 is to 9, which ratio so far can be safely neglected in our experiment; but if one now assumes the diameter of the pipe as two times the diameter of the orifice, with three-quarters of the entire aperture closed, it occurs that $n = 4$, and $\frac{c}{a}$ will be to $\frac{1}{nn}$ as 4 is to 9, which ratio will now be small enough that the experiment can be affirmed to satisfy the conditions of the theory with sufficient precision.

Therefore, it will now be appropriate to inquire further what should be stated here concerning these cases in which $\frac{c}{a}$ and $\frac{1}{nn}$ indeed have an appreciable ratio to each other, but each quantity is very small, which certainly occurs when the cylinder is submerged very deeply and at the same time the base is perforated by a very small orifice.

§11. But that case which we just treated is better deduced from the differential equation of §3, rather than the integral [equation], as was done previously; however, under these circumstances the term $-v\,dx$ can be rejected with respect to $nnv\,dx$, and thus it can be assumed that $-x\,dv + nnv\,dx = (x - b)\,dx$, in which, if it is again established that $a - b = c$ and $a - x = z$, there results $a\,dv + z\,dv + nnv\,dz = (c - z)\,dz$, the second term $z\,dv$ of which can again be neglected with respect to the first one, and thus one obtains $a\,dv + nnv\,dz = (c - z)\,dz$.

MOTION OF WATER THROUGH SUBMERGED VESSELS 147

Let it be assumed here (after e has been taken as the number of which the hyperbolic logarithm is unity) that $v = \frac{1}{nn} e^{-nnz/a} q$; in this way the last equation will be changed into this:

$$e^{-nnz/a} a\, dq = nn(c - z)\, dz,$$

or

$$a\, dq = nn e^{nnz/a} (c - z)\, dz.$$

But this must be integrated so that z and v, or as well z and q, vanish at the same time; thus there will result

$$q = \left(c + \frac{a}{nn} - z\right) e^{nnz/a} - c - \frac{a}{nn},$$

or, finally,

$$v = \frac{1}{nn}\left(c + \frac{a}{nn} - z\right) - \frac{1}{nn}\left(c + \frac{a}{nn}\right) e^{-nnz/a}.$$

But from that equation it is deduced:

I. That it again develops, as was found by another method in §10, that $v = \frac{2cz - zz}{2a}$, if indeed the number $\frac{nnz}{a}$ is again established as very small. But in order that this be evident, the exponential quantity $e^{-nnz/a}$ must be resolved into the series which is equivalent to it, $1 - \frac{nnz}{a} + \frac{n^4 zz}{2aa} - \frac{n^6 z^3}{2 \cdot 3 a^3} +$ etc., from which the first three terms are sufficient for our purpose; therefore, with this value substituted and the term to be rejected having been rejected, there is obtained, as I said,

$$v = \frac{2cz - zz}{2a}.$$

II. But if, on the other hand, $\frac{nn}{1}$ is assumed infinitely greater than $\frac{a}{z}$ or $\frac{a}{c}$, because then $e^{-nnz/a} = 0$, and also $\frac{a}{nn} = 0$, it occurs that $v = c - z$, or $v = x - b$, as in §4.

III. But it is evident that neither of the aforementioned formulas stands without noticeable error when the number $\frac{nnc}{a}$ is moderate, that is, neither infinite nor infinitely small, and nevertheless both quantities $\frac{nn}{1}$ and $\frac{a}{c}$ are infinite.

For example, let an elevation of one inch be indicated by c, let the immersion of the cylinder b be 80 inches, and let a itself be 81 inches; then let the diameter of the pipe be established as triple the diameter of the orifice, that is, $nn = 81$, and there will result $v = \dfrac{2 - z - 2e^{-z}}{nn}$; and if further it is established that $z = c = 1$, in order that the height for the velocity be obtained when both surfaces are positioned at [the same] level, one will have $v = \dfrac{e - 2}{nne}$, that is, $v = \frac{1}{307}$ inch, approximately, although according to §10 it should have developed that $v = \frac{1}{162}$ inch, and according to §4, $v = 0$. In the same example it happens that the entire space through which the surface passes is not fully eight-fifths of one inch, and the point of maximum velocity is more or less sixty-nine hundredths of the same measure below the initial height.

§12. It should not be any more difficult to extend those things which were said so far to all shapes of vessels, finally even to finite spaces by which the external water may be terminated; however, the formulas become mostly so obliging that I would consider it rather well advised to pass in silence over them and to show only by an example the particular manner in which the theory should be applied for eliciting any number of other cases.

Deserving more particular attention are those things which I indicated concerning the motion of water in pipes opened considerably at the bottom and submerged very deeply, because in these the oscillatory motion, as in pendulums, is of constant period, and the flow of waves in the sea is illustrated by them. But I thought that the backflow of water in submerged cylinders in general is to be treated first, and it is to be shown that according to this hypothesis the backflow does not differ from the preceding flow, before the entire oscillatory motion is examined. Therefore, we will now comment on the backflow, and then we will combine both motions in different cases, lest anything could be wanting in the proof.

SECOND PART: CONCERNING THE ASCENT OF WATER

§13. After the water has descended in the submerged vessel as much as the nature of the situation permits it, two things especially offer themselves for consideration: first, the excess of the height of the external surface above the internal, and second, the *live force*, or the product of the *potential ascent* and the mass of that water which was ejected from the cylinder into the surrounding water during the descent; indeed, this *live force*, which cannot return to the water in the cylinder, principally causes the water to fail considerably in approaching the original height from which it had fallen; nevertheless, this is not the only reason, even if the hindrances of tenacity, adhesion, and others of this sort interfere not at all: the other reason was indicated in §2. Indeed, the extent of that reason is to be deduced from the ascent itself, since the former pertains to the descent, and this is the only reason, disregarding the external hindrances, that the water is not elevated in ascent above the external surface as much as it had been depressed below the same. For it must be noted that, even with the water flowing in through a very small orifice, it would ascend at the same velocity as if the entire base were missing, and it would rush in through the entire orifice if only after the inflow it would exert the entire impetus which occurs in the internal water for promoting its ascent. Truly anyone who considers this matter properly sees easily that the major part of that entire impetus is expended wholly in some internal motion which provides nothing to the ascent; but I say clearly the major part (I wish it to be noted well), because, when the orifice is very large, it is not difficult to see in advance that the impetus of the entering water is produced so suitably that the internal motion is thence increased by no means slightly; but when the orifice is smaller, it is clear that the situation is otherwise. Therefore, our hypothesis is applied correctly when either the whole base is absent or is almost completely perforated (for thus the excess of the velocity of the water flowing in over the velocity of the internal water is nil or very small, and that does nothing to this impetus) or even when the orifice is very small, because thus all the impetus is overcome. But if the orifice should have a ratio to the area of the pipe such as $\sqrt{2}$ to 1, or as 2 to 1, or thereabout, the motion will be a little greater than that which follows from that hypothesis, for then the water rushing in produces a noticeable impetus, and not all of it is lost because of the nature of the matter.

Therefore it is easy to see in advance, without performing any calculation, the following relationships for backflow in water after it has fallen from a certain height.

I. Certainly no noticeable backflow will occur if the orifice is very small.

II. When the submerged portion of the cylinder remains unchanged, the water in backflow will never pass a certain limit, even if it had been elevated to infinity in the previous descent; indeed, from whatever height the descent begins, not all the water ever flows out of the cylinder, as we saw in §5 and §7.

III. If the descent is understood to begin at the height XY [Fig. 36] and the subsequent ascent to be produced right up to CD, the product of the *actual descent* of the mass of water $XYDC$ right up to TV by the mass will be a measure of the combined effect of both of those causes which, as was said in §2, make the ascent differ from the preceding descent, and when the cause reviewed in the second place vanishes, if the entire base IM is removed, that product will then be equal to the *live force* of all the water ejected during the descent, so that, without any other calculation except those things already considered so far, the ascent of the water in the entire open cylinder can be defined.

IV. The ascent will be equal to the descent if the cylinder is understood to be infinitely submerged, the previously mentioned causes of diminution then vanishing.

V. Therefore, these oscillations will be endless, because the last oscillations are always just as if infinitely small in proportion to the heights of submergence; however, the alien hindrances, of which we have taken no account up to now, soon cause all the motion to cease altogether.

§14. With these things having been generally shown in advance, let us submit the problem to more accurate calculation; however, I will give a double solution, one accommodated to the principles just explained, the other different in kind to some degree.

Therefore, with the figure [36] and the notation of §3 retained, we will consider that the water has descended from the height XY all the way to xy, and from this terminal point it begins its ascent; let My or Ix be called α, and after it has already ascended to cd or ef, let $Md = \xi$ and $df = d\xi$. After these things have been so prepared for the calculation, and with the height due to the velocity of the water at cd again designated by v, and the similar height in the adjacent position ef by $v + dv$, we will inquire into the *increment of potential ascent* of the water entering while the volume element $LONP$ goes into the cylinder and the surface ascends from cd to ef. Moreover, it is clear that when the *potential ascent* of the internal water multiplied by its proper mass is expressed by $n\xi v$ (indeed, no attention is to be paid to the internal motion), the increment of the same product will be $n\xi\, dv + nv\, d\xi$. But if in addition one were to consider the *potential ascent* $nnv - v$ (see §2) which the volume element $n\, d\xi$ flowing in loses, and which equally is due to the *actual descent* of the aqueous particle $n \, d\xi$ through the height $b - x$, it is evident that one must write

$$n\xi \, dv + nv \, d\xi + (nnv - v)n \, d\xi = (b - \xi)n \, d\xi$$

or

$$\xi \, dv + nnv \, d\xi = (b - \xi) \, d\xi.$$

But the same thing is found differently as follows.

Indeed, let it be considered that the volume element $LONP$ has almost no velocity before it begins to flow in, but that, once it begins to flow in, the same acquires a *potential ascent nnv*, although soon after its influx (according to the notation following §2) it is to be considered as continuing its motion at the common velocity $\sqrt{v}$. From this fact, the reasoning is this: before the influx of the volume element, the *potential ascent* of the water $cdMLPIc$ (the mass of which is $n\xi$) is v, and the potential ascent of the volume element $LONP$ (the mass of which is $n \, d\xi$) is 0; therefore, the *potential ascent* of all the water $cdMLONPIc$ is

$$\frac{n \, \xi v}{n \, \xi + n \, d\xi} = \frac{\xi v}{\xi + d\xi}.$$

But, indeed, after the volume element $LONP$ has flowed in and taken the position $LonP$, its *potential ascent* is nnv; moreover, the *potential ascent* of the remaining water $efMLonPIe$ (the mass of which indeed again is $n\xi$) is $v + dv$; therefore, the *potential ascent* of all the water considered here after the influx of the volume element is

$$\frac{n \, d\xi \, nnv + n\xi(v + dv)}{n\xi + n \, d\xi} = \frac{\xi v + \xi \, dv + nnv \, d\xi}{\xi + d\xi},$$

while before the same influx it was $\frac{\xi v}{\xi + d\xi}$; consequently, it took on the increment $\frac{\xi \, dv + nnv \, d\xi}{\xi + d\xi}$, or, more simply, $\frac{\xi \, dv + nnv \, d\xi}{\xi}$. But that increment is to be equated to the *actual descent* which the water makes in changing position from $cdMLONPIc$ to $efMLONPIe$, which descent is equal to the fourth proportional of the mass $n\xi$ of the internal water, the volume element $nd\xi$, and the height Vf or $b - \xi$, so that the previously mentioned descent is $\frac{(b - \xi) \, d\xi}{\xi}$; from which again the following equation results:

$$\xi \, dv + nnv \, d\xi = (b - \xi) \, d\xi.$$

But the integral of this, after the addition of the required constant, becomes the following:

$$v = \frac{b}{nn}\left[1 - \left(\frac{\alpha}{\xi}\right)^{nn}\right] - \frac{1}{nn+1}\left[\xi - \left(\frac{\alpha}{\xi}\right)^{nn}\alpha\right]$$

which we will now consider under varied circumstances.

§15. And, indeed, when the area of the pipe is an infinite number of times greater than the area of the orifice, it is evident that $v = \dfrac{b-\xi}{nn}$, and that accordingly the water rushes in at a velocity which is due to the height of the external surface above the internal, and then the ascent will not occur beyond the surface of the external water.

But when the area of the orifice has a finite ratio to the area of the pipe, the ascent occurs beyond the surface RS, for example, right up to st; but Vt will always be less than Vy, except when the entire base is missing, for then $Vt = Vy$. Just as we warned in §5 that in descent the difference between VY and Vy is proportional to and has its origin in the *potential ascent* of the water ejected during descent, so it can now be observed in ascent that the difference between Vy and Vt has its origin in the collision of the volume elements $LonP$ with the mass of water lying over it, which collision indeed does not promote the ascent but is lost in useless internal motion, just as was indicated in §2. Therefore, when the entire base IM is absent, the water flows into the pipe at the same velocity at which the water having previously entered the pipe moves, and no collision occurs, which is the reason why in that case the water ascends as much above the surface RS as it had been depressed below it, which the equation indicates, as we shall soon see.

§16. The maximum ascent st will be found by making $v = 0$. Therefore, in order to define the entire motion correctly, the formulas brought forth in §3 and §14 will have to be applied alternately, which I will now show by the single example in which $nn = 1$.

Accordingly, if $nn = 1$, it occurs that $v = b\left(1 - \dfrac{\alpha}{\xi}\right) - \dfrac{1}{2}\left(\xi - \dfrac{\alpha\alpha}{\xi}\right)$; and $v = 0$ will result when it is assumed that $\xi = 2b - x$, that is, when it is assumed that $Vt = Vy$. Therefore, if, for example, the pipe $ABMI$ is full of water, destitute of any base and immersed up to its midpoint in the exterior water, and if the entire length of the same is called a, the water will be set in motion so that at first it descends through a distance $0.297\,a$ below TV, then it is elevated through a similar distance above the same TV, and again it is depressed through

MOTION OF WATER THROUGH SUBMERGED VESSELS 153

a distance $0.240a$ below it, and it transcends that line again in the same way, and so on.

§17. It is evident as well that when $\alpha = 0$, with the pipe of course empty of all water, generally there will result $v = \dfrac{b}{nn} - \dfrac{\xi}{nn+1}$; and consequently the entire ascent will be $\dfrac{nn+1}{nn}b$, or the ascent above the exterior surface of the water will be $\dfrac{b}{nn}$.

§18. I come now to infinitely submerged pipes, the descent within which we have determined with the appropriate relationships of §10. Moreover, let us use clearly the same method which we used there for defining this case; therefore, for us the initial depression will be $Vy(= b - \alpha) = c$, and the ascent thence produced will be $yd(= \xi - \alpha) = z$. Thus $\xi = \alpha + z$, and $b = \alpha + c$, where the quantities z and c must be considered as infinitely small in proportion to the quantity α. Hence there results $\left(\dfrac{\alpha}{\xi}\right)^{nn} = \left(\dfrac{\alpha}{\alpha+z}\right)^{nn} = \left(1 + \dfrac{z}{\alpha}\right)^{nn}$, which, by applying a known series and taking the first three terms of it, one makes equal to $1 - \dfrac{nnz}{\alpha} + \dfrac{nn(nn+1)zz}{2\alpha\alpha}$. After these values have been substituted for b, ξ, and $\left(\dfrac{\alpha}{\xi}\right)^{nn}$, the last equation of §14 is changed into this:

$$v = \frac{\alpha + c}{nn}\left(\frac{nnz}{\alpha} + \frac{nn(nn+1)zz}{2\alpha\alpha}\right)$$
$$\qquad - \frac{1}{nn+1}\left(\alpha + z - \alpha + nnz - \frac{nn(nn+1)zz}{2\alpha}\right)$$
$$= (\alpha + c)\left[\frac{z}{\alpha} - \frac{(nn+1)zz}{2\alpha\alpha}\right] - \left(z - \frac{nnzz}{2\alpha}\right)$$
$$= \frac{cz}{\alpha} - \frac{zz}{2\alpha} - \frac{(nn+1)czz}{2\alpha\alpha}.$$

But that last term can be neglected, and thus one has simply

$$v = \frac{2cz - zz}{2\alpha},$$

in which equation n no longer appears. This does not differ from the equation for descent given in §10, namely, $v = \dfrac{2cz - zz}{2a}$, since indeed the quantities a and α differ only by the very small quantity $2c$.

For the rest, here as well are to be understood all those things which were said in the same §10 on the fact that a pipe must not be overly obstructed.

§19. Consequently, the descent and ascent are equal to each other; for from our equations it is evident that the liquid is balanced equally beyond the surface of the external water. But then it follows especially from those formulas that even the unequal oscillations are isochronous to each other, if only all of them can be considered infinitely small with respect to the submersion, and that moreover the simple tautochronous pendulum is of the same length as the submerged portion of the pipe.

That theorem differs from the one which was cited in §4, Chapter VI, concerning the oscillations in the cylindrical pipe composed of two vertical legs, as follows: there all oscillations are tautochronous, the oscillations of finite magnitude not having been excluded, while in the present case finite oscillations are of unequal duration; further, there the length of the pendulum is equal to half the length of the pipe, while here it is equal to the whole; however, if the matter is pondered properly, this should be considered as consistent rather than inconsistent, on account of the duplication of the pipe which occurs in the former case.

§20. In either type of oscillations the nature of waves agitated by the wind is illustrated; for they are not being moved otherwise than that the water in them continually ascends and descends again. Thus, what Newton says is evident, that the times of the oscillations are in proportion to the square roots of the lengths of the waves, because he assumes that the form of the waves is constantly similar to itself and accordingly that their length is proportional to the depth to which the water is agitated. Moreover, it is probable that the depth is that of a simple pendulum tautochronous with the waves, that is, for example, $60\frac{1}{3}$ Paris feet if the ascent or descent of the waves occurs every two seconds.

§21. Although I should not want, for the sake of avoiding the abundance of calculation, to pursue this argument to its full extent, and in view of these things I should only treat cylindrical vessels, nevertheless, since in the case of infinite submersion the propositions and theorems lose little of their elegance, let me extend the general theorem for the case of oscillations of water in an arbitrarily shaped pipe. However, the proof has been omitted, since it will be obvious to everyone from things said elsewhere, but especially from those which were presented in Chapter VI, §6, §7, etc., up to §20. However, that upper part of the vessel in which the oscillations occur must have a cylindrical form.

MOTION OF WATER THROUGH SUBMERGED VESSELS

§22. Therefore, let bd be the length of a submerged vessel (Fig. 35b). Let bF represent its area at the location of the surface, and let the vessel be so shaped that the curve FGH is the scale of the areas. Let the line bc be drawn and the curve LMN be formed, the ordinate cM of which is everywhere equal to $\dfrac{(bF)^2}{cG}$, and the length of the pendulum isochronous with the oscillations of the aqueous surface will be equal to the area $bdNL$ divided by bL.

§23. COROLLARY. It follows from the preceding paragraph that if the submerged pipe were conic, and if it had an area in the region of the surface of the water which is to the submerged orifice as m is to n, the length of the pendulum Isochronous with the vibrating water will be to the length of the submerged pipe as $\sqrt{m}$ is to $\sqrt{n}$, that is, as the roots of the previously mentioned areas, but if the same pipe is submerged not quite fully, once in the correct and then in the inverted position, the lengths of the isochronous pendulums will be in reciprocal proportion to the submerged orifices.

GENERAL SCHOLIUM

§24. Attempting the things which are contained in this chapter by experiments will be worth the effort all the more, since the majority of them arise from new hypotheses. I indeed performed several [experiments], but there was no time to execute some individual ones which I had planned; those which I did I will recount below. Meanwhile, in order that judgment can be passed more safely on the agreement of experiments with theory, first there is to be understood for the circumstances of the matter, generally, whether and how much the contraction of the stream flowing out (the nature of which I explained in Chapter IV) can disturb the calculation; this inconvenience can be removed for the most part if the walls of the final orifice form some small cylinder of barely half a line in height; concerning this let the fourth experiment pertaining to Chapter IV be brought to mind. Thence as well one has to pay attention to the resistances arising from the adhesion of the water, which indeed retard the motion little if one considers the times of the oscillations, but they detract much from the displacements, especially if rather narrow and rather long pipes are used. Therefore, more faith is to be put in experiments which are performed on the times of oscillations because these times are not altered much at all by the diminution of the displacements. With respect to the first kind of experiments, in which the displacements of the fluids in the pipes come from seeking and observing the descent as much as the ascent, I used this foresight: I wrapped a thread around the pipe at that place to which I expected the water to ascend or descend, and thus I finally located this thread, after frequent repetition of the experiment, so that the surface of the oscillating fluid would run neither above it nor below it. And I also marked the rest of the places which were to be observed in the pipe equally by wrapping a thread around it. Then, what pertains to the times of oscillations, because these decrease very quickly and become imperceptible and clearly null, I was not able to investigate otherwise than by examining the length of the simple isochronous pendulum, after the experiment had been repeated very often; while this [pendulum] was oscillating, I placed a finger over the orifice of the pipe and removed it therefrom precisely at such an instant of time that both the pendulum and the fluid would begin oscillation together.

EXPERIMENTS WHICH PERTAIN TO CHAPTER VII

EXPERIMENT 1. I acquired a cylindrical glass tube of almost four lines in diameter, entirely open below. I submerged it in water standing in a very large clear vessel to a height of 44 lines, and I put a finger over the opening above, so that in extracting a portion of the tube the water would not descend in it; then I extracted the tube to a height of 22 lines, so that the submerged portion of the tube as well as the height of the internal water above the external was 22 lines, and, as soon as the finger was removed, I observed the descent of the surface in the tube below the surface of the standing water, and I saw that it was $9\frac{1}{2}$ lines.

But according to §§7 and 17 it should have descended thirteen lines. It seems that the defect of three and a half lines is to be attributed almost solely to the adhesion of the water to the walls of the tube.

After the descent had been observed, I repeated the entire experiment in order that I might discover the next ascent also. However, it seemed to me to be 8 lines, which, according to §16, with the previous descent having been considered, should have been $9\frac{1}{2}$ lines, that is, just as much as was the preceding descent. But here the experiment failed by only one and a half lines, while in the first part of the experiment a defect of up to three lines and a half was present, because, indeed, there the displacement was greater, and this at a greater velocity, so that one finds altogether greater hindrances, which increase together with the velocities.

EXPERIMENT 2. I used the same tube, but [the end of it was] covered by a thin plate which was perforated by an orifice with an area having a ratio of $\sqrt{\frac{1}{2}}$ to the area of the tube; when the surface of the tube was elevated 18 lines above the standing water and the base submerged the same number of lines, I saw that the surface of the tube in descent fell almost five lines beneath the standing water.

However, §8 argues a descent of $7\frac{1}{2}$ lines; the defect, which was more than $2\frac{1}{2}$ lines, I again ascribe to the adhesion of the water to the walls of the tube.

Then I immersed this tube, completely empty of water, furnished with the same plate, with a finger placed over the top, to a depth of 18 lines into the water; after the finger had been removed, the surface in the tube emerged above the standing water a full eight lines, while §17 indicates nine for this case.

I attributed the fact that here the defect was altogether less than in descent to the reason which I indicated freely in §13, where I said that a slightly greater motion would develop when the orifice would have a considerable area with respect to that of the tube, such as in the ratio $\sqrt{\frac{1}{2}}$ to 1 or thereabout, than that which follows from the hypothesis; and in order that I might be made clearly certain of this matter, I applied a shorter and wider tube, so that the effect of almost all outside hindrances would be forestalled, and I performed the experiment which follows.

EXPERIMENT 3. I provided a tube the diameter of which was more than seven lines, which I took pains to have made of iron, because sufficiently cylindrical glass was not at hand; the length of this was four inches and six and a half lines; its area was in a ratio of 1.860 to that of the orifice, indicated by n, and nn was 3.458.

With that tube I performed the experiment thus:

With the upper orifice closed off, of course, I tried many times [to determine] to what depth it should be submerged in the water standing in a very large tank so that, directly after the finger had been removed which covered the orifice, the water would ascend precisely to the edge of the same orifice, and nothing would flow past. Indeed, I found that depth to be 3 inches and three lines; therefore, the ascent above the external water was one inch and three and a half lines, whereas, even with all the hindrances removed, the ascent should have been just a little beyond eleven lines, according to §17. Accordingly, one was correctly advised in §13 that the ascents cannot be a little greater in cases of this sort than the hypothesis postulates. I then applied another base to the same tube; now $n = 3.68$ and $nn = 13.54$; it was difficult to determine correctly the success of the experiment, because the surface ascending in the tube was always bubbling; nevertheless, it seemed that now the tube had to be immersed to a height of 4 inches and two or three lines, with four lines more or less thus remaining above the water, precisely as the theory indicates.

EXPERIMENT 4. I immersed a cylindrical glass tube, which had a diameter of three lines more or less, to a depth of 20 inches, and I caused the water in it to be held in equilibrium after it was first elevated to a height of almost one inch. It did not produce beyond four or five clearly noticeable departures and returns, and therefore I was not able to determine the length of the simple isochronous pendulum with all rigor; nevertheless, it seemed to me that it was 22 or 23 inches, from which I inferred that the adhesion of the water to the walls of the tube not only diminishes the displacements, but it also delays the times of the oscillations slightly; for according to §19 the previously mentioned length should have been only 20 inches. I found the same thing in the oscillations which we considered in the section above.

Finally, with the lower orifice approximately half blocked off, I was not able to observe that the displacements had been diminished or the oscillations retarded, which agrees with those things which are found in §§7 and 18.

EXPERIMENT 5. I immersed a conical tube with a length of 21 inches with the wider orifice in the water, so that one inch extended beyond the water; moreover, one orifice was a little more than twice the other. I found that the length of the pendulum isochronous with the vibrations of the water balanced in the tube was 15 inches, but according to §23 the same length should have been a little less than 14 inches. Finally, using the same tube similarly, but in an inverted position, I discovered that the length of the isochronous pendulum was a little more than double that which it had been before, just as is indicated in the cited paragraph.

EIGHTH CHAPTER

Concerning the Motion of Homogeneous as well as Heterogeneous Fluids through Vessels of Irregular and Abrupt Shape, where from the Theory of Live Forces, a Part of which is Continually Absorbed, are explained Excellently Singular Phenomena of Fluids driven through Several Orifices, after General Rules have been Set Forth for Defining the Motions of Fluids Anywhere

§1. So far, except in the chapter immediately preceding, we have not made use of any principles other than those two: *that the velocities of fluids are everywhere reciprocally proportional to the areas of the vessels*, with the aid of which is found the *potential ascent* of all the water from the given *potential ascent* of any particle whatsoever; and, further, that the *potential ascent* of all the water always remains equal to the *actual descent*. So often do these two principles apply that it is by no means to be questioned whether the motion of fluids is correctly defined by the method furnished by us. Nevertheless, I will not deny that vessels in which fluids are moved can be made of such a shape that neither one of these principles develops correctly. The former [principle] indeed rarely or never varies noticeably from the truth, because wherever it does not apply the water usually has almost no motion and can without noticeable error be considered as standing; but the other principle is regarded otherwise by far, which will be apparent from the following example, and those things that we offered in the preceding chapter on the backflow of water can serve as a splendid testimony to this matter; indeed, it is far from possible that water in a submerged vessel, having fallen from a given height, rises again to this height, as it should by virtue of this principle after external hindrances have been subtracted; rather, its ascent is often barely noticeable compared with the descent which it made before; on the contrary, indeed, the water surface certainly cannot rise as much above the water in which the pipe is immersed as it had been depressed below the same unless the entire pipe is open; but that surface is depressed much less than it had been elevated before. The reasoning on these things we gave in the preceding chapter. Because they are so, I will now give two rules for defining the motion of water everywhere, and then I will illustrate them with certain examples which could not be explained by any theory up to this point, but which conform most excellently with ours.

Rule 1

§2. One must discover what the velocity will be in the other parts of the fluid after the velocity anywhere in the vessel under consideration has been assumed as known. For thus the *potential ascent* of the entire fluid and its differential may be found. So far we have considered fluids to be divided into infinite parallel layers, or rather layers everywhere perpendicular to the walls of the vessel, and we have stated that the velocities are reciprocally proportional to these layers; certainly it is easy to fashion a vessel wherein the fluid is moved differently; I would believe, though, that the fluid in these places never has a conspicuous motion, so that a noticeable deviation from this hypothesis can hardly ever arise; nevertheless, for the sake of greater accuracy the above-mentioned rule could be applied. Indeed, this pertains especially to the contraction of jets whenever fluid is forced to go through orifices made in very thin plates, in which matter great care has to be taken. I believe that the effects of contractions of this sort will be understood in advance quite properly if what I said about them in Chapter IV is considered correctly.

Rule 2

§3. One must discover at any instant how much of the *live force* or what product of *potential ascent* and mass may develop without contributing anything to the main flow, the nature of which is sought. But this in turn is to be left to anyone's careful estimation. Then the product is to be added to the product of the mass and the *potential ascent* that the main motion contains, and the sum of the products finally is to be considered equal to the total mass of the water times its *actual descent*.

This rule is of great progress-making importance and I believe almost unique for obtaining measures of motions which occur in irregular vessels divided into several cavities connected to one another, which I will now illustrate by several examples.

Problem

4. Let there be proposed a vessel $ACRB$ (Fig. 37) of area everywhere infinite, so to speak, in proportion to [the areas of] the orifices

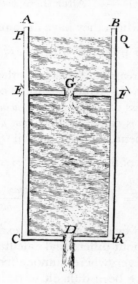

Figure 37

to be defined presently, divided into two intercommunicating cavities by some diaphragm EF, with the orifice G in the center; furthermore, let that vessel have another orifice D in its lowermost part; then let it be supposed that the vessel is full of water up to PQ, so that the lower cavity $CEFR$ will be filled completely with liquid, and that furthermore the other part $PQFE$ lies above the diaphragm. With these things having been set forth and the fluid already starting to be moved, the velocity of the water flowing out through the orifice D into the air, or the height creating this velocity is sought.

SOLUTION. Let the height of the surface PQ above the orifice D be x, the area of the orifice D be n, and that of the other one, G, be m. It is also very clear that the *potential ascent* of any drop flowing through G promotes nothing toward the efflux through D, and that all is used for exciting some internal motion which is soon absorbed without any other effect; therefore, it is necessary that in every instant a new motion be generated in the particles that will pass the orifice G, and no less in the particles flowing out through D. But if the *potential ascent* of the volume element flowing out through D is called v, that is, if the water is assumed to spring forth through D with a velocity of which the generating height is v, then the similar height, in reference to a volume element (equal to the former in bulk) flowing through G at the same time, will be $\frac{nnv}{mm}$. After these *potential ascents* have been multiplied by the mass which they have in common and which I shall call M, the sum of the products will be $Mv + \frac{Mnnv}{mm}$. And since, because of the infinite area of the vessel no other motion is generated, the aforementioned sum (by Rule 2) is to be considered equal to the product of the total mass of water and its *actual descent*. But if now the total mass of water is called μ, the *actual descent*, which occurs as long as the volume element M flows out, will (per §7, Chapter III) be $\frac{Mx}{\mu}$, so that the product of the two is Mx. Hence one obtains

$$Mv + \frac{Mnnv}{mm} = Mx, \text{ or } v = \frac{mmx}{nn + mm}. \quad \text{Q.E.F.}$$

§5. SCHOLIUM 1. It is evident from this example that the motion can be determined without differential calculus, since the shape of a vessel that is very wide everywhere cannot affect this motion. Meanwhile, it would not have been difficult to define the flow, with consideration having been given also to the areas of the vessel, and only for the sake of brevity did we avoid it and will we similarly omit it in the future, unless perhaps the motion be noticeably changed by the varying shape of the vessel, which can happen in vessels in which fluid is moved that are wide enough but very long, particularly if the motions to be determined are oscillatory. Finally, we have seen in the preceding Chapter that, if there are very small oscillations in extremely deeply submerged pipes, then it is so far from necessary that one should pay attention to the orifice in the base alone, the areas, although large enough, having been neglected, that rather one should take those [areas] alone into consideration.

§6. SCHOLIUM 2. Because in the calculation which we have performed the *live force* of any volume element flowing through G must be absorbed by the water in the lower cavity, it is evident that the proposition must not be extended to those cases which oppose the hypothesis, as, for example, when the diaphragm *EF* is close to the base *CR* and consequently the orifices lie directly opposed to one another; thus indeed it is not hard to predict that the motion will be different by far from that which the present theory indicates. However, if the distance *DG* is large, and if as well the position of the orifices is oblique and the walls of the orifices do not allow the aqueous stream to contract, then there is no doubt that the theory corresponds accurately to all phenomena.

§7. COROLLARY. If the orifice *G* is fairly large compared to the other, v becomes almost equal to x; but this height v, to which certainly the velocity of the water flowing out through *D* corresponds, decreases considerably with the orifice *D* increasing, so that if, for example, there would be twice the orifice *G*, v would be $\frac{1}{5}x$, but the total [height] almost vanishes when the orifice *G* is extremely small with respect to the orifice *D*.

Thus, with these things having been effected, anyone will now perceive the true understanding of those motions which Mariotte first observed and concerning which very *admirable* [findings] he states that he was overly pleased; and simultaneously one will understand how far this Author, most clear-sighted in other matters, diverged from the [right] path in these treatises. I believe that it will not be irrelevant to insert the observations of Mariotte here.

§8. He used a vessel such as Fig. 38 represents, which differs from the former only in this, that in the lowermost part there is inserted in

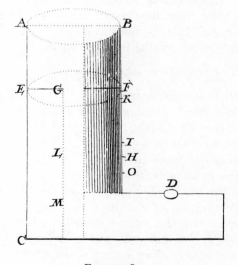

FIGURE 38

the cylinder ABC a horizontal pipe MD, perforated by an opening D through which water springs forth vertically: but the Diaphragm EF is, as before, perforated in the middle by the opening G; below this there was a very small orifice K, [placed there] in order that the lower cavity could be filled with water more easily; after this had been done, [the small orifice] was closed and the rest of the vessel was filled.

Having prepared things in this manner, with water flowing out through D, Mariotte observed that it soon ascended up to I, then, gradually, with the velocity diminished, up to H, and finally, with total depletion of the upper cavity $ABFE$ impending, up to O, and then, new forces suddenly having been added, jumps almost up to F. He also noticed, if I remember correctly, that the height of the initial thrust is the smaller, the smaller the orifice G with regard to the other, D. This may be seen in his *Traité du mouvement des eaux*, Part IV, disc. I. Moreover, he believes that the changes in these motions can be explained by imagining that to the very wide vessel $ABFE$ a rather narrow pipe $GLMD$ is connected through which the water flows. But we have certainly demonstrated and experience teaches daily that the motion of water out of the vessel $ABGLMD$ is very different from that which has just been indicated. One would be no less wrong if he believed that the water springs forth with the same velocity through the orifice D as if the latter would have been located in the diaphragm EF, for it can happen that the height of the initial thrust is larger or smaller than the height FB. And, finally, the water does not flow out in that quantity, as one might easily suspect, in which it would, at the same time, flow out of the upper vessel alone through the opened part $EFDC$, although this is approximately the case when the orifice G is so much smaller than the orifice D.

§9. In truth our equation, namely $v = \dfrac{mmx}{nn + mm}$, corresponds altogether correctly to the phenomena: it indicates indeed that the water ascends soon after the beginning of flow to a certain height, which is less, the smaller the orifice in the diaphragm is with respect to the other orifice; that this ascent is then gradually diminished until all the water has flowed out of the upper cavity; that at this very instant it immediately experiences an increase and [the thrust] reaches not quite the total height of the water lying above, because then the water is to be regarded as flowing out of a simple vessel which is infinitely wide; nevertheless, even now the water is retarded a little by the transition of the air through the orifice G, and understandably it is retarded noticeably if the upper orifice is very small, about which subject we will soon say something when the discussion will concern heterogeneous fluids. If Mariotte's figure stands in proper proportion in the argument brought forth, then it is necessary that one make the orifice *G* a little more than half of the other.

§10. Our formula indicates further that it could perhaps have seemed to be a paradox to many, when this theory had not yet been understood, that either a higher or a lower position of the diaphragm *EF* in no way changes the impact or the velocity of the water flowing out; however, I believe that the understanding of this phenomenon is now manifest to everybody.

§11. Now, however, we will examine further the motion of water when there are many diaphragms perforated by orifices through which the water is forced to flow in order that efflux through the orifice *D* may occur. This can be solved by the same method which we have used in the problem of §4. Moreover, after the calculus has been correctly applied, and with the notations applied in the same manner having been retained, there appears

$$v = x \Big/ \left(1 + \frac{nn}{\alpha\alpha} + \frac{nn}{\beta\beta} + \frac{nn}{\gamma\gamma} + \cdots \right),$$

where by α, β, γ, etc., are understood the areas of the orifices which are in the diaphragms, while n expresses, as before, the area of the orifice *D* through which the water flows out.

§12. If then in place of one diaphragm there are in the same vessel, which Fig. 39 represents, many diaphragms, let us say *B*, *C*, *R*, etc., through which the water flows as long as it flows out through the lowermost orifice *D*, then the velocity of the outflowing water will be changed and increased immediately every time some cavity is depleted: further, the proportion between the heights *AB*, *BC*, *CR*, *RE*, etc., and the areas of the orifices *D*, *G*, *F*, *H*, etc., can be such that every time when a new chamber starts to be depleted, the outflowing stream always rises to the same height *O*, or it flows out at the same velocity. But this is obtained (the areas of the orifices *D*, *G*, *F*, *H* etc., having been designated by n, α, β, γ, etc.) by setting

$$BC = \frac{nn}{\alpha\alpha} AB; \quad CR = \frac{nn}{\beta\beta} AB; \quad RE = \frac{nn}{\gamma\gamma} AB; \quad \text{etc.,}$$

so that after the orifices have been set equal to one another, the lines *AB*, *BC*, *CR*, *RE*, etc. are similarly to be made equal to one another. It will also be easy in a cylindrical vessel to ascribe such a magnitude to the orifices that the surface of the fluid descends in the same time from one diaphragm to any subsequent one, and since these diaphragms are spaced equally from each other and from the base, a uniform construction of clepsydras can be invented.

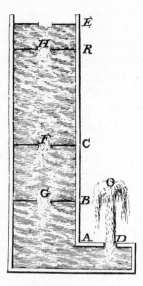

FIGURE 39

§**13.** Indeed, if all diaphragms are placed very high, it will be a pleasant hydraulic game to observe the discharging stream DO, which increases repeatedly by equal increments and in equal intervals of time, which can be done.

§**14.** Let it be proposed now to investigate the motion of the discharging fluid when different fluids flow through all the individual orifices. But evidently the successively lighter fluids have to be placed so that they are located higher in order that the motion does not become disordered, which happens when a lower fluid ascends at the same time that an upper one descends through a common orifice. In this manner one may determine what the motion is in the water flowing out of a vessel closed everywhere except for some small orifice located at the top which allows air to enter. But we will retain the hypothesis of the infinite area of the cylindrical vessel in relation to the orifices, and further we will designate the specific gravity of the fluid discharging through D by A and that of the one which flows through G we will denote by the letter B, and similarly we will indicate the specific gravities of the fluids flowing through F, H, etc., by the letters C, D, etc., respectively. Finally, since here also one must consider the heights of the different fluids, of which only the lowermost discharging one changes height, of course, because of the cylindrical shape of the vessel, we will let x be the height of the lowermost fluid above the orifice D; the heights of the remaining fluids we will designate, in that order in which they lie on top of each other, by b, c, d, etc., respectively, and we will retain the other designations of §11; with these things having been thus prepared, the computation may be performed as has been done in §4, for there is nothing else further to be observed than that the masses of the volume elements passing through the different orifices in the same small time intervals are estimated not simply from the bulk, but also from the specific gravity; the *actual descent* for the individual fluids will have to be taken separately, though. By following this path an equation is found at first in this form:

$$Av + \frac{nn}{\alpha\alpha} Bv + \frac{nn}{\beta\beta} Cv + \frac{nn}{\gamma\gamma} Dv + \cdots = Ax + Bb + Cc + Dc + \cdots$$

which, reduced, gives

$$v = (Ax + Bb + Cc + Dd + \cdots) / \left(A + \frac{nn}{\alpha\alpha} B + \frac{nn}{\beta\beta} C + \frac{nn}{\gamma\gamma} D + \cdots \right).$$

§15. If there are two liquids, two terms will have to be taken in the numerator as well as in the denominator, and three terms if there are three liquids, and so forth; if then the liquid flowing out were mercury, for example, and if water were lying on top of it, and if the specific gravities of these liquids were established as 14 to 1, it would occur that

$$v = \frac{14x + b}{14 + nn/\alpha\alpha},$$

and if the ratio of the orifices D and G should be, for example, as 3 is to 1, it would occur that

$$v = \frac{14x + b}{23}.$$

§16. It is also evident that that reasoning does not exclude those cases in which the upper fluids are specifically heavier than the lower ones, only that the lower fluids do not ascend through the same orifices through which the upper ones descend; but I presume (however, I do not confirm) that this will not happen when instead of a simple orifice there is some little pipe of small height through which the upper liquid may descend into the lower cavity, just as in Fig. 40, where indeed only two liquids are considered.

Here, though, the height CR is variable, the height AC is constant; meanwhile, nevertheless, for the sake of uniformity of nomenclature, we will set the height $AC = x$, the other one, CR equal to b; the specific gravity of the fluid going out through D we will set again equal to

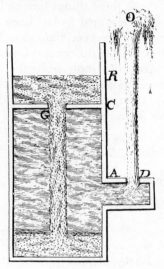

FIGURE 40

A, and that of the other fluid, passing through G equal to B; the height DO will result, or

$$v = \frac{Ax + Bb}{A + Bnn/\alpha\alpha}.$$

Therefore, if water and mercury flow through the orifices D and G, respectively, now there will be

$$v = \frac{x + 14b}{1 + 14nn/\alpha\alpha}.$$

§17. In order to understand further the motion of a simple fluid out of a vessel that admits air through a very small orifice on top, it is to be observed that here the height b is null; because the air can be considered to lie above each orifice up to the same height, hence there will be

$$v = \frac{Ax}{A + Bnn/\alpha\alpha},$$

and if $\frac{A}{B}$ were 850, which is more or less the usual proportion between the specific gravities of water and air, there will be

$$v = \frac{850x}{850 + nn/\alpha\alpha}.$$

§18. All these principles which we so far have applied are, as I have said already, easily extended to vessels which have a finite area in proportion to the orifices; but the truth of these things can be proved also in another, very different manner, as I shall show when I come to *hydraulico-statics*, because, by that other method of proof, the pressures of the fluids on the individual parts of the vessel become more clear; however, the statical rules of those fluids differ strongly from the laws which are due to standing fluids.

Otherwise, these things have their use in correctly understanding hydraulic machines; indeed the professional men seem not to have attended enough to this; occasion will also be given to elaborate on these things more copiously in the following chapter, where we will perform a calculation of how much force applied in propelling water may be lost from the passing of water through many orifices, and we will simultaneously show the remedies to be applied in order that that loss of forces be diminished as much as possible. But we will consider certain other composite vessels in this chapter before we turn to those.

§19. It happens sometimes that vessels put next to each other receive water from one another that finally will flow out of the last one. We will now illustrate those motions by an example.

Let there be proposed a vessel *AGMB* (Fig. 41) of any shape what-

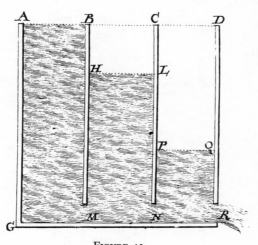

FIGURE 41

ever, which is kept constantly full up to *AB* by a new supply of water. Meanwhile, let the fluid be understood to go from this very vessel through an orifice *M* into another adjoining vessel *BMNC*, and from that again into another one, *CNRD*, through the orifice *N*, and so on, until finally the water is emitted into the air, and let the locations of the surfaces *HL, PQ*, etc., be sought after they have been reduced to a state of *permanence*. The question is solved as follows:

It is certainly clear from the fact that the surfaces *AB, HL, PQ*, etc., remain in the same position that the water goes through the orifices *M, N, R* at the velocities which are due to the heights *BH, LP, QR*, if only the transit of water through the one orifice does not accelerate its flow through the next orifice, which certainly does not occur unless an effort is made expressly in order that this happen somehow. But furthermore it is to be considered that the velocities of the water flowing through the orifices are reciprocally proportional to the orifices, because in the state of *permanence* the same amounts of water are released in the same time through the individual orifices. From this it is recognized, once the areas of the orifices *M, N, R* have been designated by *m, n, p*, that *LP* will be $\frac{mm}{nn} BH$; $QR = \frac{mm}{pp} BH$. But *BH + LP + QR* is equal to the height of the surface *AB* above the last orifice *R*, or [equal to] *DR*; therefore,

$$BH + \frac{mm}{nn} BH + \frac{mm}{pp} BH = DR$$

and thence

$$BH = DR \Big/ \Big(1 + \frac{mm}{nn} + \frac{mm}{pp}\Big);$$

and similarly

$$LP = \frac{mm}{nn} DR \Big/ \Big(1 + \frac{mm}{nn} + \frac{mm}{pp}\Big)$$

and

$$QR = \frac{mm}{pp} DR \Big/ \Big(1 + \frac{mm}{nn} + \frac{mm}{pp}\Big),$$

or

$$BH = DR \Big/ \Big(1 + \frac{mm}{nn} + \frac{mm}{pp}\Big)$$
$$LP = DR \Big/ \Big(1 + \frac{nn}{mm} + \frac{nn}{pp}\Big)$$
$$QR = DR \Big/ \Big(1 + \frac{pp}{nn} + \frac{pp}{mm}\Big)$$

and thus are determined the invariable locations of the surfaces *HL, PQ*, etc. But on the other hand we will examine below, together with other questions pertaining to this, how much time it takes for this to occur if those surfaces are located differently and in the interim some certain quantity of water flows through the individual orifices. However, we will deduce [some] outstanding relations arising from the presented values of the heights BH, LP, QR, etc.

§20. I. If the individual orifices are equally large, there will be $BH = LP = QR$, etc., and something of these heights will be contained in the height DR as long as the vessels are open.

II. But if some orifice is infinitely small in relation to the others, all surfaces which are located upstream from the orifice will be at the same height as the first surface AB; but the remaining will be close to the base GR.

III. If a continuous conduit passing through the individual orifices M, N, R, etc., is assumed, then it is recognized that the water must flow out through the orifice of the conduit at a velocity which is due to the total height DR. But in our case that velocity corresponds only to the height QR, of which matter the reason and origin is this, that the *potential ascent* of the individual volume elements flowing through the orifices—except only for the orifice of efflux—is absorbed. Therefore, the live force which is lost at any individual instant is to the live force which is generated at any individual instant as DQ is to DR. But the heights BH, LP, etc., represent *respectively* the live force which is continually withdrawn separately from the volume elements flowing through the orifices M, N. Nevertheless, if the orifices are almost equal and if their centers are located in a straight line, and if, finally, the walls BM, CN, DR are placed not very far from each other, [then] I believe it can happen that the water springs forth at some higher velocity than this theory indicates. In the remaining cases I do not doubt their accuracy, neglecting the often indicated hindrances.

IV. Finally, it is evident that every time the water surfaces HL, PQ, etc., change their position, whether many or one alone, soon all surfaces will change their locations until they have been brought back to equilibrium in the manner that has been mentioned. But to define these changes generally is [a matter] of equally intricate as well as laborious calculation, unless the vessels are taken as prismatic and as practically infinite with respect to the area of the orifices, namely in order that the increments of the *potential ascents* of the water ML, NQ, etc., which change their locations, can be neglected in comparison to the *potential ascents* which are perpetually generated in the volume elements flowing through M, N, R. And this restriction at the outset must not affect us, since we already saw in passing that in moderately wide vessels the increments of the motion of the internal masses can be left out of the calculation without noticeable error. Therefore, let me omit the general solution, which is mine, on account of its overly great involvement, and for greater elegance let me, as I have done so far in this chapter, assume the vessels as infinitely wide and certainly prismatic. But let me start with a sectioned vessel.

§21. A sectioned vessel of this kind (Fig. 42) is represented, the part AM of which is assumed full of water; the other, BN, is assumed to be

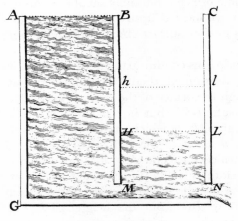

FIGURE 42

filled only up to HL when flow starts through each orifice M and N; water is also poured in at AB in order to keep the vessel constantly full, and thus it will occur that the water in BN rises (or even descends, according to the conditions). When this is so, we will seek the velocity of the surface of the water when it arrives at the position hl.

To this end we will express the area of the orifice M by m, that of the orifice N by n, and the area hl (which certainly is taken everywhere the same) by g. Then let us set $BM = a$, $HM = b$, $Bh = x$, and hence $hM = a - x$. But thus it is evident from the assumption of a practically infinite area of the vessels AM and BN that when the variable surface of the water is at hl, then the height due to the velocity of the water flowing through M will be $Bh = x$, and the velocity itself will be $\sqrt{x}$, and with respect to the orifice N the similar height will be $hM = a - x$, and the velocity of the water flowing through N will be $\sqrt{a - x}$; therefore the quantity flowing into the vessel BN through M in a given time element is to the quantity flowing out of

MOTION OF FLUIDS THROUGH IRREGULAR VESSELS 173

the vessel in the same time element as $m\sqrt{x}$ is to $n\sqrt{a-x}$, and the difference of these quantities divided by the area g gives the velocity of the surface hl; hence, this velocity, which we will call v, will be expressed by the equation:

$$v = \frac{m\sqrt{x} - n\sqrt{a-x}}{g}$$

§22. Now, in order to find the time in which the surface of the fluid rises from HL to hl, we will call this time t. But because $dt = \frac{-dx}{v}$, there will be, after the value just found has been substituted for v,

$$dt = \frac{-g\,dx}{m\sqrt{x} - n\sqrt{a-x}}.$$

Certainly this formula can be made rational at once by putting $x = \frac{4aqq}{(1+qq)^2}$ and then arranging in the required manner. But this method is slightly more favorable than that other in which the quantity to be reduced is divided into two parts that are consequently to be integrated; certainly the equation set forth does not differ from the following:

$$dt = \frac{mg\,dx\sqrt{x}}{nna - (mm+nn)x} + \frac{ng\,dx\sqrt{a-x}}{nna - (mm+nn)x}.$$

And also

$$\int \frac{mg\,dx\sqrt{x}}{nna - (mm+nn)x} = -\frac{2mg}{mm+nn}\sqrt{x}$$
$$+ \frac{mng\cdot\sqrt{a}}{(mm+nn)\sqrt{mm+nn}} \ln \frac{n\sqrt{a} + \sqrt{mm+nn}\sqrt{x}}{n\sqrt{a} - \sqrt{mm+nn}\sqrt{x}};$$

the integral of the other part, namely $\int \frac{ng\,dx\sqrt{a-x}}{nna - (mm+nn)x}$, becomes

$$\frac{-2ng}{mm+nn}\sqrt{a-x} + \frac{mng\sqrt{a}}{(mm+nn)\sqrt{mm+nn}}$$
$$\times \ln \frac{m\sqrt{a} + \sqrt{mm+nn}\sqrt{a-x}}{m\sqrt{a} - \sqrt{mm+nn}\sqrt{a-x}}.$$

HYDRODYNAMICS, CHAPTER VIII

It is evident that, after the required constant has been added, this will yield

$$t = \frac{2mg\sqrt{a-b} - 2mg\sqrt{x} + 2ng\sqrt{b} - 2ng\sqrt{a-x}}{mm+nn} + \frac{mng\sqrt{a}}{(mm+nn)\sqrt{mm+nn}}$$

$$\times \ln\frac{mna + (mm+nn)\sqrt{ax-xx} + m\sqrt{mm+nn}\sqrt{ax} + n\sqrt{mm+nn}\sqrt{aa-ax}}{mna + (mm+nn)\sqrt{ax-xx} - m\sqrt{mm+nn}\sqrt{ax} - n\sqrt{mm+nn}\sqrt{aa-ax}}$$

$$- \frac{mng\sqrt{a}}{(mm+nn)\sqrt{mm+nn}}$$

$$\times \ln\frac{mna + (mm+nn)\sqrt{ab-bb} + m\sqrt{mm+nn}\sqrt{aa-ab} + n\sqrt{mm+nn}\sqrt{ab}}{mna + (mm+nn)\sqrt{ab-bb} - m\sqrt{mm+nn}\sqrt{aa-ab} - n\sqrt{mm+nn}\sqrt{ab}}.$$

§23. From §19 it is clear that the surface hl remains in its position since $Bh(=x) = \frac{nna}{mm+nn}$. But if in the integrated equation of the preceding paragraph one puts $x = \frac{nna}{mm+nn}$, the denominator in the logarithmic quantity becomes $= 0$, and hence the quantity itself is infinite. The time of the total motion, therefore, is infinitely greater than that of any part.

But in order that we may determine another case beyond this, we will see in how much time the surface of the water would ascend from its lowermost position MN (namely by setting $b = 0$) by the quantity $\frac{1}{2}a$, but setting $m:n = 4:3$, one has

$$t = \frac{8g\sqrt{a} - 14g\sqrt{\frac{1}{2}a}}{25} + \frac{12g\sqrt{a}}{125}\ln\left(\frac{49 + 35\sqrt{2}}{49 - 35\sqrt{2}}\right) - \frac{12g\sqrt{a}}{125}\ln(-4),$$

or,

$$t = \frac{8g\sqrt{a} - 7g\sqrt{2a}}{25} + \frac{12g\sqrt{a}}{125}\ln\left(\frac{49 + 35\sqrt{2}}{140\sqrt{2} - 196}\right)$$

that is, approximately $t = \frac{15g}{100}2\sqrt{a}$, which indicates that this time is to the time during which a heavy weight falls through the height BM approximately as $15g$ is to 100: equally, the time of descent is found, if in the beginning the surface hl would have been located above the position of equilibrium. Let, for example, either one of the vessels be completely void of water, and let the orifices M and N now have a ratio 3 to 4, and let the time be determined in which the surface descends from B through half of BM: these hypotheses make $m = 3$; $n = 4$; $b = a$; and $x = \frac{1}{2}a$, so indeed one has

$$t = \frac{8g\sqrt{a} - 7g\sqrt{2a}}{25} + \frac{12g\sqrt{a}}{125}\ln\left(\frac{49 + 35\sqrt{2}}{49 - 35\sqrt{2}}\right) - \frac{12g\sqrt{a}}{125}\ln(-4).$$

From this it is apparent that the time is the same in either example.

§24. Before we get onto manifold vessels, it is convenient to have investigated what quantity of water flows through each orifice M and N while the surface of the water goes from the position HL into hl. And first of all, certainly, it is evident, as far as the orifice M is concerned, that the quantity of water flowing through it in a given time interval (dt) is proportional to the velocity $(\sqrt{x})$ multiplied by the magnitude of the orifice (m) and the same small time interval dt, so that this quantity is (on account of $dt = \dfrac{-g\,dx}{m\sqrt{x} - n\sqrt{a-x}}$ per §22)

$\dfrac{-mg\,dx\sqrt{x}}{m\sqrt{x} - n\sqrt{a-x}}$, and hence the entire quantity which has flowed out from the beginning is

$$-\int \frac{mg\,dx\sqrt{x}}{m\sqrt{x} - n\sqrt{a-x}}.$$

But

$$-\int \frac{mg\,dx\,\sqrt{x}}{m\sqrt{x} - n\sqrt{a-x}} = \frac{mnga}{(m+n)^2}\ln\left(\frac{ma - mb - nb}{mx + nx - na}\right) + \frac{mg}{m+n}(a - b - x).$$

In the same manner one evaluates the quantity of water flowing out meanwhile through the orifice N, which, of course, is

$$-\int \frac{ng\,dx\sqrt{a-x}}{m\sqrt{x} - n\sqrt{a-x}},$$

is equal to

$$\frac{mnga}{(m+n)^2}\ln\left(\frac{ma - mb - nb}{mx + nx - na}\right) - \frac{ng}{m+n}(a - b - x).$$

And from here the quantity of water which is poured into AB becomes known also, and it certainly does not differ from that which flows through M; finally, the water collected in the vessel BN is represented by $g(a - b - x)$, and when the difference of the water flowing through M and through N is determined, then that same quantity $g(a - b - x)$ appears.

§25. Just as in §21 we have determined for a two-part vessel the velocity of a surface continuously changing its position, so let us now define the velocities of the individual surfaces for manifold vessels. Certainly one may set the height of the uppermost surface above the next equal to x, the height of that one above the following equal to y, then equal to z, and again the next height equal to s, and so forth. But let the areas of the orifices be designated by m, n, p, q, etc., let the areas of the second, third, fourth, etc., vessel be M, N, P, etc. Thus it is evident that the velocity of the second surface will be $\dfrac{m\sqrt{x} - n\sqrt{y}}{M}$, the velocity of the third surface will be $\dfrac{n\sqrt{y} - p\sqrt{z}}{N}$, the velocity of the fourth surface will be $\dfrac{p\sqrt{z} - q\sqrt{s}}{P}$, etc.

Further, since the small spaces passed through by the surfaces in the same small time intervals are in proportion to the velocities, it is thus apparent that at any instant the position of these surfaces is determined, although the equations are almost intractable. This is evident by itself, or, if a single surface were put off its position of equilibrium defined in §19 above, then all the remaining are agitated by reciprocal motions, until after an infinite time they will simultaneously go back to their original position.

§26. Further, let a vessel be formed such as Fig. 43 shows, divided, of course, into two parts $ABEG$ and $LQNE$, communicating with each other through the middle orifice M; and let there be, furthermore, the orifices H and N through which water springs forth as long as the same

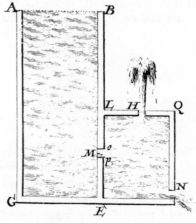

FIGURE 43

amount is poured into *AB*. Also, let the areas in either vessel be infinitely large in proportion to the orifices *M*, *H*, and *N*; after these things have been established, let it be proposed to find the velocities with which the water is ejected through *H* as well as through *N*, or the heights due to these velocities. But the velocities will be invariable, because the vessel is kept full of water and at the same time the areas of the vessel are taken as infinite with respect to the orifices.

The solution of this problem may be easily understood from the preceding, if only the orifice *M* is conceived to be divided into two parts *o* and *p*, of which the one, *o*, sends water to the orifice *H*, the other, *p*, to the orifice *N*: the parts *o* and *p*, however (because through either one the water flows at the same velocity) will have the ratio which the quantities of water flowing out through *H* and *N* at the same time have, that is, a ratio composed of the ratio of the area *H* to the area *N* and of the velocity at *H* to the velocity at *N*. After these things have been admonished, then, if the areas of the orifices *M*, *H*, and *N* are indicated by α, β, and γ and the heights due to the velocities at *H* and *N* are designated by x and y, and hence the velocities themselves by $\sqrt{x}$ and $\sqrt{y}$, it is clear that one will obtain the area

$$o = \frac{\beta\sqrt{x}}{\beta\sqrt{x} + \gamma\sqrt{y}}\alpha \quad \text{and the area } p = \frac{\gamma\sqrt{y}}{\beta\sqrt{x} + \gamma\sqrt{y}}\alpha.$$

Now let the height of the surface *AB* above the orifice *H* be given equal to *a*, and *x* will result, as it was proven in §4, if the square of the orifice *o* is divided by the sum of the squares of the orifices *o* and *H*, and if the quotient is multiplied by *a*; and so it will occur that

$$x = \frac{\alpha\alpha a x}{\alpha\alpha x + (\beta\sqrt{x} + \gamma\sqrt{y})^2},$$ from which this equation results:

(A) $$\alpha\alpha x + (\beta\sqrt{x} + \gamma\sqrt{y})^2 = \alpha\alpha a.$$

In the same manner, from the ratio of the orifices *p* and *N*, after the height *AB* above *N* has been set equal to $a + b$, this other equation is obtained:

(B) $$\alpha\alpha y + (\beta\sqrt{x} + \gamma\sqrt{y})^2 = \alpha\alpha(a + b).$$

After equation (B) has been subtracted from equation (A), there results $y = x + b$, from which it follows that, if both streams are directed vertically upwards, each one springs up to the same position. Hence, if in equation (A) the value of $x + b$ is substituted for y, then

(C) $$\alpha\alpha x + (\beta\sqrt{x} + \gamma\sqrt{x+b})^2 = \alpha\alpha a,$$

from which the value of *x* itself is deduced from the quadratic equation.

§27. From the equation of the preceding paragraph the following conclusions result.

I. Because the velocity of the water flowing through M is $\dfrac{\beta\sqrt{x}+\gamma\sqrt{y}}{\alpha}$, the height generating this velocity will be $\left(\dfrac{\beta\sqrt{x}+\gamma\sqrt{y}}{\alpha}\right)^2$ but if the equations (A) and (B) are added, there results:

$$\left(\frac{\beta\sqrt{x}+\gamma\sqrt{y}}{\alpha}\right)^2 = \frac{2a+b-x-y}{2} = \text{(since } y = x+b\text{)} \; a-x.$$

II. If the orifice H is very small in proportion to the orifices M and N, that is, if β can be assumed as zero in proportion to α and γ, then equation (C) changes into this:

$$\alpha\alpha x + \gamma\gamma x + \gamma\gamma b = \alpha\alpha a, \quad \text{or} \quad x = \frac{\alpha\alpha a - \gamma\gamma b}{\alpha\alpha + \gamma\gamma}.$$

But this agrees splendidly with §19, since it is manifest that the water springs forth through a very small orifice to the same height which the water would have if it pressed the section LQ as much downward as it is pressed upwards by the internal water; but this mentioned height is, by virtue of §19, $\dfrac{\alpha\alpha a - \gamma\gamma b}{\alpha\alpha + \gamma\gamma}$. Further, in this hypothesis one finds the height of the velocity of the water at N, or

$$x + b = \frac{\alpha\alpha a + \alpha\alpha b}{\alpha\alpha + \gamma\gamma},$$

and finally the height of the velocity of the water at M, or

$$a - x = \frac{\gamma\gamma a + \gamma\gamma b}{\alpha\alpha + \gamma\gamma},$$

which latter equations could have been immediately understood or predicted in this particular case from §19 as well.

III. But if now another orifice N, sufficiently small, is placed in front of the remaining two, there will be, after one has set $\gamma = 0$,

$$x = \frac{\alpha\alpha a}{\alpha\alpha + \beta\beta}; \quad \text{then } x + b = \frac{\alpha\alpha a + \alpha\alpha b + \beta\beta b}{\alpha\alpha + \beta\beta}, \quad \text{and } a - x = \frac{\beta\beta a}{\alpha\alpha + \beta\beta}.$$

IV. If $\gamma\gamma b = \alpha\alpha a$, x becomes null. Therefore, in this case the various portions of the section LQ sustain no pressure: in fact it is pressed downward if γ is larger than $\dfrac{\alpha\alpha a}{b}$ and the section is not perforated anywhere.

But, similarly, all these things are understood easily from §19.

V. Thus, also, by means of the same paragraph it could have been predicted without new calculations what should happen if, the orifices H and N having been located at the same height, the sum of these orifices or a unique one of area $\beta + \gamma$ can be considered. Certainly §19 as well as §26 indicates that

$$x = \frac{\alpha\alpha a}{\alpha\alpha + (\beta + \gamma)^2}.$$

VI. It can also be noted that, when the value of x itself becomes imaginary, it happens not only that does the water not flow out through H in certain cases, but also that the surface LQ descends; whence it can happen that it descends below the orifice M, in which case the continuity of the water ceases, contrary to the hypothesis of the proposition. Moreover, if the value x is real, then it is doubly expressed, but the other value is to be considered useless; accordingly, therefore, care has to be taken lest the absurd root be taken as useful.

VII. Finally, in order that we may treat a very special case, let us set all orifices equal to one another, and there will result

$$5xx + (2b - 6a)x = -aa + 2ab - bb$$

or

$$x = \frac{3a - b - 2\sqrt{(aa + ab - bb)}}{5};$$

and if, furthermore, $a = 3b$, then x will be (approximately) $\frac{4}{15}b$, hence the height of the velocity at the orifice N or $x + b = \frac{19}{15}b$ and the height due to the velocity at M or $a - x = \frac{41}{15}b$. And so the velocities or even, since the orifices are equal, the quantities of water flowing in the same time through the orifices M, H and N, are approximately as $\sqrt{41} : 2 : \sqrt{19}$.

§28. From all this the method is evident for determining the motion of fluids, even when the amount of *live forces* is not conserved; and the computation is always performed in a similar manner as long as it can be presumed from the nature of the subject of investigation (as could be done accurately in the investigations in this chapter), how much of the *live force* vanishes that is useless for determining the motion at any instant. Certainly the cases are not singular which we have examined so far; it is pleasing, therefore, to add another one which treats oscillations of fluids, in order that one may know for this how much the displacements of the fluid decrease.

Let there be two pipes, equal in size and cylindrical, AL and BH

(Fig. 44), inserted vertically into a very large horizontal vessel *ABOP*. Let that vessel be completely full of water, but let the pipes contain water up to *C* and *F*; then, with equilibrium having been disturbed, let one surface stay at *G* and the other at *E*; and let the water, left to itself, soon begin to move. These things having been

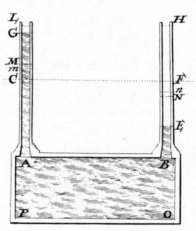

FIGURE 44

set forth, the surface *G* should descend as far below the position *C* and the other surface *E* should ascend as far above *F* as the height *CG* or *EF* is, if the entire *live force* were conserved (we disregard the hindrance of frictions and other similar things); in truth it is evident that the *live force* of all the water flowing through *A* in the horizontal vessel is absorbed without any other effect from the water standing there, and hence it follows that the descent of the surface *G* and the ascent of the other will be less than was just mentioned; therefore, we will now explore this decrement.

To this end let it be assumed that the surface from *G* has reached *M*, and let $GM = x$, $GC = b$, $CA = a$; it will occur that $BE = a - b$, $EN = x$, $MC = FN = b - x$. Further, let the height due to the velocity of the surface at *M* be v, and at the next position let it be $m = v + dv$; and the increment of the *live force* of the water (while the surfaces run through the elements *Mm*, *Nn*, or *dx*) will be $2a\,dv$, to which is to be added the *live force* of the volume element which is absorbed by the water in the horizontal vessel, namely $v\,dx$, and the sum $2a\,dv + v\,dx$ will be equal to the *actual descent* of the water multiplied by the mass of water, which product is equal to the *actual descent* of the volume element dx, multiplied by $2b - 2x$. Therefore,

MOTION OF FLUIDS THROUGH IRREGULAR VESSELS

$2a\,dv + v\,dx = 2b\,dx - 2x\,dx$. But this equation, integrated correctly, transforms into the following:

$$v = 4a + 2b - 2x - e^{-x/2a}(2b + 4a),$$

whence, if one sets $4a + 2b - 2x - e^{-x/2a}(2b + 4a) = 0$, the value of x itself will give the total displacement; if b is subtracted therefrom, the residual will indicate the descent below the point C of equilibrium.

§29. But in order that by a certain example one may show how much the oscillations are diminished by this reasoning, let us set $a = b$, having made, of course, $CA = GC$ and $BE = 0$.

Thus arises

$$3a - x = e^{-x/2a}(3a) \quad \text{or} \quad e^{x/2a} = \frac{3a}{3a - x}, \quad \text{or} \quad x = 2a \ln \frac{3a}{3a - x},$$

in which equation the value $x = \tfrac{7}{4}a$ is almost fully satisfactory. Therefore, the decrement of the displacement, or $a - b$, is equal to the fourth part of the elevation of the fluid above the middle point; if it is observed to be greater in the experiment, the balance will have to be attributed to the adhesion of the water to the walls of the pipes.

§30. This reasoning of the diminished displacements clearly should not be withdrawn, as I suspect, if the horizontal pipe becomes equal in area to the vertical ones, on account of the changed direction of the fluid at the points A and B.

Furthermore, infinitely many other cases could be invented to be solved by these principles, just as the nature of the oscillations is to be investigated in the vessel of Fig. 44 when in the diaphragmatic horizontal section it is split into two parts communicating with one another through the single opening which the diaphragm has, and other cases of that sort. But I believe that this suffices already, so that anybody can easily form for himself the general rules for solving questions of this type.

EXPERIMENTS WHICH PERTAIN TO CHAPTER VIII

EXPERIMENT I. The fourth paragraph, in which it is said that the height for the velocity of the water flowing out through the orifice D (Fig. 37) is $\dfrac{mmx}{nn + mm}$, I confirmed in this manner, that either of the orifices G and D has an edge like a little belt, very slightly elevated, so that there be no place for the contraction of the stream, and a safe judgment could be made of the velocities from the quantity of water flowing out in a given time. Then, having taken the measurements accurately and having observed the time in which the surface descended through the given space AP, I saw that this time corresponds correctly to the velocities defined in said paragraph; I also observed that the motion is not at all changed by an elevation or depression of the diaphragm. The remaining matters pertaining to the experiment have slipped my mind, and I did not keep a record of them; however, it seemed superfluous to me to repeat the experiment, since it will be easy for anyone to imitate it; but it is the basis for the remaining matters, which therefore hardly need any further experimental investigation; nevertheless, I wanted to try the following things in addition.

EXPERIMENT 2. I used a vessel exactly of the kind which Mariotte applied (see Fig. 38) and I confirmed our equation again in this manner: I made the water flow out of the orifice D horizontally, and then I took measurements of the height of the orifice D above the floor and the distance of the spot where the stream struck the floor from the point on the same floor vertically above which the orifice D was located; from this I found the height due to the velocity of the water flowing out at D; moreover, this very same height I found by a related experiment, which [height] the theory of this chapter indicates in §4. Similar experiments I may add at the end of the experiments pertaining to Chapter XII, which at the same time will confirm our *hydraulico-static* theory.

Finally, since there are many things in §§26 and 27 which would have to be evaluated by individual calculation, it will be worthwhile also to perform experiments concerning them, particularly since in the same effort the other experiments which will be enumerated in Chapter XII could be performed also, if, to this end, one would care to make a vessel such as Fig. 43 shows.

Furthermore, this theory is also confirmed by the experiments listed in Chapter VII which I performed concerning the oscillations of fluids flowing into pipes through openings.

NINTH CHAPTER

Concerning the Motion of Fluids that are Pushed forth not by their own Weight but by an Outside Force, and particularly concerning Hydraulic Machines and their Ultimate Grade of Perfection that can be Attained, and how this could be Perfected further through the Mechanics of Solids as well as of Fluids

§1. In this chapter, in which I have chosen especially to examine hydraulic Machines and to perfect the use of them as much as this can be done, let us disregard the variations of the motion which take their origin from the force or inertia of the internal fluid because, as we have seen, the motion of the internal water is as much not uniform from practically the first instant of flow, if the orifice is small in proportion to the internal areas, as is the case in most hydraulic Machines. It would be ridiculous to be concerned in practical cases about the changes which occur in the first instants of flow, and which we have already determined in Chapter IV, since there it had been worth the effort in order that the whole force of the theory might hence be brought into the light. Therefore, for the sake of brevity, let us assume that during the entire motion the water is constantly expelled with a velocity that is proportional to the root of the internal pressing force, after that force will have been reduced to the weight of an aqueous cylinder lying over the opening; because, whatever that force should be, there will have to be considered the weight of a vertical aqueous cylinder lying over the internal water surface, and the height of that cylinder will give the height due to the velocity of the water springing forth, if only there are no extrinsic obstacles and the water is emitted from a very wide vessel. This is to be understood in such manner that, if the lid *AB* loaded with the weight *P* (Fig. 45)

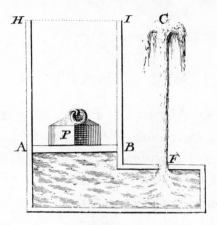

FIGURE 45

expels the water through the orifice *F*, and, moreover, if the weight *P* is equal to the weight of the aqueous cylinder *HABI*, then the aqueous jet *FG* ought to attain the height *HI*.

DEFINITIONS

§2. By *moving potential* then let me understand that acting principle which consists of a weight, an activated pressure, or other so-called dead forces of this kind.

Moreover, the product which arises from the multiplication of this *moving potential* by its velocity and also by the time during which it exerts its pressure I shall designate by *absolute potential*. Or, because the product of velocity and time is simply proportional to the distance covered, it will be permitted also to understand the *absolute potential* as the *moving potential* multiplied by the distance which the same moves through. But this very product I call *absolute potential*, because from that finally is to be estimated work endured by day laborers elevating water, which I shall soon show, proven in rules which were observed by me in this matter. Meanwhile hydraulic Machines seemed to me apt to be conveniently reduced to two types, of which the one emits water with impetus, and the other transports it, so to speak, smoothly from one place to another. I will treat an example of each in the proper order, and, finally, before the end I may add something about the diverse moving potentials.

FIRST PART: CONCERNING MACHINES EXPELLING WATER UPWARD WITH IMPETUS

Rule I

§3. The work of day laborers which is applied to hydraulic Machines for elevating water is to be estimated from the *absolute potential*, that is, from the *moving potential* or pressure which they exert, from the time, and from the velocity of the point to which the *moving potential* is applied.

PROOF. (α) Concerning the *moving potential* the matter is clear: namely, everything else being equal, the work is in any case proportional to the number of laborers or to the *moving potential*. (β) With respect to time the matter is no less manifest from the reproduction of all circumstances which arises from a duplication of the time. (γ) Finally, the matter that pertains to the velocity is to be deduced from the fact that, whether one doubles the moving force or its velocity, the effect is no different from twice [the effect] of either part. Imagine that the weight P [Fig. 45] by its descent ejects water through the orifice F to the height FG; then, the rest remaining the same, imagine the orifice F to be doubled, and one sees that twice the quantity of water will be ejected to the same height FG in the same time from the same *moving potential* P, but with the latter descending twice as fast. Equally, the quantity of water will be doubled, the rest remaining the same, if one doubles both the orifice F and the area AB and the weight or the *moving potential* P, but then the velocity of this doubled potential remains unchanged. Therefore, in either way the effect is doubled. Q.E.D.

Scholium

§4. The preceding proposition is not to be interpreted in a physiological, but in a moral sense: morally I estimate neither more nor less the work of a man who exerts at some velocity a double effort than that of one who in the same effort doubles the velocity, because certainly either one achieves the same effect, although it may happen that the work of the one, despite being no less strong than the other, is very much greater in a physiological sense. If someone advances in an effort of 20 pounds a distance of 200 feet in the first minute, he will easily be able to double the effort, but with great difficulty double the velocity. From this it follows that for every kind of machine it is to be considered particularly how it should be constituted in order that for the minimum fatiguing of the men at the same time the product of their effort by the velocity of all be a maximum: and hence it will be evident what length should be prescribed to the levers in windlasses, how large the radius should be made in wheels or rollers for treadmills, how great a length should be considered for oars, and so forth regarding other machines.

Moreover, by the reasoning of the use of treadmill rollers, which are very frequently applied in order that the moment become clearer to us, let this experiment be considered.

Let us suppose in Fig. 46 a vertical height of many thousands of feet, to which a man ought to ascend in a given time; further, let us

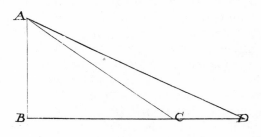

FIGURE 46

take a time of ten hours, because such is usually the limit of a day for workers; finally, let us consider several paths, AC, AD, etc., inclined differently to the horizontal BD; having supposed all this, we understand that a walker must progress the faster, the less inclined a path he will have chosen, so that he reaches the top of the mountain A in the same time, and it is evident that there will be some path, as, for example, AC, along which he travels the way with the least fatigue, insofar as nobody can either proceed up a vertical plane or travel in a given time an infinite distance; let us state that this path of least fatigue makes an angle ACB or 30 degrees with the horizontal.

If this is so, the treadmill roller will have to be fabricated such that the weight is raised with the desired velocity when the man in the treadmill is constantly 30 degrees away from the lowermost point of the roller.

According to the same principle a selection is to be made between machines of a different type: thus, for example, if on a windlass the operator exerts a potential or a horizontal pressure which has the effect of the fourth part of his own weight, and by this effort travels a distance of 200 feet in the first minute, he will, as I believe, hardly be fatigued in the same manner as if he were treading on a rotating roller with the same velocity at an angle of 30 degrees; meanwhile the man in the treadmill will nevertheless in this manner carry double the weight in the same time to the same height, because he exerts double the pressure, other things being equal.

RULE 2

§5. With the same *absolute potential* existing, I say that all machines which suffer no friction and generate no motions useless to the proposed end maintain the same effect, and that one is therefore not to be preferred to the other.

PROOF. From mechanics it is certain that any composite machine can be reduced to a simple lever: therefore, it will be pleasing to represent all hydraulic machinery by the simple pump supplied with a lever (Fig. 47), where, for example, by aid of the lever *MN* movable

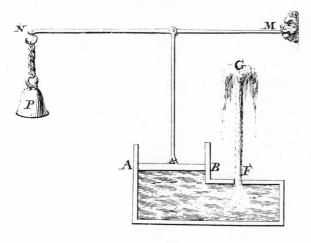

FIGURE 47

around the point *M*, a piston is pushed down, and thus water is expelled through the orifice *F*. But if the moving potential *P* applied to the lever is understood [to be] at *N*, we may see from the preceding proposition that no benefit comes to the *absolute potential* from an increased or diminished length of the lever *MN*; and certainly, whatever this length may be, it can occur that the same *moving potential*, moved at unchanged velocity, expels the same quantity of water with the same impetus as long as the area *AB* of the pump has a constant ratio to the length *MN* of the pole. From this it is very clear that all machines maintain the same effect under the same *absolute potential* as long as one disregards friction and motions which are useless to the destined end.

Scholium

§6. There are some who believe that a machine can be invented by aid of which a maximum quantity of water can be elevated to any height with a minimum of work, and they torture their minds with investigating wheels, levers, and weights to be applied: but they waste their effort, and proponents of this kind ought not to be heard, since what do these great [men] seem to have found for themselves? The best machine is, if we consider its effect alone, that which suffers the least friction and creates no useless motions, the precepts concerning the avoiding of either one of which we shall treat below.

Rule 3

§7. In pumps such as are represented in Figs. 45 and 47, in which the internal surface AB of the water is at approximately the same height as the orifice F, the *absolute potentials* for the same instants are in a threefold ratio to the velocities of the water springing forth.

Proof. The *moving potentials* are certainly in a twofold ratio to the velocities at which the waters flow out through the orifice F, and the velocities of the *moving potentials* follow the same ratio as the velocities of the water springing forth; but for the same instants the *absolute potentials* are as the moving forces multiplied by their velocities, hence the proposition is evident.

Scholium

§8. It follows from this rule that, if it be our will to elevate water through the orifice F to the height FG, a large part of the *absolute potential* is wasted fruitlessly, since the water springs forth with a greater impetus than corresponds to the height FG; for example, arrange for water to be expelled at twice the velocity, and an eightfold *absolute potential* is required, and, nevertheless, according to reason the limit of the proposed effect is not to be considered [to be] more than double, because certainly at the same time twice the quantity of water is elevated: and this effect could have been obtained with a quarter of the *absolute potential* by expelling the water at the simple velocity through double the orifice; therefore, on this account three quarters of that potential must be said to have been wasted uselessly.

MOTION OF FLUIDS PUSHED BY AN OUTSIDE FORCE 189

I have indicated the origin of this loss in §5, and it consists of the motion which is generated that is useless to the proposed end: namely, the entire motion which remains in the water after it has attained the height G is to be called superfluous in our case.

RULE 4

§9. When water is expelled through the conduit DF (Fig. 48) and has at the orifice F a velocity which is due to the vertical height GF,

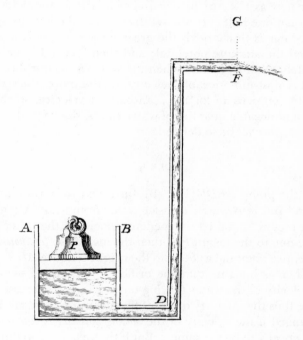

FIGURE 48

the *absolute potential* applied at the same time is proportional to the velocity of the water at F multiplied by the height G above AB.

PROOF. The moving potential P is certainly proportional to the indicated height, and the velocity of that potential is proportional to the velocity of the water at F.

SCHOLIUM

§10. The *absolute potentials* increase at a greater rate than the velocities of the outflowing water, that is, than the quantities ejected in the same time; but, nevertheless, the difference in ratios is almost unnoticeable, since the height FG is very small in proportion to the height FD of the conduit. For example, let FG be equal to $\frac{1}{4}FD$ (neglecting the height BD); then let the water be emitted at twice the velocity, such that now $FD = FG$; thus the *absolute potentials* will be as $1 \times \frac{5}{4}$ to 2×2 or as 5 to 16, so that a more-than-threefold *absolute potential* is required for emitting twice the quantity of water. But if the former FG is set equal to $\frac{1}{100}FD$, and then again the water is assumed to be expelled at twice the velocity, [then] the absolute potentials will be now as 1×101 to 2×104 or as 101 to 208, which ratio is just less than one half. It follows thence that the less the speed at which the water is discharged, the greater the success with which I have applied the absolute potential; and then finally I have applied approximately all of it usefully when the water flows out through the orifice F at almost unnoticeable velocity; furthermore, the size of the orifice could compensate for the scantiness of velocity, so that in a given time a noticeable quantity of water can be discharged. Let the loss of *absolute potential* be so defined.

Rule 5

§**11.** Let the pump $ABDF$ [Fig. 48], furnished with a little valve at the base and put into water, transfer water from a lower region AD to a higher region F, and let the median velocity of the water flowing out at F be due to the height FG; then the loss of *absolute potential* will be to that entire potential as FG is to the height G above AB.

Proof. Let us imagine that the orifice F is enlarged very much, with the velocity of the water flowing out through F decreased in the same ratio; thus the quantity of water flowing out in a given time will not be changed if the velocity of the *moving potential* is the same, and thence the effect will be the same. But if the velocity is so diminished that the height due to it is unnoticeable, the *moving potential* may be expressed by the height F above AB, since previously the *moving potential* was equal to the height of G above AB; and since in either case the velocity of the *moving potentials* is the same, the *absolute potentials* for the same times will be as the height G is to the height F above the common [base] AB. Therefore, the difference of the heights G and F will express the loss, since the entire height G above AB represents the total *absolute potential*.

§**12.** The same reasoning is valid for every kind of machinery: Indeed, whenever water, having been conveyed to the location to which it is to be elevated, has a noticeable velocity, the loss of *absolute*

MOTION OF FLUIDS PUSHED BY AN OUTSIDE FORCE

potential becomes great; for if one sets the height of the elevation equal to A, the height due to the velocity of the water at the place at which it is emitted equal to B, and the entire absolute potential equal to P, the quantity $\dfrac{B}{A+B} \cdot P$ will be lost.

It can also be noted that when water has to be conveyed over some height the culmination of which is located at F by means of a pump attached to a pipe, the pipe DF is to be continued downward as much as it may please and is not to be discontinued at F, just as it appears in Fig. 49. Because if, let us say, the point F is located twice as high as

FIGURE 49

the extremity G of the pipe, twice as large an *absolute potential* is required for transferring water through the conduit discontinued at F than through that continued to G, even if in either case it flows out at very low velocity, [and] its generating height is indeed small in proportion to the heights FD or GD.

RULE 6

§**13.** When in pumps which we have considered so far the covers AB, or rather the pistons, do not correspond well to the sides of the machines, an opening is left, and from this arises another kind of loss in absolute potentials, which is determined thus in pumps in which the height of the orifice above the piston can be neglected. As the sum of the orifice of efflux and the aforementioned opening is to this very opening, so the *absolute potential* which is exerted is to that part of it which is useless, or to its loss.

PROOF. The water is indeed pressed equally through the orifice and the opening, and it flows at an equal velocity; but the entire *absolute potential* that forces the water through the opening is lost, and this is to the complete *absolute potential* as the opening is to the sum of the orifice and the opening.

SCHOLIUM

§14. It certainly suits the piston to be well formed and smooth; it is also necessary that the cavity of the pump be exactly cylindrical and its sides be very smooth as well. But I should hardly believe, unless it is done for another purpose, that it is of importance that the pistons fill the cavities with ultimate accuracy, because thus perhaps a greater loss of forces arises from friction than if a more or less very small opening would have been left. For if that opening amounts to, let us say, the hundredth part of the orifice of efflux, there will hardly be any place for friction, and thence only approximately the hundredth part of the *absolute potential* is lost, and perhaps a larger loss arises from the friction of a piston occupying the cavity of the pump exactly. Therefore, it is not in this respect that we only too carefully avoid the transit of water through an opening left by the piston. But this consideration does not refer to those machines in which the water is to be drawn into the pump by the retraction of the piston. For here the correct and full size of the piston is entirely necessary.

RULE 7

§15. In machines which have several orifices transmitting water from one cavity to another, something of the *absolute potential* is lost, the reason for which we said in the preceding chapter is that the *potential ascent* of the individual volume elements flowing from one cavity into another through a common orifice vanishes.

The more and the smaller the orifices of this type are, the greater a loss of *absolute potential* arises, which usually is of great importance, and this perhaps apart from the common opinion, in the machines which Vitruvius names after their inventor, Ctesibius. Indeed, I speak of

MOTION OF FLUIDS PUSHED BY AN OUTSIDE FORCE 193

orifices located such that all the water that will flow out must go through them. That type of loss may now be determined by the following calculation.

Let the area of the last orifice emitting water into the air be n, but the areas of the remaining orifices through which the water is driven inside the machine be designated by the letters α, β, γ, etc., and when the same *moving potential* has been assumed in either case, the height due to the velocity of the outflowing water will be to the similar height with no internal orifices obstructing as 1 is to $1 + \frac{nn}{\alpha\alpha} + \frac{nn}{\beta\beta} + \frac{nn}{\gamma\gamma} +$ etc. (by §11, Chapter VIII), and hence it follows that with these heights having been made equal to one another, the *moving potential* will be as $1 + \frac{nn}{\alpha\alpha} + \frac{nn}{\beta\beta} + \frac{nn}{\gamma\gamma} +$ etc. is to 1, and because in either case the velocities of the moving potentials are the same, the *absolute potentials* will also have an equal ratio for these instants. Therefore the portion $\frac{nn}{\alpha\alpha} + \frac{nn}{\beta\beta} + \frac{nn}{\gamma\gamma} +$ etc. is superfluous, whence the loss of *absolute potential* will be to that entire potential as $\frac{nn}{\alpha\alpha} + \frac{nn}{\beta\beta} + \frac{nn}{\gamma\gamma} +$ etc. is to $1 + \frac{nn}{\alpha\alpha} + \frac{nn}{\beta\beta} + \frac{nn}{\gamma\gamma} +$ etc.

Scholium

§16. Whenever the idea of a machine requires orifices through which water flows from one small container into another (which happens in every kind of pump, such as aspirating ones, *aspirantes* in French, or pressing ones, *foulantes*, etc.) those orifices are to be made very large, as much as the remaining circumstances permit, so that the area of the orifice of efflux is very small with respect to those internal orifices. But, in order that the use of the rule be more clearly evident, we will consider examples of other, no less useful machines.

EXAMPLE 1. Let a machine be proposed (which Fig. 50 represents) in which the pistons C and F are alternately depressed, and by which water is introduced into the small container BEH through the passage AB and DE, in order that a continuous jet may thus discharge through the orifice H. Since here the pistons act alternately, we will consider one or the other alone, so to speak, but acting continuously; and so one must consider the orifice of efflux H, of area n, and one or the other of the orifices o and p, each one of which let have the area α;

so the loss of *absolute potential* will be equal to $\frac{nn}{\alpha\alpha}$, the whole potential having been set equal to $1 + \frac{nn}{\alpha\alpha}$, which quantities are as nn is to $nn + \alpha\alpha$. Certainly that loss is considerable, if one may trust the representations of those machines in which often the orifices o and p

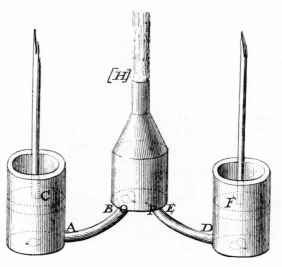

FIGURE 50

are smaller than the orifice of efflux H, because if this were so, more than half of the *absolute potential* would be lost. The conduits AB and DE will also have to be enlarged throughout their entire extents, as much as this is permissible, in order that the machine may lose little of its excellence.

As for the rest, this machine was thought up in order that a continuous jet emerge through H. Nevertheless, because it cannot happen but that some interval of time occurs between the last point of the elevation of the piston and the beginning of the instant of its depression, it will not be possible for the jet to be completely continuous and steady. However, the inventor of that machine presents an optimal remedy for this inconvenience, which Mr. Perrault mentions in *Commentarii ad Vitruvium*, p. 318, edition 2, Paris, which he says is kept in the Royal Library in Paris; this machine will serve us as another example: also, let me take the figure together with its description from Perrault himself.

MOTION OF FLUIDS PUSHED BY AN OUTSIDE FORCE 195

EXAMPLE 2. "There is a machine," according to the aforementioned Perrault, "in which water is expelled from the small container A (Fig. 51) by means of the piston B into the jar FG, out of which the

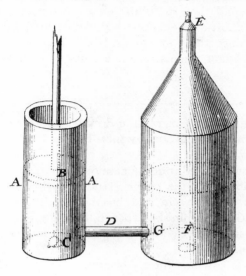

FIGURE 51

air cannot discharge as long as there is already some water present, because the pipe EF descends almost down to the bottom: indeed, it happens thus that the water, propelled from the small container A through the passage D and occupying the lowermost portion of the jar, closes the orifice of the pipe at F and prevents transit of the air. Therefore, when the piston brings new water into the small container, filled partly with air, partly with water, this newly supplied water exerts a force on either fluid, and since the water cannot spring forth through the pipe FE at the same velocity at which it flows in from the pump through the passage D, because naturally (these are Perrault's words) the pipe FE is perforated at its extremity E by an orifice much smaller than the orifice of the pipe D, the water accumulated in the vessel compresses the air and, pressed reciprocally by the latter, springs forth through the pipe FE even while the piston is raised."

In this machine a large part of the *absolute potential* is lost by the transition of the water through the passage D, and that loss will be greater, the narrower that little pipe is; therefore, it should be made wide, or even several pipes transmitting water may be constructed; this annotation is of greater importance in the present case, since a much greater loss arises from a narrow passage D than in other machines; indeed, make the area of this passage the same as the orifice E, and assume furthermore that the piston is depressed and retracted in equal time intervals: now not only is one-half of the *absolute potential* lost as previously, but clearly four-fifths will become useless. But since there are many [things] in this machine which postulate individual calculation, it is suitable to illustrate that one separately.

A Digression Containing Some Comments on the Hydraulic Machine which Fig. 51 Represents

(α) The aqueous jet through E cannot be completely steady during the entire agitation of the piston. Indeed, while the piston is elevated, no new water flows in, and thus the quantity of water contained in the vessel GE is diminished: hence the water discharges also at a continuously smaller velocity until it is accelerated again by the intruded piston.

But if the space which the air occupies in the vessel is set [to be] much larger than that space occupied by the water which is ejected during a single elevation of the piston, this entire inequality almost ceases, it having been assumed that the piston is agitated uniformly and has been agitated for a long time previously, which latter hypothesis is necessary insofar as the first ones differ very much in agitation from the following. Therefore, for the sake of brevity let us satisfy all these hypotheses, that is, let us assume everywhere what is called the *state of permanence*.

(β) Therefore, since the velocity of the water flowing out through E is increased noticeably by the first agitations of the piston, it happens soon that the aqueous jet attains almost the entire velocity; with this state of the matter having been assumed, it is evident that during the depression of the piston as much water is pushed into the vessel as is ejected out of the same during the total agitation of the piston.

During the first agitations, however, more is pushed in than is ejected, and this not for the reason, as Mr. Perrault believed, that the orifice at E is less than the other one at G (for the same would happen if it were larger), but that the generating cause does not immediately exert its total effect in ejecting water.

(γ) It would seem perhaps that it will not be sufficient for investigators that, with everything assumed in the permanent state already and no outside obstacles being present, the water springs forth from

MOTION OF FLUIDS PUSHED BY AN OUTSIDE FORCE   197

the orifice E at a velocity which can ascend to the height of an aqueous column assumed in equilibrium with the pressure of the piston; and it would be reasonably so if the pressure of the piston were present without interruption, and if no *potential ascent* were lost in the water; but because in both [cases] the situation is different, it is not possible that in the aqueous jet no other estimation of the velocity arises; hence everyone sees not obscurely that attention has to be paid to the consideration of the time in which the piston is depressed and retracted, [and] then also to the consideration of the areas in the small conduit D and of the orifice E.

(δ) Let us therefore set the time in which the piston is depressed equal to θ; the time of one entire agitation equal to t, the area of the orifice E equal to μ; and [that of] the passage D equal to m; then, after the force pushing down the piston has been compared with the aqueous column lying over it, let us make the height of this column equal to a, but the height due to the velocity of the water springing forth equal to x. After these [things] have been prepared thus for the calculation, it will be permitted to investigate in two ways the ratio which will prevail between the velocities of the water at the orifice E and at the passage D, and from here to elicit the value of the unknown x. *First*, namely, it is evident that in the time θ (in which certainly the piston is pushed down) as much water flows through the passage D as flows out through E in the time t (in which the piston is depressed and retracted). The velocity at D is therefore to the velocity at E as $\frac{1}{m\theta}$ to $\frac{1}{\mu t}$; and since this latter velocity is equal to $\sqrt{x}$, the other one will be equal to $\frac{\mu t}{m\theta}\sqrt{x}$. *Second*, because the velocity of the outflowing water is due to the pressure of the air in the jar, it follows that this pressure is equivalent to the weight of an aqueous column of height x; but if one subtracts the pressure of the air from the pressure of the piston, one will have the pressure which generates the velocity of the water at D; hence, because the difference of pressures is expressed by $a - x$, the velocity of the water at D will be represented by $\sqrt{a - x}$; therefore, the velocity of the water at D is now to the velocity of the water at the orifice E as $\sqrt{a - x}$ to $\sqrt{x}$. After combining the ratios found by either method,

$$\sqrt{a - x} : \sqrt{x} = \frac{1}{m\theta} : \frac{1}{\mu t} \quad \text{or} \quad x = \frac{mm\theta\theta}{mm\theta\theta + \mu\mu tt} \cdot a.$$

It is evident from this equation that the height of the jet is deficient for a double reason from the height a of the pressing column; indeed, it is deficient by more when the piston is depressed faster or elevated more slowly, and then also when the orifice E increases in proportion to the area of the small conduit D. For example, let the area of this orifice be equal to the area of the small pipe D, and let the piston be depressed and elevated at an equal velocity, and there will result $x = \frac{1}{5}a$, such that the outflowing stream rises only to the fifth part of the height a.

(ϵ) The loss of *absolute potential* is now evaluated in the following manner after it has been assumed beforehand that no work is done in elevating the piston. Let the velocity at which the piston is depressed be v, and the *absolute potential* expended in the time of one entire agitation will be $av\theta$ (by §3), but because the effect consists in the fact that efflux occurs through E during the time t and the water itself is elevated to the height $\dfrac{mm\theta\theta}{mm\theta\theta + \mu\mu tt} a$, the simple pump of Fig. 45 could have managed this, if in the latter as the *pressing potential* an aqueous cylinder of height $\dfrac{mm\theta\theta}{mm\theta\theta + \mu\mu tt} a$ had been taken, and this potential had acted during the time t at the velocity $\dfrac{\theta}{t} v$; whence the required *absolute potential* in this simple machine in which nothing of the former is lost would have been

$$\frac{mm\theta\theta}{mm\theta\theta + \mu\mu tt} \cdot a \cdot \frac{\theta}{t} v \cdot t = \frac{mm\theta\theta}{mm\theta\theta + \mu\mu tt} \cdot av\theta.$$

The total *absolute potential* is therefore to the uselessly wasted part of it as $av\theta$ is to $av\theta - \dfrac{mm\theta\theta}{mm\theta\theta - \mu\mu tt} \cdot av\theta$, or as $mm\theta\theta + \mu\mu tt$ is to $\mu\mu tt$. Therefore, if the entire *absolute potential* is designated by P, the loss of it will be $\dfrac{\mu\mu tt}{mm\theta\theta + \mu\mu tt} P$.

Therefore, it is necessary in this rather than in other pumps that the passage at least exceed the orifice E in area, or that it be many times as great. Indeed, if there is a single one, and this is equal in area to the orifice E, and at the same time the piston is assumed to be agitated upward and downward at uniform velocity, a loss of four-fifths of the total will arise; and if it were made twice as large, then still half of the *absolute potential* would be lost.

(ξ) Finally it is clear that the sides of the jar GE sustain a lesser pressure than [those of] the small container AA; indeed, these pressures are as x is to a, that is, as $mm\theta\theta + \mu\mu tt$ is to $mm\theta\theta$, from which ratio engineers will judge the strength of the sides which is required for either one.

──────────────── [*End of Digression*] ────────────────

Rule 8

§17. When the piston in pumps is extracted and the water flows into the small container, not only excited by its own weight but for the most part drawn by the piston, then all the *absolute potential* expended in this attraction comes into the problem in addition, because a pump placed under water, as it happens, would be filled on its own if sufficient time for filling were allowed; thus that attraction does not especially pertain to the ejecting of water with a certain velocity, so the entire [attraction] could be avoided, and on this account the work expended in that [attraction] is called useless by me.

But as the inflow occurs partly by the water's own weight, partly also by the lifting of the piston, the loss of *absolute potential* cannot be estimated from the effect; indeed, the calculation is to be set forth rather so that, after the force elevating the piston to a certain position has been set equal to π, the velocity of the piston equal to v, and the small time interval corresponding to the quantities π and v equal to dt, the entire *absolute potential* expended in elevating the piston is called $\int \pi v \, dt$ or $\int \pi \, dx$, if by dx is understood an element of space traversed in the small time interval dt. It follows hence that if the effort by which the piston is raised is of constant magnitude, as it is almost, the *absolute potential* will be equal to the *moving potential* multiplied by the traversed space; but since a similar consideration is valid also for the depression of the piston, and also the piston is as much raised as it is depressed, it is apparent that the *absolute potentials* which are exerted in alternately attracting and expelling the water are approximately in proportion to the *moving potentials* in either case; whence a loss arises which is equal to $\dfrac{\pi}{\pi + p} P$, after one has set, of course, the elevating potential equal to π, the depressing potential equal to p, and the *absolute potential* exerted in the elevation and depression of the piston equal to P.

The loss of *absolute potential* can be estimated approximately in a different way from the fact that the whole *potential ascent* of the water flowing into the pump must be thought of as generated uselessly. But if the piston is moved upward and downward in the same time intervals, or at the same velocity, the velocity at which the water is taken in will be to the velocity at which it is expelled reciprocally as the corresponding orifices, and the *potential ascents* themselves in either case will be in the inverse-square ratio of the corresponding orifices; if, further, the elevation and depression of the piston occur in different time intervals, the velocities are reciprocally as the time intervals, and the *potential ascents* reciprocally as the squares of the time intervals. Therefore, the *potential ascent* generated by the inflow of water is to the *potential ascent* which arises from the efflux, and which alone is intended, in an inverse-square ratio composed of the ratio of the orifice of inflow to the orifice of efflux and of the time in which the water is drawn in to the time in which it is expelled.

Scholium

§18. From either means of estimating, it follows that the piston is to be raised slowly; for thus the *moving potential* becomes small, by reasoning of the first method, or the time of elevation becomes large, by reasoning of the second, and thus the laborers may recover from the exhausting effort of the preceding depression during the individual intervals of the elevation of the piston. The latter method indicates further that the orifices through which the water is drawn are to be made larger and more numerous; but this is also in accordance with the former method because thus an almost sufficient quantity of water flows in on its own, and so less *moving potential* is needed.

Rule 9

§19. Finally, it is to be observed that the aqueous stream, rising vertically, never attains that height which would be due to the initial velocity of the water; that is, if the stream of fluid would start to rise vertically from its origin with as great a velocity as a weight falling freely from the height a would acquire, the fluid could not ascend to the total height a, even if one were to remove the resistance of the air or whatever one may think might retard the motion in this case. Indeed, the very nature of the matter inevitably demands some defect, the physical reason of which is this: certainly any volume element whatsoever, even though beginning a vertical ascent, can nevertheless not help but be deflected noticeably to the sides, and finally, when it reaches the summit, it is carried by a horizontal motion, which must be noticeable, because through the uppermost limb or section of the aqueous stream all the water passes, which has flowed out through the orifice; assume, therefore, that the velocity of

MOTION OF FLUIDS PUSHED BY AN OUTSIDE FORCE

any volume element at the instant of time at which it is moved horizontally is that which a weight acquires by free fall through the height b; thus one sees that the stream cannot ascend beyond the height $a - b$. And for this reason a loss arises in proportion to the total *absolute potential* as b is to a.

Scholium

§20. It has been observed that among quantities of water ejected at a common velocity from differently formed small pipes, some rise higher than others; therefore, attention is to be paid here to the most apt configuration of the final pipes emitting water (*des ajutages*).

Mr. Mariotte set up experiments on this matter in his *Traité du mouvement des eaux*.

General Scholium

§21. So far we have examined the hindrances which appear in the case of hydraulic machines ejecting water with impetus; I consider those which I exposed to be the outstanding ones; nevertheless, still others could be considered, but, as I believe, only of less importance. Almost everywhere we gave completely geometric measures and have indicated simultaneously the extent to which these hindrances could be counteracted for the most part. He who reaches for greater ones, believing that the effect expected in elevating water by the least work, or (which I have shown in §3 to come back to the same thing) by the smallest absolute force, can be exceeded, is tricked by his opinion and wastes [lamp] oil and effort. For if one disregards the indicated hindrances and other similar ones that might perhaps be considered, in the nature of things the most perfect machine will be the simple pump of Fig. 45, and if water projected upwards by means of it is collected at G, [then] I say that it could not have happened that the same amount of water was elevated to the same height FG with less work.

There is, furthermore, another kind of machine which differs from the machinery treated so far in that the latter ejects the water with impetus, while the former transfers it quietly without noticeable motion. But also in the latter the ultimate degree of perfection which can be reached comes back to the same thing. But most [of them] are subject to many hindrances of very great importance. Therefore, these will have to be treated by us directly.

SECOND PART: CONCERNING HYDRAULIC MACHINES TRANSPORTING WATER WITHOUT NOTICEABLE IMPETUS FROM A LOWER POSITION TO A HIGHER

RULE 10

§22. If a given weight is elevated through a given vertical height a by a *moving potential* [that is] arbitrarily variable but applied directly, and if the body retains no motion at the summit of the proposed height, the *absolute potential* expended in the lifting of the weight will always be equal to the product of the weight of the elevated body and the height a of elevation.

PROOF. Indeed if a weight, which I shall call A, ascends through the height y and is assumed to be moved with the velocity v, and animated by a variable moving potential P directly applied in this position, the small time interval in which the weight is elevated through the element dy will be $\frac{dy}{v}$, which multiplied by the *moving potential* P and its velocity v, gives the element of the *absolute potential* (by the definition of §2) equal to $P\,dy$, therefore $\int P\,dy$ will give the total *absolute potential*, if after the integration one sets $y = a$; but during the entire motion the increment of velocity dv is equal to the exciting or moving potential, which here is $\frac{P-A}{A}$ multiplied by the small time interval which is now $\frac{dy}{v}$; therefore, we have $dv = \left(\frac{P-A}{A}\right)\frac{dy}{v}$ or $Av\,dv = P\,dy - A\,dy$, that is, $\frac{1}{2}Avv = \int P\,dy - Ay$, or $\int P\,dy = \frac{1}{2}Avv + Ay$, where one is to set $y = a$ and $v = 0$ (by hypothesis) so that $\int P\,dy = Aa$.

Furthermore, because $\int P\,dy$ expresses, as we have seen, the entire *absolute potential* expended in elevating the weight, this very potential will constantly be the same and expressly equal to the product of the weight A and the height a, just as the proposition states. Q.E.D.

§23. COROLLARY. From the proof it appears to us also that the *absolute potential* is the same whenever the velocity at the summit is the same, that is, whenever the height to which a body can ascend at its residual velocity, namely $\frac{1}{2}vv$, is constant; and if this height is called b, the *absolute potential* will be equal to $A(a + b)$. Therefore, it is now evident how large a portion of the *absolute potential* is lost when one intends to elevate the weight A to the height a, and when the same has

MOTION OF FLUIDS PUSHED BY AN OUTSIDE FORCE

at the summit a residual velocity due to the height b; certainly the loss of force will be to the entire force as b is to $b + a$.

SCHOLIUM 1

§24. And so precautions have to be taken lest the machines are constructed such that the water is transported to the determined location with a violent motion. Usually, however, this kind of loss is small in most machines.

SCHOLIUM 2

§25. Everything behaves similarly if the body is not elevated vertically but along a plane however inclined or even curved in any manner whatever; indeed, the total *absolute potential* will always be equal to $A(a + b)$, that is, to the product of the weight by the height of elevation augmented by the height due to the residual velocity of the body at the summit, the proof of which matter I omit since it differs little from the preceding proof.

GENERAL SCHOLIUM

§26. Because the effects of all machines, however composite, can be reduced to the nature of the inclined plane, it is evident that if we disregard frictions and these losses of *absolute potentials* which we have dealt with so far, all machines come back to the same, because the absolute potential simply depends on the height to which a body is to be elevated and its weight. The *absolute potential* has this in common with the *live force* or with the *actual ascent* or *descent*. And this is the ultimate level of perfection of machines, which cannot be exceeded, on the contrary, not even be reached, for a larger weight could always be elevated to the same height by the same *absolute potential* when the frictions and losses have been removed. In order that some comparison can be made of the loss in those machines which, so to speak, project water to a desired height as well as in those which transport it, we will now indicate also the most greatly noticeable losses in the latter ones.

I. In most machines of this kind friction is of so great a hindrance that it alone absorbs the largest part of the *potential*, particularly when square blades or oval bowls connected to a chain moving around in a circle elevate the water passing through the conduit to which they are fixed.

II. Most machines, but particularly those which we have just described, usually designated by the name of water wheels, are so joined together that continually, while the water is elevated, part of it trickles down, or plainly runs back to the place from which it was drawn or at least from a higher position to a lower, just as in water wheels; if in these the bowls or blades are well adapted to the conduit, the friction becomes almost unsurmountable, but if they are less [well adapted], a very great amount of water drips through the openings that are left, from the higher portions into the lower, so that a very small part of that quantity of water which they received in their entire traverse is left when they reach the culmination point. Therefore, or for this reason only, it seems that these machines are to be strongly condemned, and particularly if clear water is to be elevated which could be drawn by pumps.

III. Machines are also customarily of such a nature that they lift up the water beyond the proposed height; but the potential which corresponds to the excess is wasted, and if the water is to be raised through labor, [then] that which I indicated in §12 is obtained with difficulty.

IV. There are also machines which do not allow direct application of a moving potential, from which drawback again some loss arises.

§27. These are the obstacles, more or less, which seemed to me of notable importance; I do not know, however, whether those can be counteracted so much as we have shown regarding the first kind of machines; the mechanics know certain tricks of diminishing frictions; I would prefer, to water wheels, machines which draw and lift the water in buckets; but the buckets are to be constructed so that, if this can only be done, they are filled immediately in the lowermost position and emit nothing before they have reached the uppermost position. Since the water is to be transferred through the higher location to another one, less high, an effort has to be made that the impetus of the falling water promotes the motion of the roller or wheel acting in a circular course, although thus the *entire absolute potential* is far from being expended usefully, as we have shown to happen in the pump of Fig. 49 (§12). The principle of action exists, if I judge correctly, most aptly in the treadmill: for these men are best accustomed to work. That which I advised in §4 on occasion of the first rule about the angle of inclination according to which a walker can attain a certain vertical height in a given time with the least fatigue pertains here. I would believe that a man of ordinary stature, healthy and robust, marching on a path inclined at 30 degrees will accomplish 3600 feet in a single hour without difficulty, and therefore

MOTION OF FLUIDS PUSHED BY AN OUTSIDE FORCE 205

he will elevate to a vertical height of 1800 feet the weight of his body, which I may assume [to be] 144 pounds or two cubic feet of water. Such a man, therefore, could by means of a treadmill machine, acting in a circle and being most perfect (in which of course nothing of the *absolute potential* is wasted) elevate in a single hour two cubic feet of water to a vertical height of 1,800 feet, or, which is the same, in a single second one cubic foot to the height of one foot; machines which are of much lesser effect, doing a favor to the workers, I believe to have little to recommend them; meanwhile, having set up an experiment with a pump in the house of the illustrious General de Coulon, which I shall account for at the end of the chapter, I experienced an effect by no means less, by which I am confirmed in my statement that workers usually accomplish more with a treadmill: I easily foresee, moreover, that in very composite machines a far lesser effect is achieved, because in these the greatest part of the *absolute potential* is expended uselessly. Notably, I shall now contribute to this matter the example of the very well-known *machine de Marly* showing what an almost incredible loss of *absolute potential* arises from all the collected hindrances.

Weidler published a treatise about *hydraulic machines* in which he gives a full description of the *machine de Marly*, and reports that all the water is elevated by the motion of 14 wheels, the blades of which are propelled by the impetus of the Seine; this makes the impetus for all wheels equal to a weight of 1,000,594 pounds, and this is what we have designated by the name of *moving potential*. Furthermore, I could understand from some circumstances that the blades are carried by a motion by which they travel $3\frac{3}{4}$ feet in a single second, and this velocity is to be taken for the velocity of the *moving potential*. Then he adds that in a single day 11,700,000 pounds of water are elevated by means of this machine to a height of 500 feet. These things having been so assumed, let us see now how great a potential P, similarly moved at a velocity of $3\frac{3}{4}$, would be required for this effect in the very simple machine of Fig. 45, in which it is assumed that none of the *absolute potential* is lost. Indeed, the height FG will be 500 feet, and since now in the time of 24 hours 11,700,000 pounds must be ejected through the opening F, that is, 162,500 cubic feet, the size of this opening will have to be taken as 0.0108 part of one square foot. The velocity of the water is so great that it travels 173 feet in a single second. Therefore it contains the velocity $3\frac{3}{4}$, which the weight P is assumed to have, 46 times, and the area AB of the pump has to exceed the area of the opening F just as many times. Consequently, the area AB will have to be taken as 0.4968 part of a square foot, from which follows that the weight P will be equal to the weight of an aqueous cylinder constructed over the base AB to a height of 500 feet, or to the weight of 248.4 cubic feet of water, that is, to a weight of 17,885 pounds, which brings about only the fifty-sixth part of the *moving potential* which Weidler shows to be applied to the [water] moved at the same velocity. Thus, therefore, a loss occurs in the entire machine which equals $\frac{55}{56}$ of the entire *absolute potential*.

After we have so examined the nature of hydraulic machines, as much as it can be done in general, by no means will it be irrelevant to treat some special example more accurately, and because the waterscrew of Archimedes possesses many outstanding properties which, as far as I know, no one has exposed sufficiently, I want to take the example from it, and this all the more willingly since there are many who believe contrary to our rules that this waterscrew has a singular virtue for elevating a large quantity of water in a short time and by a small force; but those who think so are deceived; for if no account of accidental hindrances is taken, this vouches for the same *absolute potential* as all other machines.

Special Comments on the Waterscrew of Archimedes

I. There are various authors who taught a method of constructing this waterscrew: the summary comes back to the fact that some conduit or several cylindrical surfaces are bent around, and certainly so that the conduit has everywhere the same inclination in relation to the axis of the cylinder, which Vitruvius, beyond necessity, orders to be made at a half right angle [45°] in all waterscrews. Therefore, it is required first of all that on the surface of the cylinder a spiral line be drawn, to the normal of which the conduit is to be put, which can be done most easily, in my judgment, on a very smooth surface (particularly since the helices have to be no little distance from one another) by winding a string around the same several times. For here the tension will produce the desired line on its own, and the spiral indeed cannot be everywhere similar to itself or have a constant inclination to the axis of the cylinder unless the arc spanned between two points is the smallest of all arcs having the terminal points, which is shown to be the case with a stretched string; but if friction is a hindrance, the string need only be extended to smaller intervals. But this is not why we are hesitant in a matter that is intrinsically very simple in many respects.

A primary law of the spiral is that it is everywhere equally inclined to the axis of the cylinder, following which law the construction is undertaken, which I shall add for the benefit of the things to be said below.

Imagine a right cylinder $MafN$ (Fig. 52–1), on the surface of which is to be drawn the desired spiral $a1b2c3d$, etc., and consider the same surface to be laid out in the plane given by the shape of the rectangular parallelogram $AafF$ (Fig. 52–2), [and] let here be taken from the one part AB, BC, CD, DE, and EF, from the other ab, bc, cd, de, and ef, each one to be equal to the corresponding one; the points B, C, D, E, and F are joined by straight lines with the points a, b, c, d, and e: if, after these things have been done in this way, the plane surface is again rolled up into a cylindrical one, with the lines AF and af joined, and the points A and a, B and b, etc., coinciding, it will happen that the lines aB, bC, cD, etc., form a continuous line on the cylindrical surface which will be the desired spiral itself. For easier understanding I marked homologous points in either figure with common letters.

II. The cylinder $MafN$ (Fig. 52–1) was already proposed, having as a duct the curved conduit of the spiral just described, the diameter of which we shall assume as infinitely small in proportion to the diameter pertaining to the cylinder; and thus the waterscrew of Archimedes will be obtained; if we want to use this for elevating water from M to N, the cylinder will have to be inclined with respect to the horizontal, and certainly so that the angle aMH (the intercept between diameter Ma of the base, which is in the vertical plane, and the horizontal MH) is greater than the angle sao, which the tangents to the circle and the spiral form at the common point a. Then, after the cylinder has been turned around its axis in the direction $aghMs$, the water will flow in through the lower orifice of the bent conduit and flow out through the upper one.

III. So that we may understand the nature of this elevation correctly, three points in any arbitrary helix of the spiral are to be examined by us, namely the points o, p, and q, the first of which, o, is the farthest away from the horizontal, the other, p, is closest to it, and q is located at the same height as the point o taken in the next lower helix; through the individual points o is drawn the straight line gn, through the points p the straight line hm, and through the points q the straight line st. But the locations of these lines will be determined below.

IV. Let the radius that pertains to the base of the cylinder be 1 and let it be taken as the total sine; the sine of the angle sao equals m; its cosine equals M; the sine of the angle aMH equals n, its cosine equals

208 HYDRODYNAMICS, CHAPTER IX

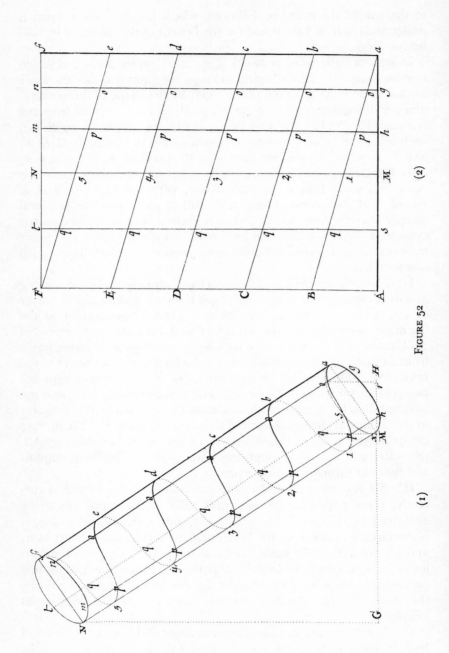

FIGURE 52

N; the arc ag equals X, the cosine of that arc equals x, [and] the perpendicular drawn from o down to the horizontal will certainly be $or = \frac{mNX}{M} + n(1 + x)$. But because or is a maximum, it happens that $\frac{mN\,dX}{M} + n\,dx = 0$, and since from the nature of a circle there is $dX = \frac{-dx}{\sqrt{1-xx}}$, there will be $\frac{-mN\,dx}{m\sqrt{1-xx}} + n\,dx = 0$, therefore $\sqrt{1-xx} = \frac{mN}{Mn}$. Therefore, the sine of the required arc ag is $\frac{mN}{Mn}$, or cosine $x = \pm \frac{\sqrt{nn-mm}}{Mn}$; the upper sign gives the arc ag, the lower the arc ah determining the lowest points p.

And so we have determined both the uppermost points o and the lowermost p, and it is evident that the arcs Mh and ag are equal to one another, but simultaneously it is understood from the irrational quantity $\sqrt{nn-mm}$ determining the value of the letter x that it cannot occur that m is larger than n; and, indeed, in this case the lowermost point is not given, since the entire spiral ascends continuously everywhere. Indeed, the waterscrew will not serve thus in elevating water, hence the reasoning is now evident which I pointed out in the second article of this digression concerning the required excess of the angle aMH over the angle sao.

V. Let us suppose now that a sphere is located somewhere in the cavity of the conduit and that the waterscrew is fixed in its position; thus the sphere is certainly not at rest unless it is located at some point p. But if the waterscrew is assumed not to be held back, the sphere will descend, and by its descent it will drive the waterscrew around, and if, furthermore, it is imagined that the waterscrew is of no weight and that the motion of the sphere occurs very freely with no hindering friction, [then] the sphere descends on the straight line mh by no other law than a sphere descending freely on an inclined plane. And so it is evident that a potential is required for holding back the descent of the sphere and fixing the waterscrew. Let us assume that that potential is applied at the point f in the plane of the circle and perpendicularly to the radius in question in the ratio which it has to the weight of the sphere resting at some point p.

Let the weight of the sphere be p; but, because the action of the sphere is vertical, it will have to be resolved into two others remaining perpendicular to each other, the one of which let have a common direction with the axis of the waterscrew, the other let be perpendicular to it; the former will have to be rejected, since it contributes nothing to turning the waterscrew, and the latter will have to be considered alone; but that residual action equals np, and it acts on a lever which is equal to the sine of the arc Mh or the arc ag, and this sine (by Art. IV) is $\frac{mN}{Mn}$. Therefore, the moment of the action is $\frac{mN}{Mn} \cdot np = \frac{mNp}{M}$; if one divides this by the radius of the base, which is the lever pertaining to the potential applied at f set in equilibrium with the action of the sphere, one will have the required potential equal to $\frac{mNp}{M}$. Therefore, others customarily derive from a foreign principle that which can thus be deduced directly from the nature of the lever. With these things having been set forth which were to be set forth, let us now begin to consider the use of the machine for elevating water.

Problem

VI. It is asked what the maximum quantity of water is that a given waterscrew can discharge in a revolution.

SOLUTION. Let us consider an entire helix $a1b$, and let the quantity of water which it contains [when it is] full be q: it is also to be noted that the helix cannot be entirely filled with water; for if the entire conduit were full, water would flow out through the lower orifice: therefore some branch, which is $a1b$, is occupied partly by air, partly by water; also one extremity of the water will be at o, or the uppermost point, the other at q, or a point situated at a level with the former; therefore, the part full of water is opq, and if this part is assumed to be in proportion to the length of the entire helix $a1b$ as g is to h, the maximum quantity of water to be discharged in one revolution will be equal to $\frac{gq}{h}$. Q.E.I.

Scholium I

VII. Because, as we have said, it cannot happen that the water is continuous through the full extent of the conduit, care has to be taken that no separation is imparted to the water, which can easily be accomplished if the entire base of the cylinder is immersed in the water, because thus air cannot enter through the lower orifice of the conduit; neither must it happen that too large a part of the base projects from the water, because then the waterscrew does not draw

MOTION OF FLUIDS PUSHED BY AN OUTSIDE FORCE

all the water that it could otherwise [draw] in one revolution; on the contrary, it draws nothing if the immersion does not reach the point h; but due immersion should occur up to the point g, because thus the arc opq of the helix, which is capable of retaining water, becomes largest. Although indeed I never conducted an experiment on the matter, and most authors seem to speak differently about it, I would rather trust in reasoning than in the authority of those who did not pay attention to this immersion.

Therefore, this *rule for the ratio of immersion* will be observed, namely that the base is submerged until the chord of the arc projecting from the water is $\frac{2mN}{Mn}$, where the letters m, N, M, and n signify the same [variables] as in the fourth article.

Scholium 2

VIII. It is apparent, indeed, after light contemplation of the matter, that the ratio between the arc opq of the helix and the entire helix $a1b$, that is, between g and h, is greater, and hence a larger quantity of water is discharged in a single revolution, the rest being the same, the smaller the angle sao and the larger the angle aMH, or, the smaller the distance between two adjoining helices and the more the waterscrew is inclined towards the horizontal; but it is not possible to express that true ratio algebraically; nevertheless, in every particular case this is obtained by an easy approach.

Let me select *an example of the preceding rule* from a waterscrew which Vitruvius shows how to construct and apply. He makes sao a semi-right angle, and thus $m = M = \sqrt{\frac{1}{2}} = 0.70710$; then he sets a ratio between NG and MG, which is as 3 is to 4; whence one deduces the angle GNM or aMH is equal to $53° 8'$, and the sine n of it equals 0.80000 and the cosine $N = 0.60000$; therefore (by Art. III), the sine of the arc ag defining the highest point o equals $\frac{mN}{Mn} = \frac{3}{4}$, and the arc ag itself equals $48° 35'$. And thus, by virtue of the rule of Art. VII, the arc projecting from the water at the base must be $97° 10'$, and an arc of $262° 50'$ is immersed.

Furthermore, in order that we may now define the ratio between the arc opq of the helix and the entire helix $a1b$, it is to be noted that the ratio is the same as that which exists between the circular arc $ghMs$ and the circumference of the circle, which is manifest from the accompanying figure. But the arc $ghMs$ may be determined in the following manner: for example, the arc $ghMs$ = arc $aghMs$ − arc ag.

But we have seen in the third article that, if from any point of the spiral, such as *o* and *q*, perpendiculars are drawn down to the horizontal going through *M*, such as *or* and *qx*, that perpendicular will be $\frac{mNX}{M} + n(1 + x)$, or in our case $0.60000X + 0.80000(1 + x)$, with *X* denoting the circular arc corresponding to the assumed point on the spiral, namely the arc *ag* or the arc *aghMs*, and *x* denoting the cosine of that arc. But the arc $ag = 48° 35' = 0.84794$ (because the radius is expressed by unity), and the cosine of it equals 0.66153; therefore, in our case *or* becomes $0.50878 + 1.32922 = 1.83800$. Further, because the points *o* and *q* are located at the same height, and the lines *or* and *qx* are equal to each other, it is apparent that the question is now reduced to this: that the other arc *aghMs* corresponding to the point *q* be found so that, if it is called *X* and its cosine *x*, then $0.60000X + 0.80000(1 + x) = or = 1.83800$; for this condition the arc *aghMs* is found [to be] approximately $175\frac{1}{2}$ degrees, intersecting the cut *agM* at the point *s*. And since the arc *ag* will be $48° 35'$, the arc *ghMs* will finally be $126° 55'$, which thence will be to the circumference of the circle approximately as 10 to 29: the same ratio prevails between the arc *opq* of the helix and the entire helix.

From this follows that in a single revolution there is discharged by the waterscrew described by Vitruvius approximately $\frac{10}{29}$ of that quantity which the full helix contains, or very little more than one third.

Scholium 3

IX. It is nevertheless to be noted that whatever be the quantity of water which enters the conduit at the bottom at any revolution of the waterscrew and flows out of the same at the top, it imparts neither a loss nor a gain to the *absolute potential* if no consideration is given to friction, because the *moving potential*, the rest being equal, is proportional to that quantity. But because friction always hinders and is almost the same on account of the very weight of the machine whether a larger or smaller quantity of water is pumped, and certainly an effort is to be made that that quantity becomes great, the rest being equal, this matter I shall now treat a little more expressly.

Scholium 4

X. I already hinted above that the ratio of the arc *ghMs* to the circumference of the circle increases with decreasing angles *sao* and

MOTION OF FLUIDS PUSHED BY AN OUTSIDE FORCE   213

NMG; either one should therefore be built very small unless other inconveniences interfere, particularly in the consideration of the angle NMG. As far as the angle sao is concerned, it can be diminished almost arbitrarily, and thence no other inconvenience results apart from the fact that the sides of the conduit to be curved cannot come too close to one another: on the contrary, from the diminishing of that angle another benefit is obtained, namely, that then the machine can be erected more nearly vertical, and the water itself is elevated higher, for truly the angle aMH must always be larger than the angle sao; from the more nearly vertical position of the waterscrew, moreover, it occurs simultaneously that the very weight of the machine is of less inconvenience, and that the latter is more easily supported.

Considering these things accordingly, I should believe that it usually suffices for the conduit to make an angle of 5 degrees with the base of the center. Cardano also made that angle smaller than Vitruvius, and since the fewer conduits can be wound around the same center, the more obliquely they are attached, Vitruvius stated that eight are to be placed, Cardano only three; but the conduits are longer in the waterscrew of Cardano, so that it contributes in the lengths what it lacks in the number of conduits. In the consideration of the other angle NMG it merits being observed that the water can be elevated higher, the larger the angle becomes, but, on the contrary, the quantity of water discharged in a single revolution is less. Probably those who make that angle 60 degrees will reach a just median.

XI. Now we will also perform the calculation of our waterscrew constructed to the norm of the preceding article as we have done for the waterscrew constructed according to the concept of Vitruvius in Art. VIII. But because by hypothesis the angle sao is $5°$ and the angle $NMG = 60°$, the arc ag will, by Art. IV, be found to be $8° \ 43'$, and the vertical line $or = 1.00574$, to which the other vertical qx will be equal if $284° \ 57'$ is assigned to the arc $aghMs$; hence, if the arc ag is subtracted, the arc $ghMs$ remains as $276° \ 14'$, which corresponds to the arc of the helix capable of retaining water; therefore, this part is to the entire helix as $16{,}574$ to $21{,}600$ or as 8287 to $10{,}800$, such that in a single revolution more than four-fifths of the capacity of the entire helix can be discharged, and two and one-third times as much is accomplished by this machine as is obtained from similar machinery constructed according to the understanding of Vitruvius; also, the water is elevated higher from the same center in the ratio as $\sqrt{3}$ is to $\sqrt{2}$. I come now to the *moving* as well as the *absolute potential* that is expended in elevating water.

Problem

XII. Given the weight of the water resting in the helix, find the tangential potential located at f in equilibrium with that weight.

Solution. We have seen how this problem may be solved geometrically by reasoning of a sphere resting at the lowermost point p. But in the present case the situation is slightly different, since the weight of the water is distributed through a large arc of the helix and not concentrated at some given point. It is certainly easy to foresee that in either case the potentials will be the same from the indirect rules of mechanics; it pleases, nevertheless, to present the desired proof of this matter from the nature of the lever, because the mechanics love to reduce everything to that.

We shall consider the helix $a1b$ taken separately from Fig. 52 in order to avoid confusion of the lines, the notations applied in Art. IV having been preserved. Thus, therefore, in Fig. 53 the angle NMG

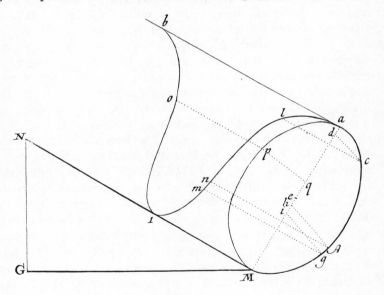

Figure 53

will again be the angle which the center makes with the horizontal, its sine equals N, and the sine of the angle aMH equals n; $a1b$ is one revolution of the spiral. The circle $acMpa$ is the base of the center; the sine of the angle pal, as before, equals m, and its cosine equals M; but the points l and o are the extremities of the water resting in the spiral and located at the same height from the horizontal; from these points are drawn to the periphery of the base the straight lines lc and op, perpendicular to the base. In the part of the helix which the water occupies, two infinitely close points m and n are assumed, and through these the straight lines nf and mg are drawn, again perpendicular to the base. Finally, from the points c, f, g, and p the perpendiculars cd, fh, gi, and pq are drawn to the diameter aM; and the center of the base is assumed at e, and the radius $ea = 1$. Now let the arc $l\mathbf{1}o$ of the spiral, full of water, be equal to c, and consequently the circular arc cMp corresponding to the same be equal to Mc : $al = e$; $ac = Me$; ad (or the sine with respect to the arc ac) equals f; $aq = g$; the weight of water in lso equals p; the arc $aln = x$; $nm = dx$; $acf = Mx$; $fg = M\,dx$; $ah = y$; $hi = dy$; $hf = \sqrt{2y - yy}$, [and] the weight of the volume element at nm equals $\dfrac{p\,dx}{c}$; but if the line hf is multiplied by the sine of the angle aMH and divided by the entire sine, there results the lever arm by which the particle nm attempts to turn the waterscrew; therefore, this lever arm is equal to $n\sqrt{2y - yy}$, which, multiplied by the weight of the volume element given above, $\dfrac{p\,dx}{c}$, gives its moment $\dfrac{np\,dx}{c}\sqrt{2y - yy}$. But from the nature of a circle $M\,dx = \dfrac{dy}{\sqrt{2y - yy}}$; therefore, after this value has been substituted for dx, the moment of that same volume element nm becomes $\dfrac{np\,dy}{Mc}$, the integral of which, after subtraction of the proper constant, is $\dfrac{np(y - f)}{Mc}$ and denotes the moment of the water in the arc ln; whence, therefore, the moment of all water in $l\mathbf{1}o$ is $\dfrac{np(g - f)}{Mc}$. This divided by the lever arm of the potential applied at f, or by 1, yields in a like manner the desired potential $\dfrac{np(g - f)}{Mc}$.   Q.E.I.

Scholium I

XIII. In order that it be apparent that the value of this potential does not differ from that which we found in Art. V for a sphere of the same weight p, namely $\dfrac{mNp}{M}$, the equality between $\dfrac{np(g - f)}{Mc}$ and $\dfrac{mNp}{M}$ or between $n(g - f)$ and mNc is to be shown; but this equality is to be deduced from the fact that the extremities l and o of the water are located at the same height above the horizontal; for hence it follows, as we have shown in Art. IV, that the sum of the arc ac multiplied by $\frac{mN}{M}$ and the line Md multiplied by n equals the sum of the arc $acMp$ multiplied in the same way by $\frac{mN}{M}$ and the line Mq multiplied by n. And so, with the notations of the preceding article having been applied, there results

$$Me \cdot \frac{mN}{M} + (2-f) \cdot n = (Me + Mc) \cdot \frac{mN}{M} + (2-g) \cdot n,$$

or $n(g-f) = mNc$, which equality has to be shown for demonstrating the equality of the potentials to be applied for the sphere as well as for the water at f.

Scholium 2

XIV. Because the potential $\frac{np(g-f)}{Mc}$ does not differ from $\frac{mNp}{M}$ and the quantity $\frac{mN}{M}$ remains the same, whatever quantity of water is drawn in or discharged in one revolution, that potential will be proportional to that very quantity of water discharged in a single revolution, or to the weight p. Also, it is easy to prove that, if the same quantity of water is elevated by the same *moving potential* and at the same velocity to an equal vertical height above the base plane, which to this end must be appropriately inclined towards the horizontal, it will happen that the time of elevation is also the same.

Therefore, the same *absolute potential* is required in the waterscrews of Archimedes as [is required] on an inclined plane, to which all machines can be reduced, and this waterscrew does not have any prerogative over other machines viewed in the theory. Perhaps in practice it is less exposed to the inconveniences indicated in §26; by no means do I reject its use, but neither do I prefer it to the pumps of Ctesibius.

——————————— [*End of Digression*] ———————————

§28. From what has been said so far, one understands under what conditions one machine ought to be preferred to another: namely, what degree of perfection of the machine these [conditions] permit; to what one should pay most particular attention in their construction and use; how large a part of the *absolute potential* is lost; and other similar things. Of course, we have considered only machines driven by *animated potentials*, as they are called; but it is readily apparent that those machines that are to be driven by the impetus of water, by wind, or by the gravitation of water and other principles of this kind are subjected to the same laws; always, indeed, the *moving potential* multiplied by the time and the velocity of the point to which the potential is applied will give the product of the quantity of water and the height to which that quantity can be elevated in a given time by means of the proposed machine, other hindrances having been set aside. However, I am speaking about machines which lose none of the *absolute potential*; it can happen, indeed, that the greatest part is lost, which we have shown often enough above.

§29. Hence it is apparent that water elevated to a certain height can by its descent produce the same effect again; but the effect will have to be estimated from the quantity of water to be elevated and the height of elevation; for example, by the descent of 8 cubic feet from a height of one foot, it is wholly possible for 8 cubic feet to be elevated again to the same height, or 4 cubic feet to a height of two feet, or one cubic foot to a height of 8 feet, and thus however one would please. A specimen of a machine which can elevate water to any height whatever by a minimum descent of water is found among [the works of] Mr. Perrault in the *Commentarii ad Vitruvium*, Book 10, Chapter 12, which machine he introduces as an almost incredible paradox, and he makes the Italian Mr. Francini the inventor of it, by whose industry and planning it was constructed successfully in the garden of the Royal Library. The basis of the machine consists in the fact that buckets chained together and moving around in a circle take up water and transport it to the lowermost point where they discharge it, while another series of buckets take up water, although less in quantity, and carry it to a much higher location and discharge it. It is very clear that if all descending buckets are heavier than all ascending buckets, the former series will activate the other perpetually in a circle. There exist also machines which produce the same through simple pipes by means of flaps that are to be reversed at regular time intervals, in which conversion certainly no potential is expended. Carlo Fontana describes machinery of this kind.

But if anyone believes that the same can be obtained from the impetus of water falling from a certain height and impinging on the blades of the machine, he will be far off. Machinery of this kind would pertain to that class in which the largest part of the *absolute potential* vanishes without benefit.

It will not be beside the point to follow this argument more accurately and to show how great an effect can be obtained from the impetus of water or wind, and under which circumstances this effect may be considered the greatest of all.

THIRD PART: CONCERNING MACHINES WHICH ARE DRIVEN BY THE IMPETUS OF A FLUID, SUCH AS BY THE FORCE OF THE WIND

§30. After water elevated to a certain height falls down again from the same and impinges continuously on the blades of a wheel to be turned, it cannot happen differently than that the *absolute potential* required for so turning the wheel is much less than that which was expended in the elevation of the water, the foremost reason for which matter is that the water falling down after the impulse on the sides still preserves a velocity which contributes nothing to the rotation of the wheel. Therefore, a large part of the *absolute potential* of elevated water would become useless if a machine were driven by the impetus of this water and finally, by this in turn other water were elevated to a certain height; and indeed a larger or smaller part is lost because of different circumstances, but never, as I shall show, is lost less than $\frac{23}{27}$ of the total if a computation of the ordinary impulse of water is made according to the norm.

§31. Furthermore, it is commonly stated that if water flows out of a very wide cylinder through a simple orifice at its total velocity, that is, that which would be due to the total height of the water above the orifice, and the stream immediately in front of the orifice impinges directly on a plane, [then] it will occur that the impetus of the fluid against the plane is in equilibrium with the weight of the aqueous cylinder erected above the orifice to the height of the water. Authors certainly misled by a false experiment have supported this completely false theory. I nevertheless did not want to withdraw here from the latter, because I have not yet shown the true theory, and then, after our theory has been explained, it will be easy to correct the calculation. May one therefore be allowed to adhere to the common, although erroneous, statement until we consider the matter more correctly in its proper place. The greater the impetus of a fluid, the greater the ratio by which the *absolute potential*, which we shall give, will have to be increased.

§32. Consider now (Fig. 54) a vessel *ABC* or a pump which expels water through the orifice *C* in a not quite vertical direction; but the water is taken up by another vessel *EFD* when it has reached the summit. At the base of this other vessel picture an orifice *D*, equal

MOTION OF FLUIDS PUSHED BY AN OUTSIDE FORCE

to the former C and located at the same elevation, so that as large an amount of water flows out through D as is poured in above, and the vessel EDF is kept constantly full. Assume further that the water flowing out through D impinges continuously on the blades of some

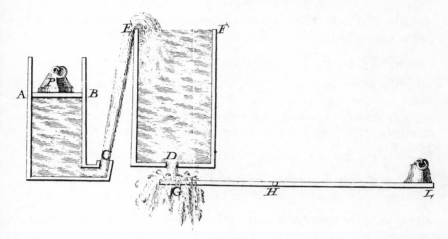

FIGURE 54

wheel, which, turned in this manner, elevates other water; in place of this machine is described in the simple figure a lever arm rotating around H, by assuming that continually some one or another different lever arm of this kind is present in front of the orifice D, which receives the water and draws water at its other extremity and elevates it to the given height.

After these [things] have so been assumed, I shall first inquire about the *absolute potential* that elevates the water flowing through the orifice C to the height CE; then also about the *absolute potential* that is required at G for moving the lever arm at the same velocity at which it is moved by the impulse of water DG.

§**33.** Let the area of the orifice C or D be n, the area AB be m, the velocity of water at C or D be v, the weight of the cylinder erected above the orifice C or D to the height CE be p, the time of flow be t; then the weight P will be $\frac{m}{n}p$; the velocity at which the weight descends while water is expelled equals $\frac{n}{m}v$; therefore (by §3) the *absolute potential* expended in ejecting water through C is

$$\frac{m}{n}p \cdot \frac{n}{m}v \cdot t = pvt.$$

§34. Now, in order that the *absolute potential* expended in the gyration of the lever arm GL around the point H may be determined, it is to be noted that the former is the least consistent with itself; for it is changed by the changed velocity at which the lever arm is rotated. Therefore, let us make the velocity at which its extremity at G is moved equal to V. But in this manner the water is to be considered as impinging at G with the velocity $v - V$, and thus it exerts a pressure which is $\left(\dfrac{v-V}{v}\right)^2 p$ (for the pressures are in a square ratio to the velocities of the impinging fluid, and a pressure equal to p is substituted for the velocity v). But this pressure exists in place of the *moving potential*; in place of the pressure of the fluid we can certainly substitute the weight lying above the lever arm at G, which is $\left(\dfrac{v-V}{v}\right)^2 p$. But that weight will be moved at the same velocity as the point G, namely, at the velocity V, and it acts during the time t. Therefore, the *absolute potential* required for the rotation of the lever arm during the time t and at the velocity V is $\left(\dfrac{v-V}{v}\right)^2 p \cdot V \cdot t$.

§35. Thus, if the lever arm LG is not rotated immediately, but the fluid is elevated to the height CE with the intention that the stream of fluid, by its impulse at G for rotating the lever, elevates water from the other part, the entire *absolute potential* will be to the useful *absolute potential* as pvt is to $\left(\dfrac{v-V}{v}\right)^2 pVt$, or as v^3 is to $(v-V)^2 V$, and it will be to its useless part as v^3 is to $v^3 - vvV + 2vVV - V^3$.

§36. In almost all machines in which the principle of motion consists of the impulse of fluid, it usually happens that the velocity V of the lever where it sustains the impetus of the fluid is very small in proportion to the velocity v of the fluid; but in these [machines] the largest part of the effect that could be obtained from the same quantity of fluid moved at equal velocity is lost.

§37. The greatest effect from the impulse of fluid develops, or, which is the same thing, the *absolute potential* defined in §34 becomes greatest, if $V = \tfrac{1}{3}v$, and then this *absolute potential* is $\tfrac{4}{27}pvt$; and even then it falls short by twenty-three twenty-sevenths of the similar potential that is expended in elevating water from C to EF.

If a natural descent of water exists and is to be used for elevating water or for accomplishing anything else, it must be arranged that the machine is moved at that place where the impulse occurs at a velocity of one third of the velocity of the impinging fluid. But this condition can always be satisfied, which is evident from the cited example of the

MOTION OF FLUIDS PUSHED BY AN OUTSIDE FORCE   221

lever. For if the point G is moved at a greater velocity, one must diminish the part *HG*, the rest remaining the same, or one must increase it if the point G is moved at a lower velocity. Or even with the length *HG* retained, one must arrange that the water is drawn in at the extremity *L* in a larger or smaller quantity.

§38. Truly this is the reasoning about fluids impinging perpendicularly on blades; the computation is different for fluids attacking obliquely the arms of mills agitated by the force of the wind, and other similar machines. About these let me now add some few things, and with these I will bring this chapter to an end.

Since the fluid impinges on the surface of the entire blade, arbitrarily located and to be rotated in the direction perpendicular to the motion of the fluid, writers show that the fluid exerts the greatest pressure on the blade for promoting the rotation when the blade makes an angle with the direction of the wind the sine of which is to the total sine as $\sqrt{2}$ is to $\sqrt{3}$; but if the same entire stream of fluid is received by the blade, whether [it is] thus or inclined differently to the direction of the fluid, then that blade which makes a half right angle with the direction of the fluid will sustain the greatest pressure in the direction of rotation.

The first Rule pertains to machines which are driven around by a wind surrounding everything; the other to those which are moved by a solitary stream and by a certain determined quantity of fluid. But either hypothesis depends on the fact that the motion of the blades is very small with respect to the motion of the fluid; for if one refers to the motion of the blades, both rules are false; and in the outset this motion is not to be neglected; indeed, I have often observed on mills that the tips of the arms are carried at a velocity which almost equals the velocity of the wind itself.

Since these [things] are so, let us perform a calculation so that we obtain an understanding of either motion.

§39. Therefore, let there exist the fluid *DEBA* (Fig. 55) which impinges on the entire plane *AB* in the direction *EB*; moreover, the plane is assumed to be moved in parallel motion in the direction *Bb* perpendicular to *EB*. Further, let the velocities be of the kind that, while a particle of fluid moves through the line *EB*, the point *B* of the plane travels the line *Bb*. Under these assumptions one may consider that the entire system, namely, the fluid and the plane, is moved from *b* towards *B*, and certainly with the velocity *bB*. But so it will occur that the plane *AB* is at rest, yet the particle of fluid striking at the point *B* is to be considered as having come from the point *e*, *Ee* = *Bb* having been assumed, and so accordingly for all volume elements. Therefore, instead of the fluid *DEBA* striking the moving plane *AB* with the velocity *EB*, one will have to consider a fluid *deBA* striking the same but immobile plane *AB* with the velocity *eB*.

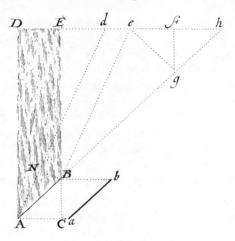

FIGURE 55

Let *AB* now be extended up to *h*, and let *DEdeh* be moved perpendicular to *EB*; the motion represented by *eB* of the particle of fluid will have to be resolved into *eg* and *gB*, remaining perpendicular to each other, the latter of which does not act upon the plane *AB*; but the other, *eg*, is again composed of two motions, *ef* and *fg*, the latter, *fg*, of which tries uselessly to propel the plane *AB* in the direction *EB*, while the former, *ef*, alone propels this plane in the direction *Bb*. It is therefore shown that any arbitrary particle causes an impulse proportional to the line *ef*; then it is also evident that if the line *AB* represents the entire plane, the number of particles impinging in a given time on the plane is to be represented by the line *BN*, perpendicular to *Ad* or *Be*. Whence finally the pressure of water for moving the plane in the direction *Bb* is proportional to the line *ef* multiplied by *BN*.

In order that now the inclination of the plane to the fluid be determined that is most favorable under these circumstances for promoting the movement of the plane in the direction *Bb*, let us set $AB = 1$, *DE* or $AC = x$, $BD = \sqrt{1 - xx}$, the line *EB*, which represents the motion of the fluid, equal to v, and *Bb* or the measure of the motion of the plane equal to V; and thus, after the calculation has been performed, one finds

$$ef = xv\sqrt{1 - xx} - (1 - xx)V,$$

and
$$BN = [xv - V\sqrt{1-xx}]/\sqrt{vv+VV};$$
whence
$$ef \cdot BN = [xv - V\sqrt{1-xx}]^2 \frac{\sqrt{1-xx}}{\sqrt{vv+VV}},$$

which quantity will be largest when this occurs:
$$(9v^4 + 18vvVV + 9V^4)x^6 - (12v^4 + 30vvVV + 18V^4)x^4$$
$$+ (4v^4 + 16vvVV + 9V^4)xx - 4vvVV = 0.$$

§40. The calculation in consideration of the inclination of the arms in mills is different, because the velocities are different in different locations on the arms; they are, indeed, proportional to the distances from the center. But now it will be easy for anyone to perform a computation for mills. I do not wish to pursue this case any further, so let it suffice to have noted that it is stated by authors not accurately enough [that] $xx = \frac{2}{3}$, and that the true value of x itself is always less than $\sqrt{\frac{2}{3}}$. For example, if V were equal to v, and all points of the arm were thought of as being moved at similar velocity, x would become $\sqrt{\frac{1}{2}}$, which indicates that the arm is to be inclined to the direction of the wind at a half right angle. The best construction of arms would be if they were curved so that the wind impinges on them higher up at a smaller angle than lower down, or if it were made that the arms receive the wind everywhere at a mean angle of approximately fifty degrees.

§41. I pass on to the other case in which all fluid is assumed to be received by the plane, whichever way it be inclined. Here, however, it is evident that because the number of particles impinging in a given time is always the same, no attention must be paid to the line BN, and that thus the pressure that the water exerts for moving the plane AB in the direction Bb is represented simply by ef or $xv\sqrt{1-xx} - (1-xx)V$. Therefore, this pressure will be made the greatest by taking $xx = \frac{1}{2} + \frac{V}{2\sqrt{vv+VV}}$, and then the pressure itself will be $\frac{1}{2}\sqrt{vv+VV} - \frac{1}{2}V$, if by v one understands the direct pressure which the stream exerts on a plane which it strikes perpendicularly.

§42. Let us consider now the stream $DEBA$ as if immediately discharged from the orifice D in Fig. 54, and let us call p again the direct pressure of the stream thus considered, just as in §33; and the pressure of this water, by which it tries to propel in the direction perpendicular to the stream the plane inclined in such a manner that the pressure becomes greatest, will be $\frac{p}{2v} \cdot (\sqrt{vv + VV} - V)$; and if, furthermore, this pressure is multiplied by the velocity V of the plane and by the time, the *absolute potential* is obtained by which the plane can be moved at the same velocity through the same time interval; so, therefore, the aforementioned *absolute potential* will be $\frac{pVt}{2v} \cdot (\sqrt{vv + VV} - V.)$

§43. The *absolute potential* which we have just defined is so constituted that it increases continuously with increasing V, and if the velocity V is assumed infinite, that same potential becomes $\frac{1}{4} \cdot pvt$. If we therefore want to use the stream DG in Fig. 54 for rotating a machine by an oblique impulse, there can never be obtained more than the fourth part of that *absolute potential* which is expended in the elevation of the water from C to EF. But we have seen in §37 that by direct impulse more than $\frac{4}{27}$ is never obtained. Therefore, by an oblique impulse or by horizontal motion of the wheel an effect can be obtained almost two times as great as by vertical motion of the wheel.

But if the impulse of fluids is estimated differently than was indicated in §31, one will have to change the value of the letter p everywhere in the same ratio in which the estimation of the impulse was changed.

The *experiment* of which I made mention in §27, Chapter IX, is this: namely, by means of a pump one worker lifted sixteen and a half cubic feet to a height of fourteen feet within seven and a half minutes.

But this effect, equally distributed, is equivalent to that action by which approximately half a cubic foot is elevated in a single second to a height of one foot. Here, therefore, the effect is only half of that which I deduced from other principles in §17, that a healthy and robust man can produce on a treadmill. I would not believe that the entire defect is to be sought in the losses which can occur to the *absolute potential* from the different causes exposed in this chapter, but rather in the fact that the men become more tired from the agitation of the piston in the pump than from the tread in the treadmill.

Some months ago at Geneva I finally performed an obviously similar experiment, but with a far more excellent machine constructed by a singular craftsman, and with these Most Famous Gentlemen present: Messrs. De La Rive, Calandrini, Cramer, and

Jallabert, Professors of the Academy of Geneva. The success of the experiment was such that I found out that one worker elevated four-fifths of one cubic foot to a height of one foot in a single second, or, rather, that he achieved an equivalent effect. The experiment is noteworthy, and I do not believe that an effect greater than this can be obtained by any other machine. Also curious is the fact that it thus appears that machines of all kinds, animated by any potential whatever, achieve, if you remove hindrances, an effect not greatly dissimilar. Having thought over the matter well, I state that by a most excellent machine a man can elevate a cubic foot of water in a single second to the height of one foot, or produce a similar effect.

Here as well, particularly in consideration of §31, would pertain the experiments that I most accurately performed for estimating the impetus of a fluid stream impinging on a plane, by which was confirmed the new theory which I had established about this matter, and simultaneously I learned that in Mariotte's time a common error was committed. But since at the end of that chapter there was no eloquent discussion on this subject, and since the intention is to treat it expressly in Chapter XIII, let us therefore delay until then these discussions brought forth from mechanical principles not yet observed.

TENTH CHAPTER

Concerning Properties and Motions of Elastic Fluids, but especially of Air

1. Now being about to consider elastic fluids, we may ascribe to them a constitution that coincides with all properties known so far, in order that thus also a path be provided to the remaining, still insufficiently explored properties. But the outstanding properties of elastic fluids are stated as follows: (1) they are heavy, (2) they extend in all directions, unless they are confined, and (3) they allow themselves to be compressed continuously more and more as the compressing forces increase. Air, to which our present considerations pertain mostly, is composed in this way.

§2. And so consider a cylindrical vessel *ACDB*, placed vertically (Fig. 56), and in it a movable lid *EF*, on top of which lies the weight *P*. Let the cavity *ECDF* contain extremely small bodies agitated in a

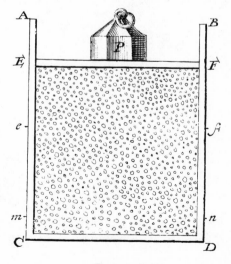

Figure 56

very rapid motion; thus the small bodies, while they impinge on the lid *EF* and also support the same by their continually repeated impacts, compose an elastic fluid which expands if the weight *P* is removed or diminished; this is compressed if the same is increased, and it gravitates on the horizontal base *CD* not at all differently than if it existed with no elastic property. Indeed, whether the small bodies are at rest or agitated, they do not change gravity, so that the base sustains either the weight or the elasticity of the fluid. Therefore, let us substitute for air a fluid that is consistent with the primary properties of elastic fluids, and thus we will explain some properties which have been already detected in air, and we will illustrate further some others [that are] not yet sufficiently investigated.

§3. We shall consider the small bodies enclosed by the cavity of the cylinder as infinite in number, and since they occupy the space *ECDF*, let us say that the latter forms the natural air, to the measures of which all [other measures] are to be referred; and thus the weight *P* holding the lid in the position *EF* does not differ from the pressure of the Atmosphere lying above it, which we therefore shall designate henceforth by *P*.

But let it be noted that this pressure is not at all equal to the absolute weight of the vertical cylinder of air lying above the lid *EF* in the atmosphere, which authors so far have affirmed inconsiderately; but that pressure is equal to the fourth proportional to the surface of the earth, the size of the lid *EF*, and the weight of the entire atmosphere on the surface of the earth.

§4. Now the weight π is sought which can compress the air *ECDF* into the space *eCDf*, the velocities of the particles in either air (the natural and the compressed), of course, having been assumed the same; moreover, let $EC = 1$, and $eC = s$; but since the lid *EF* is transferred to *ef*, it suffers a greater pressure from the fluid in two ways: *firstly*, because the number of particles is now greater in proportion to the space in which they are contained, and *secondly*, because any particle repeats the impetus more often. In order to perform correctly the calculation of the increment which depends on the *first cause*, we shall consider the particles as resting, and we shall make *n* the number of those which are adjacent to the lid in the position *EF*, and the equivalent number for the location of the lid at *ef* will be $n \bigg/ \left(\dfrac{eC}{EC}\right)^{2/3}$, or $n/s^{2/3}$.

But let it be noted that the fluid is considered by us not more compressed in the lower part than in the upper part, which is so because the weight *P* is just as infinitely much larger than the very weight of the fluid. Hence it is clear that with this designation the force of the fluid varies in proportion to the numbers n and $n/s^{2/3}$, that is, as $s^{2/3}$ is to 1. But what pertains to the other increment arising from the *second cause* is found by observing the motion of the particles; and thus it is apparent that the impulse occurs the more often, the closer the particles are located to each other; of course, the number of impulses will be reciprocal to the median distance between the surfaces of the particles, and these median distances will be determined as follows:

We assume the particles to be spherical, and we shall call D the median distance between the centers of the small spheres for the position EF of the lid, and the diameter of a small sphere we shall designate by d; so the median distance between the surfaces of the small spheres will be $D - d$; but it is evident that at the position ef of the lid the median distance between the centers of the small spheres will be $D\sqrt[3]{s}$, and therefore the median distance between the surfaces of the small spheres is $D\sqrt[3]{s} - d$. Therefore, with respect to the second cause the force of the natural air $ECDF$ will be to the force of the compressed air $eCDf$ as $\dfrac{1}{D-d}$ is to $\dfrac{1}{D\sqrt[3]{s}-d}$, or as $D\sqrt[3]{s} - d$ is to $D - d$; for both causes together, however, the aforementioned forces will be as $s^{2/3} \cdot (D\sqrt[3]{s} - d)$ is to $D - d$.

For the ratio D to d we can substitute another, more intelligible one: namely, if we consider that the lid EF, [when] depressed by an infinite weight, descends to the position mn at which all particles touch each other, and if we designate the line mC by m, D will be to d as 1 is to $\sqrt[3]{m}$, which ratio being substituted, finally the forces of the natural air $ECDF$ and of the compressed, $eCDf$, will be as $s^{2/3} \cdot (\sqrt[3]{s} - \sqrt[3]{m})$ is to $1 - \sqrt[3]{m}$, or as $s - \sqrt[3]{mss}$ to $1 - \sqrt[3]{m}$. Therefore,

$$\pi = \frac{1 - \sqrt[3]{m}}{s - \sqrt[3]{mss}} \cdot P.$$

§5. From all phenomena we can judge that natural air can be compressed into an almost infinitely small space; therefore, with $m = 0$ having been assumed, π becomes $\dfrac{P}{s}$, so that the compressing weights are almost in inverse proportion to the space which the air occupies when compressed differently; manifold experience has confirmed this. This rule can also certainly be accepted safely for air rarer than natural; but I have not explored sufficiently whether it can be also [accepted] for very much denser air; and indeed experiments have not yet been performed with that accuracy which is required here; for defining the value of the letter m, there is need of only one, but of one to be formed most accurately and certainly with violently compressed air; however, let the degree of heat in the air, while it is being compressed, be carefully kept unchanged.

§6. Meanwhile, the elasticity of air is increased not only by compression but also by increased heat, and since it is established that heat is spread out everywhere by increasing internal motion of the particles, it follows that an increased elasticity of air not changing volume discloses a more intensive motion in the air particles, which agrees correctly with our hypothesis; it is indeed evident that the greater the weight P required for keeping the air in the position $ECDF$, the greater the velocity at which the air particles are agitated. By all means it is not difficult to see that the weight P will follow the ratio of the square of that velocity, for the reason that the number of impacts as well as their intensity increases equally with an increased velocity; but [each one] separately is proportional to the weight P.

If, therefore, the velocity of the air particles is called v, the weight which it can sustain at the position EF of the lid will be vvP and in the position ef it will be $\dfrac{1 - \sqrt[3]{m}}{s - \sqrt[3]{mss}} \cdot vvP$, or approximately $\dfrac{vvP}{s}$, because, as we have seen the number m is extremely small with respect to unity and the number s.

§7. That theorem which I presented in the preceding paragraph, by which, namely, it is indicated that *in all air of any density whatever but of the same prescribed degree of heat, the elasticities are as the densities, and therefore also the increments of the elasticities, which are created by equally increased heat, are proportional to the densities,* that theorem, I say, Mr. Amontons was taught by experience, and he recorded it in *the memoirs of the Royal Academy of Science of Paris for the year 1702.* The sense of this theorem is that if, for example, natural air of moderate heat sustains a weight of 100 pounds imposed on a given surface, and then its heat is increased until it can carry 120 pounds on the same surface and at the same volume, then it will occur that the same air, compressed to half the space and possessing the same degrees of heat, can carry, respectively, 200 pounds and 240 pounds, so that increments of 20 pounds and 40 pounds, proportional to the densities, are generated in either case by the increased heat. He further affirms that the expansion of air, which he calls tempered, is to the expansion of the same air with the heat of boiling water approximately as 3 to 4 or, more accurately as 55 to 73. However, I have learned from performed experiments that very hot air, such as it is in the hottest summer in this country, is not yet of such an expansion as Mr. Amontons attributes to tempered air; I believe that not at the equator itself is the air ever of this heat. But I believe that my experiments are to be trusted more than the Amontonian ones for the reason that in the latter the air will not preserve its volume, and to this variation no consideration was given by the Author in the calculation. I learned that the expansion of air which was very cold here in Petersburg on 25 December 1731 (old statutes) was to the expansion of similar air attributed with the same heat as boiling water as 523 is to 1000.

But in the year 1733, on the 21st day of January the cold was much more intense, and for this I observed that the elasticity of the air corresponds to within half of that which the same air has when heated to boiling water. But when the heat of air in a shadowed place was greatest in the year 1731, it had an elasticity of approximately $\frac{4}{3}$ and more accurately $\frac{100}{76}$ of that which the coldest air had and $\frac{2}{3}$ of that which air of the same heat as boiling water has; therefore, the greatest variations in the air in these places are contained within the limits 3 and 4; I have read that in England they do not exceed the limits 7 and 8. But I believe that the heat of air, the elasticity of which equals three quarters of the elasticity of air as hot as boiling water, is almost intolerable for an animal body.

§**8.** From the known relation between the different elasticities of the same air contained in the same space, it is easy to deduce a measure of the heat which pertains to the air if only we agree in defining twice, three times, etc., the heat which definition is arbitrary and not fixed in the nature of things; to me it seems indeed not incongruous that the heat of the air, if it is of common density, is proportional to the state of its elasticity. But let the first degree of heat, from which the others obtain their measure, be taken from boiling rain water, because this has, without doubt, approximately the same degree of heat everywhere on earth.

If these [things] are so accepted, the temperatures of boiling water, of air in the hottest time of summer, and of air in the coldest time of winter in this country will be approximately as 6, 4, and 3. Let me tell now how I found those numbers, so that judgment can be passed on the accuracy of the experiments, the success of which is so greatly different from the Amontonian.

§**9.** Indeed, I made use of an ordinary barometer $ACBE$ (Fig. 57), and I took care that it was sealed hermetically at m; in this manner I changed the instrument into an air thermometer not subjected to barometric changes, for with increasing heat an expansion of the air

PROPERTIES AND MOTIONS OF ELASTIC FLUIDS   231

AmF is achieved, and the column of mercury *BD* which the captured air sustains becomes higher, and if the space *AmF* could be assumed as practically infinite, the heat would be in proportion to the height *BD* (by §§7 and 8), and by means of this thermometer the measure of heat

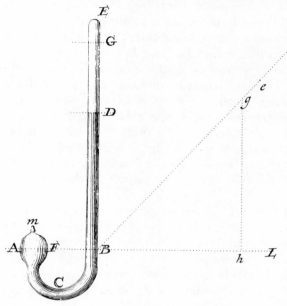

FIGURE 57

could be defined specifically everywhere. For if the instrument is immersed in boiling rainwater in a vertical position and the point *G* is observed to which the surface of the mercury ascends, then any other degree of heat whatsoever which has been observed to have sustained the mercury up to the point *D* will have to be defined, and in any case that heat will be to the heat of boiling water as *BD* is to *BG*. And since the ratio *BD* to *BG* is constant, whatever the height *BG* should be, the same degree of heat that we are discussing can be easily imitated in any location. Furthermore, *BG* could be divided into a hundred or a thousand parts, and the height *BD* could be defined by partitions of this kind.

I say nothing about the methods of obtaining more sensitive thermometers of this sort; whoever wants to will easily think up many of them. Care should be taken, however, that the height *BE* be not below 4 feet; on the contrary, it should be larger if one intends to determine the degree of heat of other boiling fluids, which is often greater than in water. If smaller thermometers of this kind are desired, these could be made so that at the time of sealing at *m* a glass ampule *AF* is put into the fire of the torch for rarefying the air contained in it, and then let the sealing be made immediately, and lest any delay be introduced in the sealing, the glass ampule could be drawn previously into a capillary tube which melts together easily when it is brought close to the flame. In this manner I have obtained thermometers not more than four or six inches long, but of little value. Furthermore, it is of great importance that the space *ED* be vacant of all air, as much as this can be done, and we shall not be sure enough of this vacuum when we shall have seen that at a horizontal position of the instrument the mercury reaches the extremity *E*, because it can happen that the air which was previously in the space *ED* retracts into the pores of the mercury and again occupies the original space when the mercury descends; the test will be safer by moving the part *DE* towards a flame; for if the surface *D* does not change its position from the heat of the flame, this will be a certain proof that the space *ED* is vacant of air.

§10. In the preceding paragraph we have considered the space *AmF* occupied by the air as practically infinite in proportion to the space *DG* or *DE*; but since it will be only eight or ten times larger, it will not yet be permissible to consider it as infinite without noticeable error; and from here I suppose that some error has arisen in defining the expansion of moderately warm air in the Amontonian experiments.

Therefore, in order that the experiment be performed most accurately, one will have to proceed as follows: Let the lower surface of the mercury be at *AF*, and let a horizontal [line] be drawn at *AL*; then, for defining any degree of heat whatsoever, let the instrument be inclined until the surface of the mercury is at the point *g* (which is the same one at which the mercury remained from the degree of heat of boiling water at the vertical position of the thermometer), and then let the measurement of the vertical height *gh* be taken, which will be in fact to the height *GB* as the expansion of the air, the heat of which is to be defined, is to the expansion of air as hot as boiling water. Thus, therefore, the heats will be accurately in the ratio of the height *gh*. Before I discontinue this argument, it will be convenient to have noted (in case perhaps to some it will seem that the *first degree of heat*, which was defined by us, taken from boiling water is not always and everywhere completely consistent with itself) that instead of the heat of boiling water also a thermometer of certain and fixed measures can be made if in the experiment the density of the air is measured or its specific gravity is noted together with the barometric height. For if the thermometer is inclined until the surface of the mercury is at g, and at that time the height of the barometer is 28 Paris inches, and a cubic foot of the air in which the thermometer is located has a weight of 600 Nuremberg grains, the vertical height gh could be considered as the *first* degree of heat. But if at a different location and time the height of the barometer is 29 Paris inches and the weight of a cubic foot of the air surrounding the other thermometer (in which it is the intention to define the *first degree* of heat) is 500 Nuremberg grains, and finally the surface of the mercury in the thermometer is again at g, the vertical height corresponding to the first degree of heat will be $\frac{29(600)}{28(500)} \cdot gh$. In using the thermometer, let the instrument always be inclined until the surface of the mercury is at g; I wanted to add this method so that it be apparent how easy it is in theory to give a fixed measure of heat; but in practice I shall prefer another much easier and sufficiently accurate one to this.

§11. Let us come now to considering the atmosphere of air that is influenced not by an overlying foreign weight, but by its own mass. But first we shall examine the pressures of vertical air columns and their equilibria between one another as well as with the mercury columns in barometers. Second, we shall investigate the elasticities of air at different heights of the atmosphere above the sea and the corresponding barometric heights; and these [things] having been set forth, we shall satisfy many other phenomena pertaining to the changes of the atmosphere.

12. Let there be two vertical pipes AC and BD (Fig. 58) of equal area, and each one of indefinite height. Then imagine narrower horizontal pipes ab, cd, ef, gh, lm, etc., practically infinite in number, open on both sides and connecting to the vertical pipes. Assume furthermore that the air particles occupying these pipes are everywhere agitated at the same velocity and so have the same degree of heat; thus there is no doubt that the bases A and B are equally pressed and simultaneously the same weight (which, of course, is the very weight of the indefinite air column AC or BD) lies on top of them.

One also understands that if one assumes diaphragms at equal heights such as g and h and imagines the lower air gA and hB to be absent, even now these diaphragms are pressed equally on both sides, and that the weights of the air columns gC and hD lying above the diaphragms are equal. Therefore, if the weight of the entire air column AC or BD is called A and the weight of the air column gC or hD is taken as B, the weight of the air contained between A and g or B and h will be $A - B$, the weight lying above the base A or B

equals *A*, and the weight lying above the diaphragm at *g* or *h* equals *B*.

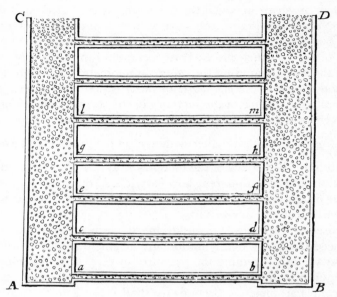

FIGURE 58

§**13.** But if the particles in the pipes *AC* and *BD* are agitated at unequal velocity, the situation will be different; nevertheless, whatever difference of velocities and temperatures be assumed in the individual locations, it is evident that nonetheless the parts of a pipe located at the same height will be pressed equally on both sides, such as at *g* and *h*, and that hence the diaphragms, if they are assumed to be located at the same height on both sides, will sustain equal pressure. For if one says that the pressure at *g* is less than at *h*, there will be nothing which may hinder the flow of air from *BD* into *AC* through the small transverse pipe *hg*, and thus this statement will contradict the state of *permanence* which we have supposed.

Therefore, since places located at the same height are equally pressed by the air lying above, the densities in arbitrary homologous places, such as at *g* and *h*, will be (by §6) approximately in the inverse-square ratio of the velocities at which the particles are agitated in these locations.

§**14.** It follows from the preceding paragraph that in every place the pressure of air is the same at equal heights above the surface of the sea if the atmosphere is assumed to be in a permanent state of equilibrium and agitated by no winds, whichever be the difference in heat between the different parts of the atmosphere. Therefore, it is obvious that everywhere on earth, at the equator and at the pole, the height of mercury is the same in barometers which are located at the surface of the sea or at equal heights above the latter if the atmosphere is not exposed to changes. I state further that the water terminated by the surface of the sea is located according to a common equilibrium, not because this is completely necessary, but because so far no difference has been observed; in reality, the currents (*les courans*) of water in many places of the ocean, which are perpetually directed toward the same region, show that this hypothesis is not to be accepted with all rigor.

§15. I have already noted that the density of air at any position of the vertical pipes depends upon the corresponding heat; and since the degrees of heat can be different, equilibrium being retained, the densities can also be different. Therefore, let the densities be taken equal to D at g and equal to δ at h; and let there be assumed on both sides two strata of equal and infinitely small height dx, with the height Ag or Bh taken equal to x. So the weight of the air column Ag will be $\int D\,dx$ and [that] of the column Bh will be $\int \delta\,dx$, and in this manner one can define the weight of both the entire column and any part [of it]. Meanwhile it is apparent that the nature of the matter requires least of all that the weights of the columns AC and BD, or Ag and Bh, or finally gC and hD are equal to one another, although (by §13) the pressures at the bases A and B as well as at the diaphragms g and h are equal to one another. It will perhaps be amazing to some at first consideration that it can happen that the base A sustains another pressure than that which is the weight of the indefinite air column AC lying above it, since indeed, with everything remaining in its state, as we have just seen, the individual orifices a, c, e, g, etc., can be conceived as closed, in which case there is certainly no doubt that the pressure at the base A is the very weight of the overlying air column. But anyone may examine this point for himself in the following manner: let us assume either column to be of finite height (for although they rise without end as long as the particles maintain some motion, they will nevertheless be terminated if those particles in the uppermost part of the columns have no motion and thus form a simple heavy fluid without any elasticity). With this assumption it is apparent, first, that either column rises to a common height after the transverse tubes which are everywhere present have been opened; second, that the uppermost layers are equally dense on both sides because they are positioned at equilibrium and have a common height. For this is now obvious why it would not be possible to consider the transverse tubes as closed, which I undertook to show. Also, it is evident by itself that the pressures are everywhere proportional to the weight of the uppermost layer, from which it follows, as was already shown in §13, that the pressures in either part are equal to one another at equal heights. If now the columns are not terminated anywhere, it will be allowable to conclude that the last layers are loaded on either side by an equal weight, or to assume [that] diaphragms [exist] at equal heights, so that hence nothing of the power of the demonstration is lost.

§16. If, therefore, the mercury descends in a barometer transported from a lower point such as A to a higher one, g, it does not follow that the weight of the mercury column which descends in the barometer is equal to the weight of an air column of the same diameter and the height Ag, which is so asserted by others. And actually, the rest being equal, the descending column of mercury will be the same in wintertime as in summertime, since from that statement it should be less in a warm season than in a cold season. It will also be the same in southern and northern regions.

Hence it is evident what ought to be thought of that method which Mr. Duhamel, in the *History of the Royal Academy of Science of Paris*, reports was used at some time or other in England for investigating the ratio between the specific gravities of air and mercury. The height of the mercury having been observed, of course, at a lower point, then also at a higher one, they announced the specific gravities in air and mercury to be as the difference of heights of the mercury in the barometer was to the height contained between the points of observation. Even if the air is assumed as of the same density from the lowermost point of observation up to the other one, it is hence not permissible to pass judgment on its specific gravity in proportion to mercury. This is all that one may conclude from the experiment.

Let us consider indeed the entire air shell surrounding the earth and contained between the two points of observation, and the weight of this shell will be to the surface of the earth as the weight of the mercury column which descends in the barometer is to its base; this is manifest from the fact that the sum of the bases A and B sustain certainly the sum of the weights which the air columns AC and BD have, and that nevertheless neither base is pressed separately by the weight of its column; and this must also be understood for the columns gC and hD lying above the diaphragms located at g and h after the columns Ag and Bh have been cut off. Therefore, the experiment does not indicate that specific gravity of the air in which it is performed, but rather it determines the *mean* specific gravity of all air close to the earth; the former is greatly variable, the other without doubt remains almost constantly the same.

Let us compute the *mean specific gravity* of all the air that surrounds the earth. Indeed, from many experiments which have been performed in different places elevated slightly above the sea, it is shown that a descent of one line in the barometer corresponds to an elevation of approximately 66 feet. It follows hence that the average specific gravity of the air is in proportion to the mercury as the height of one line is to the height of 66 feet, that is, as 1 is to 9504; therefore, with the specific gravity of mercury taken as 1, the *mean* specific gravity of air will be 0.000105. It is indeed noteworthy that this average gravity of air is so large; for I am sure that even in the most raging coldness in this country the specific gravity of the air is hardly yet as large as we have just shown for the mean state of all air surrounding the earth; and at the equator it will be much less, and, everything having been thought over correctly, I should not believe that the *average gravity* of the air which is contained between the two latitudes of 60 degrees extends beyond 0.000090; this having been assumed, the *average gravity* of the air encompassing the earth from either pole to the 30th degree (which space makes up a little more than an eighth part of the total surface of the earth) will be 0.000210, which is twice that of the most dense air in this country; but at the pole itself, particularly the Antarctic, the air will be very much heavier and perhaps almost 10 times lighter than water, since it is very cold and very dense.

§17. Let us come now to changes of both the atmosphere and the barometer. We shall therefore consider two barometers, both located at the lowest point of air, the one at A, the other at B, and let us assume that in either one the mercury is suspended at the same height. Next let us imagine the air at A to be greatly heated; thus we see that this very air will be rarefied; nevertheless, no change of the barometer would be produced, if the air had no inertia against the motion, even if all air were driven from AC over to BD; but, this inertia having been assumed, a certain pressure develops in all regions, and this is most noticeable in the region A. Therefore, the height of the mercury in either barometer increases with time, and it increases more at A than at B. The contrary will exist, if at once some great mass of air close to the barometer A or B were compressed by cold.

§18. This seems to be the unique cause which can effect some change in the barometers located at A or B, because, if it is removed, the bases A and B are always equally pressed, certainly either one by a weight which is one half of the air columns AC and BD added together, which sum of weights is indeed constant. If we want to apply this to the atmosphere, it is to be noted that the bases A and B represent the lowest points of the atmosphere, which certainly would be located at the surface of the earth if the air could not penetrate the inner parts of the earth; but because the situation is different, the locations analogous to the bases A and B will have to be considered below the surface of the earth.

§19. Let it now be assumed that the barometers are located at g and h, and let the mercury in both be suspended at the same height; with these things established, let a cause be imagined to develop by which the column Ag either alone or together with the other, Bh, is heated up and expands. From this it is evident that if the inertia of the air is practically null, the pressures of the air at g and h will increase, because a larger quantity of air than before now lies over these places; of course, the weight of all the air which was pushed upward from Ag and Bh by the heat was involved. And in order to indicate this by symbols we will set A equal to the weight of the column Ag before a new degree of heat will have developed, α to that of the other, Bh, B to the weight of the column gC, β to that of the column hD, C to the weight of the rarefied column Ag, γ to the weight of the likewise rarefied column Bh, l to the height of the mercury at g before the expansion of the air Ag and Bh, and x to the same height after that expansion; we will then have this analogy:

$$B + \beta : l = B + A - C + \beta + \alpha - \gamma : x;$$

from this

$$x = \frac{B + A - C + \beta + \alpha - \gamma}{B + \beta} \cdot l.$$

Therefore, the mercury ascends, by having rarefied the lower air, through the height $x - l = \dfrac{A - C + \alpha - \gamma}{B + \beta} \cdot l = \dfrac{A - C}{B} \cdot l$ (setting everything equal in either pipe).

But with the air again cooling in Ag and Bh, the mercury descends again in either barometer.

Here is to be noted that in this manner from a very small change of heat at Ag and Bh a noticeable variation in the barometer can develop on account of the tremendous density of air in the lower parts, by which it can happen that much more air is contained in the part Ag (finally, an infinite number of times as much if air pressed by an infinite force is assumed to be compressed into an infinitely small space) than in the remaining gC, although it is infinite in length. Whence, if the weight A is very much larger than the weight B, and simultaneously the cause rarefying the air is maintained, the weight C follows a given ratio to A; since this usually happens, it is apparent that the ascent of the mercury on account of the least degree of heat being added at Ag can be arbitrarily large.

Equally, if it is assumed that the parts Ag and Bh are very much narrower than the areas at gC and hD, it is recognized that the variations of the barometer due to the increased or decreased degree of heat at Ag and Bh thus become less noticeable, because the weights A and α and C and γ, proportional to the former, decrease in this manner; nevertheless, the barometric variations which originate from this cause can still be understood to be arbitrarily large.

§20. When this is thus considered, it happens indeed that the barometric variations are for the most part to be sought from quick changes of temperature in underground caves. It has been known for a long time that there are many caves of this kind, and that they are immense; even in solid ground, pores can make something like a cave. If one collects all cavities (both those which are formed by caverns and those formed by air-containing pores) up to a height of 20,000 or 30,000 feet below the surface of the earth and compares their capacity with the solid part of the earth's crust of the same height, and if one assumes the latter to be a thousand or a hundred thousand times larger than the former, then this will now indeed be reason enough to explain the very large changes of the barometer. From the preceding paragraph I believe that these things will be clear to anyone.

By the way, places which are closer to caves will be the more exposed to winds and changes of the barometer because of the inertia of the air to motion, which is perhaps the reason why toward the equator, where almost everything is deep sea, smaller variations are observed on a barometer than in these northern places.

§21. From the same source it is deduced that aqueous exhalations from the pores of the earth can also contribute something to barometric changes, but this will certainly be little. For if the vapors would give as much water as can fall in the most severe rain, then the mercury will hardly ascend a single line in the barometer, apart from the fact that this matter does not occur so fast but that the effect of it is distributed almost evenly over the entire atmosphere, and thus vanishes completely for a certain definite location. For if we consider the entire atmosphere which surrounds the earth, I have noticed that it certainly cannot be loaded unevenly with vapor. Undoubtedly, I would prefer the reasoning given in §20 to all others; indeed, motions of the earth, which are often noticed up to a hundred miles at the same time, and other phenomena of this kind, indicate that large and quick changes can occur in the interior of the earth.

In order to explain barometric changes, there is first of all some sudden cause required; indeed, I have already mentioned that the slow ones, which I have distributed over the entire mass of air, are of no effect, and I have demonstrated this in §14. And for this reason the changes which occur immediately in the atmosphere above the surface of the earth are to be considered as of little importance.

§22. And this also seems to be the cause of why the moon, which has such a great effectiveness in agitating the waters of the ocean, exerts no effect on a barometer that anyone has been able to notice during very careful observations. And if the remaining causes which may produce some change somewhere in the atmosphere would also act gradually, one would notice constantly, without doubt, the same height of mercury at all points equally distant from the surface of the sea. This height can be called the *mean* and will be determined approximately in the manner which Johann Jacob Scheuchzer used, by observing daily the barometric height through a long period of time and taking the mean of all of them.

And, having used this procedure, that most famous Author stated the mean height from the many observations which had been transmitted to him from many places.

At Padua	27 inches	$11\frac{1}{2}$	Paris lines
Paris	27 ,,	$9\frac{1}{2}$	,,
Turin	27 ,,	$1\frac{1}{4}$	,,
Basel	26 ,,	$10\frac{1}{8}$	,,
Zurich	26 ,,	$6\frac{1}{2}$	,,
On the mountain (St. Gotthard)	21 ,,	$27\frac{1}{2}$	,,

§23. It is known that the diversity of these *mean* heights stems from the unequal elevation of the places above the sea. Indeed, in Pascal's time experiments had already been performed on the descent of mercury in a barometer which is carried from a lower position to a higher one. Thence the philosophers inquired into the mutual proportion between cause and effect. Several rules were produced on this matter from various authors. The foremost of them, to which many people still cling, is that the heights of the locations are proportional to the logarithms which correspond to the heights of the barometer. This rule is established principally on the fact that the density of air is everywhere proportional to the weight of the air lying above it; but this principle is wrongly applied here, since it is valid only for air of constant temperature and is not a definite matter at every height of air, although it exists in the same vertical column; but if it were so that the heat would be equal, then it must be acknowledged that thus the rule behaves correctly enough.

But experiments are clearly contradictory to the rule; therefore, the same degree of heat does not exist everywhere in the entire height of the vertical air column. In order to make this plain, let me add now some experiments, performed accurately, as I believe, but nonetheless at different times and locations, which I regret; experiments performed at the same time on the same mountain, only at different elevations, would aid our undertaking more; such experiments, however, exist, as far as I know, only for moderate heights of places, with all the circumstances which one has to know.

I. At a height of 1070 Paris feet above the surface of the sea the barometer descends $16\frac{1}{3}$ lines, when at the surface of the sea it would hold 28 inches $4\frac{2}{3}$ lines (others set it simply equal to 28 inches, but in the papers which Mr. Delisle exchanged with me it was obtained as 28 inches $4\frac{2}{3}$ lines). Therefore, the elasticity of the air at the surface of the sea having been set equal to unity, as I will always do from now on, the elasticity at the higher point which I shall designate by E was found equal to 0.9520.

II. At a height of 1542 Paris feet above sea level the mercury, which at sea level was clinging to a height of 28 inches 2 lines, descended $21\frac{1}{2}$ lines in the barometer; here, therefore, $E = 0.9364$.

III. On top of the Peak on the Island of Tenerife, 13,158 Paris feet above sea level, the mercury reached a height of 17 inches 5 lines, while at the surface of the sea it maintained a height of 27 inches 10 lines, whence at that place $E = 0.6257$.

IV. If the descents of mercury are observed accurately at smaller heights, it is found that a descent of one line corresponds to a height of 65 or 66 feet. Therefore, at a height of 65 feet $E = 0.9970$. These observations are very widespread; indeed, I have a third from Mr. Delisle, and it was performed by Mr. Feuillée and presented before the *Royal Society of Science in Paris*; and that one is the stumbling block against which all theories collide that have been brought forth so far.

§**24.** In order that it now be evident to what extent these theories agree with the position of the logarithmic, or of the scale of heights corresponding to the elasticities, let us set the height of a place above sea level, to be defined by a certain number of Paris feet, equal to x; the expansion of the air at the surface of the sea we shall designate by 1, and the expansion of the air at the height x we set equal to E. Let it be noted further that the atmosphere is now considered unchanged by us, or at least always similar to itself, so that the expansions of the air at the surface of the sea and at any height x stand in a constant proportion to each other. For if only the expansions, not constant with time, would change unequally at different heights of the atmosphere, no rule could reasonably be invented. This having been set forth, let us now take the equation $\alpha \ln E = x$, where the coefficient will be determined from a single observation; let us use the first observation and there will be $\alpha \ln 0.9520 = 1070$, and hence (due to Vlacqian logarithms) $\alpha = -50{,}194$. Therefore, for this treatment, if the logarithmic [method] is to be satisfied, one must set $-50{,}194 \ln E = x$, or $\ln \dfrac{1}{E} = \dfrac{x}{50{,}194}$; but, according to the norm of this equation it is found that $E = 0.9317$ if $x = 1542$ for the second observation; the observation itself, however, indicates $E = 0.9364$; the difference between hypothesis and observation is more than a line and a half, which is clearly noticeable in respect, as usual, to the small difference in the vertical heights.

Furthermore, if now, for the third observation one sets $x = 13{,}158$, there results from the hypothesis $E = 0.5469$, while the experiment indicated $E = 0.6257$; this difference is too large to be supported by any logarithmic method, for it amounts to more than two inches and two lines.

§**25.** The logarithmic law having been rejected, it follows that the elasticities at different heights of the atmosphere are not at all proportional to the densities, or, which amounts to the same thing, that the mean degree of heat is different at different heights. Therefore, different rules were thought up by others, to whom this defect was well known; nevertheless, none of those rules can be said to be suited satisfactorily to Experiment 3 (§23). I think one should hardly hope to find the true law which nature follows; this, indeed, would provide a ratio of the mean velocities of the air particles different from [that provided] by simple interferences. Nevertheless, I have agreed strongly with a certain hypothesis that corresponds not badly to the phenomena; but before I proceed to this special hypothesis, I shall give the curve for any law of velocities.

§**26.** Let the line AD (Fig. 59) be vertical; the horizontal QF touches the surface of the sea; BF denotes the mean velocity of the air particles at the surface of the sea, *BM* the mean density, and *BQ* the elasticity, which is the same over the entire location of equal elevation. Then let there be conceived the curves *EFH*, *LMO*, *PQS*, drawn through the points *F*, *M*, *Q*, or the scales which at all elevations such

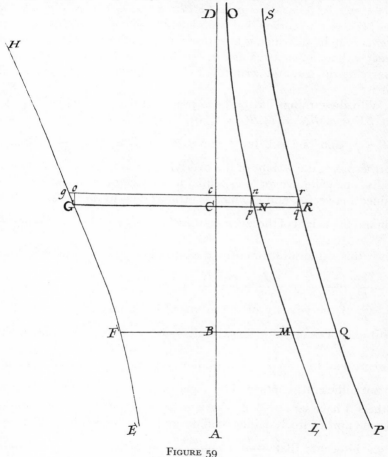

FIGURE 59

as *BC* denote by the ordinates *CG*, *CN*, and *CR* the mean velocities of the air particles, the mean densities, and the mean elasticities. Now with two curves given, one can determine the third from the fact that the elasticities (as experience has also shown and was explained in §§3, 4, 5, and 6) are approximately in proportion to the product of the square of the velocities just mentioned and the first power of the densities.

I myself have indeed advised in the above-mentioned place that this proportion cannot be exactly true, because air can certainly have an infinite expansion or can be compressed by an infinite force, but it can plainly not be compressed into an infinitely small space; nevertheless, since this property, *namely that the elasticities are in proportion to the product of the square of the velocities of the particles and the first power of the densities*, has been seen to correspond completely to the observations in experiments even in air which is four times denser than natural air, we may use it without any noticeable error for the natural air of the atmosphere overlying the sea, and certainly it will be the more true the rarer the air is.

With these things having been prepared for the calculation, let us set $BF = a$, $BM = b$, $BQ = c$, $BC = x$, $Cc = dx$, $CG = v$, $CN = z$, $CR = y$, and one will have $y : c = vvz : aab$, or $y = \dfrac{cvvz}{aab}$. Because, furthermore, the weight of the overlying air is a measure of the elasticity, one will have $qR(-dy)$ equal to the weight of the air layer contained between C and c, which is proportional to the density z of the air and the height of the layer dx; therefore, $-dy = \dfrac{z\,dx}{n}$ or $z = \dfrac{-n\,dy}{dx}$; with this value substituted in the equation $\left(y = \dfrac{cvvz}{aab}\right)$, one obtains $y = \dfrac{cvv}{aab} \cdot \dfrac{-n\,dy}{dx}$ or $\dfrac{-dy}{y} = \dfrac{aab\,dx}{ncvv}$.

§27. If the velocity of the air particles is taken the same at any altitude, for example, a, there results $\dfrac{-dy}{y} = \dfrac{b\,dx}{nc}$, or, after the required integration, $\ln \dfrac{c}{y} = \dfrac{bx}{nc}$. But we have seen in §24 that that hypothesis is not sufficiently confirmed by experiments. Therefore, having tried others, I have set $v = \sqrt{aa + mx}$, or $vv = aa + mx$, which is the law for motion of freely falling bodies, and this not without success; so thus it occurs that $\dfrac{-dy}{y} = \dfrac{aab\,dx}{naac + mncx}$ or $\ln \dfrac{c}{y} = \dfrac{aab}{mnc} \ln \dfrac{aa + mx}{aa}$. In this slightly more general equation in which m and n are still arbitrary, I further made an attempt to see whether one could not set $\dfrac{aab}{mnc} = 1$, and I saw also that this can be done aptly; truly, thus I obtained $\ln \dfrac{c}{y} = \ln \dfrac{aa + mx}{aa}$, or $\dfrac{c}{y} = \dfrac{aa + mx}{aa}$, or $\dfrac{y}{c} = \dfrac{aa}{aa + mx}$. This hypothesis indicates that the elasticities of air are everywhere in inverse proportion to the square of the velocities at which the air particles are agitated, or that CR is to BQ as $(BF)^2$ is to $(CG)^2$, and since by hypothesis EFH is a parabola above the axis AD having a vertex at a distance $\frac{aa}{m}$ below the point B, it follows that the curve PQS is a hyperbola. But I have noticed that said distance $\frac{aa}{m}$ has to be taken equal to 22,000 feet in order to satisfy approximately the observations of §23. Hence this specific equation now results:

$$\frac{y}{c} = \frac{22{,}000}{22{,}000 + x}.$$

But for the curve LMO one finds (by §26) $\frac{z}{b} = \frac{aay}{cvv}$, or, because $\frac{aa}{vv} = \frac{22{,}000}{22{,}000 + x} = \frac{y}{c}$, one obtains, after this substitution,

$$\frac{z}{b} = \left(\frac{22{,}000}{22{,}000 + x}\right)^2.$$

§28. In order to show to what extent our hypothesis coincides with the experiments of §23, in the equation for the elasticities let us set successively $x = 1070, 1542, 13{,}158$, and 65; thus one finds, respectively, $\frac{y}{c} = 0.9536, \frac{y}{c} = 0.9345, \frac{y}{c} = 0.6257$, and $\frac{y}{c} = 0.99705$; the observations, however, indicate $\frac{y}{c} = 0.9520, \frac{y}{c} = 0.9364, \frac{y}{c} = 0.6257$, and $\frac{y}{c} = 0.9970$. The third observation, so very unfavorable for the other hypotheses, agrees clearly with ours, and the others deviate not more than 0.0019 division, which constitutes three-fifths of a line in the height of the barometer. But nobody who has experienced how vague and how little consistent with one another the barometric observations are will even care about such a small difference. I myself, meanwhile, consider this no different from a precarious hypothesis, and I have presented the calculation of §§26 and 27 for no other purpose than to give the reason by which it can happen that the vertical heights do not correspond to the logarithms of the barometric heights, as it should occur if the temperature were uniform throughout the entire atmosphere; indeed, after the calculation has been performed and a comparison of it has been made with the experiments, it seemed to me that this matter cannot be sufficiently explained by the different gravitation of the air particles at different distances from the center of the earth, such as Newton has attempted by stating that the gravitations of these particles decrease with the square ratio of the distances from the center of the earth, which hypothesis affects no noticeable difference from the hypothesis of uniform gravitation at heights that do not exceed 13,000 Paris feet. Similarly, I once came across the opinion that the increased centrifugal force of the air particles in higher altitudes can contribute something here; but, similarly, after performing a calculation I did not adhere to this opinion any further. Meanwhile, I do not believe that it is absurd if we say that the mean temperature of the air is greater, the further it is from the surface of the sea. But let me wish that it be properly noted that here we are discussing the *mean* temperature in the free atmosphere; for thus it can happen that the real temperature in the mountains certainly does not rise for other reasons; nevertheless, the hypothesis is not overthrown thereby, since indeed it has already been shown in §15 and §16 that the weight of the column of mercury in the barometer is not to be understood as being precisely equal to the weight of the air column taken in that region, but equal to the mean weight of all columns surrounding the earth; therefore, I think accordingly about the different densities.

§29. If the temperature were everywhere the same, the densities would be proportional to the elasticities, as far as can be noticed, and the vertical heights would correspond to the logarithms of the barometric heights. But I state this to oppose the experiments; nevertheless, I would not believe that at two places spaced only a little apart from each other a noticeable difference of temperature can occur, because heat is quickly distributed uniformly in a body of rather small density, such as air, unless a perpetual cause exists which cools the air in the vicinity.

But the situation is different in more remote locations, and indeed I think it is not absurd to assume the air at the pole as ten times denser than at the equator, if only the air is accepted in either case as being close to the surface of the earth; but at great heights the difference will certainly be less between the density of the air which corresponds to the pole and that which corresponds to the equator, other things being equal, and therefore the densities of air decrease altogether differently away from the surface of the earth, and they decrease much more at the pole than at the equator; therefore, in this way it could happen that the real densities of the air at the pole at small altitudes, let us say, decrease in proportion as $(22{,}000 + x)^4$ is to $(22{,}000)^4$ on account of the increased temperature, and at the equator they decrease hardly noticeably, because of the decreased temperature, which decrease in temperature close to the equator is confirmed by the fact that the top of the mountain Pico is covered with snow through a period of almost ten months, while on the island of Tenerife it never snows, as they say. Therefore, the mean densities can be thought of, not absurdly, as being diminished in the ratio of $(22{,}000 + x)^2$ to $(22{,}000)^2$ as it has been assumed in §27, while the elasticities decrease everywhere in the ratio of $(22{,}000 + x)$ to $22{,}000$; and certainly these cannot differ at the same height above the surface of the earth, unless due to causes brought forth by chance and lasting only shortly.

§**30.** In countries which lie between the 40th and 60th degree of latitude, it is probably that the densities decrease in approximately the same ratio as the elasticities; and for this reason I wanted to perform an experiment [to find out] what theory of refractions would hence arise, about which subject I shall now add something.

Digression Concerning the Refraction of Rays Passing Through the Atmosphere

(α) It is a well-known property of rays passing from one medium into another, confirmed by innumerable experiments, that the angle of incidence maintains a constant ratio to the angle of refraction. Furthermore, it is also obvious that if the refraction becomes infinitely small, that is, if the difference of the two sines has an infinitely small ratio to either sine, the sine of the angle which is contained between the prolonged ray of incidence and the refracted ray will have the same ratio to the total sine as the difference of the sines of the angles of incidence and refraction has to the cosine of the angle of incidence. But from now on I shall call that angle which I just defined, contained between the prolonged ray of incidence and the refracted ray, *the differential angle of refraction*. Thence it follows that, everything else being equal, the sine of the *differential angle of refraction* is proportional to the sine of the angle of incidence divided by the cosine of the same angle.

(β) Experiments show further that if a ray passes from air into air of a density different from the former, the *differential angle of refraction* is, other things being equal, proportional to the difference in densities.

But experiments pertaining to this matter, as much as it is possible, have been performed most accurately by Mr. Hawkbee on greatly compressed air as much as on very rare air also, which finally could be assumed for null. The manner in which they had been performed is described in the *Philosophical Transactions*; but the success of all experiments reduces here to the fact that they argue that the sine of the differential angle of refraction was to the total sine as $5\frac{1}{8}$ inches is to 2588 feet when a ray passes out of the natural air into a space empty of air at an angle of 32 degrees, that is, as 1 is to 6060, and under the same conditions, with the angle changed from 32 degrees into a half right angle, as 1 to 3787 (by §α). Hence it is deduced that, if a ray hits a vacuum from *natural air* at an angle, the sine of the angle of incidence is to the sine of the angle of refraction as 3787 is to 3786.

In his *Treatise on Optics* Newton assumes, instead of this, the ratio of 3201 to 3200, and he deduces it from magnitudes of the refractions observed by Astronomers; he states, moreover, that the amount of refraction is the same if the layers refracting the ray are parallel, no matter in what ratio the densities of the medium decrease, if only the difference of densities in the first and in the last layer remains the same (see Newton's *Treatise on Optics*, page 321, French edition). Concerning the remainder, the refraction cannot be but greatly variable under diverse circumstances, since the air that we call natural is exposed to many changes, as much from heat and cold as from the pressure of the atmosphere, which both act together in forming the density of the air, to which density, other things being equal, the refractions of rays entering a vacuum are proportional. Mr. Hawkbee has mentioned the same thing in the report of the experiments which we just discussed, and for this reason he has defined properly the state of the air that existed when he did the experiments.

(γ) Now let AC (Fig. 60) be the arc of a terrestrial circle drawn around the center B, in the plane of which lies the ray of light AG; but this curved ray AG will be of such a nature that it converges to an asymptote; let AH be assumed parallel to this asymptote; let the horizontal AE be drawn, and the straight line AF which touches the curve AG at A. Thus we may see that the angle HAE will be a measure of the true height of the star, and the angle FAE will be a measure of the apparent height, and the angle FAH will be the angle of refraction; moreover, the angle FAH is the same as the sum of all *differential angles of refraction* or of all angles of contact such as the angle cbo.

Let two elements ab and bo of the curve be considered, and let it be understood that the arcs $\alpha\alpha$, $\beta\beta$, and $\gamma\gamma$ with the common center B are drawn through the points a, b, and o; let the density of the air $\alpha\alpha\beta\beta$ be D, and the density of the air $\beta\beta\gamma\gamma$ be $D - dD$; then (through §§α, β) the sine of the angle of contact at b divided by the total sine, or

PROPERTIES AND MOTIONS OF ELASTIC FLUIDS 249

the angle of contact itself, will be proportional to the difference in densities dD multiplied by the ratio of the sines of the angles of incidence and refraction, that is, multiplied by $\frac{be}{eo}$. But if BD is drawn perpendicular to FA extended, it is evident that $\frac{be}{eo}$ and $\frac{BD}{Do}$ barely differ, because the ray is almost straight and thus the triangle BDo can be taken for rectilinear and similar to the triangle beo.

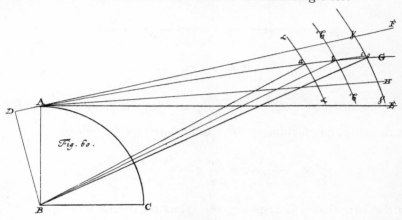

FIGURE 60

Therefore, the required angle FAH will be proportional to $\int \frac{BD}{Do} \cdot dD$.

(δ) Following these paths, and assuming that the density D is everywhere $\frac{22{,}000}{22{,}000 + x} G$, where x expresses the line na in Paris feet and G denotes the density of the air at the point of observation, I found the following: Let the sine of the apparent height of the star be f, the cosine be F, the radius of the earth be r, to be expressed in Paris feet; let the number 22,000 be indicated by a; assume further that the total sine is 1, and that the *differential angle of refraction* for a ray entering a vacuum from natural air at a half right angle is g; finally, for the sake of brevity, let $2r - 2a = \alpha$, and $-FFrr + 2ar - aa = \beta$; and β will be either a positive or a negative number; it will be positive if the apparent height of the star is small, and indeed below 2° 44′, otherwise it will be negative. In the former case the required angle FAH will be obtained in this manner: namely, let the semicircle MLF (Fig. 61) be drawn, the radius AM of which

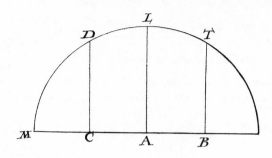

FIGURE 61

is 1; let $AC = \dfrac{\alpha}{2fr}$, $AB = \dfrac{2\beta - \alpha a}{2afr}$, and let the lines CD and BT be drawn perpendicular to MC; and the angle FAH will be

$$\dfrac{-fFrr}{2\beta}g + \dfrac{far}{\beta}g + \dfrac{far\alpha \cdot DT}{2\beta\sqrt{\beta}}g.$$

In the case that β is negative, the same angle FAH will be

$$\dfrac{-far}{\beta}g + \dfrac{fFrr}{\beta}g + \dfrac{far\alpha}{2\beta\sqrt{\beta}}g \cdot \ln \dfrac{(\alpha - 2\sqrt{\beta}) \cdot (Fr - a + \sqrt{\beta})}{(\alpha + 2\sqrt{\beta}) \cdot (Fr - a - \sqrt{\beta})}.$$

(ϵ) By means of these hypotheses, by assuming 19,600,000 for the radius of the earth, for any apparent height of a star one can determine its astronomic refraction if the value of the angle g has been found correctly by experiment; but, because it is very difficult to define this value with sufficient accuracy, it will be wiser to define the refraction in some particular astronomical case and from this to determine the other items by calculation. Let us assume, for example, that at a height of ten degrees the refraction is 5 minutes 28 seconds, which hypothesis most of the Astronomers of Paris follow. We will find the following table of refraction [page 251].

But since the refractions follow the proportion of the letter g, that is, of the *differential angle of refraction* of a ray entering at a half right angle from natural air into a vacuum, and because this angle is proportional to the density of natural air, or the air which an observer breathes, it is evident that even if the air were constantly loaded uniformly with

Apparent height of the star	Refraction		Apparent height of the star	Refraction	
0 degrees	34 minutes	53 seconds	50 degrees	0 minutes	53 seconds
5	9	59	55	–	$44\tfrac{1}{3}$
10	5	28	60	–	$36\tfrac{1}{2}$
15	3	44	65	–	$29\tfrac{1}{2}$
20	2	52	70	–	23
25	2	12	75	–	17
30	1	47	80	–	$11\tfrac{1}{4}$
35	1	29	85	–	$5\tfrac{1}{2}$
40	1	15	90	–	0
45	1	3			

vapors (which we have neglected so far), nevertheless, it cannot happen that astronomical refractions are so variable. They will indeed be larger at the surface of the sea than in the mountains, and the differences will be noticeable even at moderate heights of mountains; furthermore, they will be greater in a cold season than in a warm one, and this cause alone can increase refractions at least by a quarter in this country; finally, the refractions will also be larger when the barometer is high than when it is low. But if vapors were of no hindrance, the refractions could be defined correctly at any time if the instrument which has been described in §9 and which Fig. 57 represents were applied simultaneously with a barometer; for if one divides the height of the mercury in the other instrument, one will have the density of the air, to which, other things being equal, the refraction is to be made proportional. And I do not doubt but that the refraction of the sun is less than the refractions of the other stars, since the heat of the sun expands the air considerably and diminishes the density of the air.

———————————— [*End of Digression*] ————————————

§31. It appears from what has been mentioned about the agitation of air particles, on which in turn the heat of the air depends, but particularly from that which has been mentioned in §10, that the air possesses the same degree of heat whenever the same ratio prevails between the elasticity and the density of the former; the barometer indicates the elasticity; the density we conclude from the specific gravity of the air; and hence, as we have seen in §10, a fixed degree of heat can be obtained, if the heat of boiling water seems uncertain, just as Mr. Fahrenheit observed that it depends upon the weight of the atmosphere lying over it. Instruments which at any single instant indicate the density of the air can be invented easily and have been described by many.

It is to be noted here that that ratio just mentioned between the elasticity of air and its density simultaneously shows the height of homogeneous air, and since we discuss that height from now on, it is in order to define the latter correctly before we proceed to other things.

§32. If we consider a vertical air column of uniform density and brought to equilibrium with the mercury of the barometer, then the height of that column will be what I call the *height of homogeneous air* for the given density.

And since the specific gravity of moderately dense air is to the specific gravity of mercury as 1 is to 11,000, and since the mean height itself of the mercury in the barometer for locations slightly elevated above the surface of the sea is $2\frac{1}{3}$ Paris feet, the height of homogeneous moderately dense air will be 25,666 feet.

It is evident from this definition that those heights about which we are talking now are smaller, the denser the air to which the height must correspond and the smaller the height of the mercury in the barometer. Therefore, if the degree of heat is the same in the mountains and at the surface of the sea, the height of the homogeneous air will also be the same in either case, because for the same degree of heat of the air the density is in proportion to the elasticity of the air or to the height of the mercury in the barometer. Further, it is apparent that the height of the homogeneous air at the surface of the sea decreases considerably from the equator towards the poles, because the cold is intensified and the density of the air is increased, with the elasticity remaining the same, and in the same regions the height is less in wintertime than in summer.

§33. There are many things which pertain to defining the motion of air, the solution of which depends upon the height of homogeneous air; among these also is the propagation of sound and its celerity, for although the celerity of sound is defined differently by different methods, we can understand those concerning its propagation in this way: that at first the celerity seems to be that which is due to the height of the homogeneous air, then that which corresponds to half the height, or even to half the height multiplied by the ratio of the square circumscribed on a circle to the area of the circle; nevertheless, all opinions agree on the fact that the celerity of sound is proportional to the root of the height of the air homogeneous with that in which it is propagated. If the situation is thus, then sound is propagated faster in warm air than in cold, with a high barometer rather than low (to say nothing about favorable or contrary winds); many experiments have been performed on this matter, partly in Italy, partly in England, and the latter have shown us that the mean celerity of sound corresponds to 1140 English feet to be traveled in one second. But since in one and the same place the height of the homogeneous atmosphere is variable, and since especially in this area it rises, from barometric changes combined with changes of heat, [relatively] from 3 up to 4, the celerity of sound will be everywhere variable, even if the winds do not change at all, and in these regions that celerity will be contained [relatively] within the limits $\sqrt{3}$ and $\sqrt{4}$, or 173 and 200.

§34. I come now to solving various questions that can be proposed concerning the motion of air similar to those which we presented previously on the motion of nonelastic fluids.

Problem

The motion of air discharging from a vessel through a small orifice into infinite space empty of air, is to be defined.

SOLUTION. From the nature of the question it is apparent that the local motion of the internal air, which expands itself while a certain quantity of it flows out through the orifice, is not noticeable; therefore, only the *potential ascent* which an air particle acquires while it is expelled is to be considered here, and it is to be compared with the *actual descent* or rather with the decrease in elasticity which the internal air has. But in order to reduce the entire matter to our method applied to nonelastic fluids, we shall consider a vertical cylinder of common area with the proposed vessel and of the same height as the height of the air homogeneous with the internal air; but this cylinder, if it is considered full of similar but nonelastic air, expels the lowermost air through the orifice by its own weight at the same velocity at which the air in the proposed vessel expels itself by its own elasticity. But in the former case it is ejected at a velocity which is due to the height of that very cylinder, and hence also in the latter case. It is to be noted further that the height which we assumed for the cylinder is always the same, because the elasticity and density of the air are diminished in the same proportion, but we assume that the temperature is not changed. Therefore, if the height of the homogeneous air (which depends on the temperature of the internal air) is called A, then the air will flow out constantly with the velocity $\sqrt{A}$. And nevertheless the vessel is never emptied, which the calculation shows, because the air flowing out becomes continually thinner. In order to include this in the equation, we will set the density or the quantity of air at the beginning of flow equal to unity; the residual density or quantity of air after a definite time is x, and the time itself is t; then, since the velocity is constant, $-dx = ax\, dt$, where by a one understands a constant quantity to be defined from the size of the vessel, the area of the orifice and the height A; hence $\dfrac{-dx}{x} = a\, dt$, and $\ln \dfrac{1}{x} = at$; moreover, the value of the coefficient a is found in this manner. Because we have set $-dx = ax\, dt$, at the beginning of the efflux $-dx$ will be equal to $a\, dt$. The first element $(-dx)$ is now changed into a cylinder sitting above the orifice as a base; moreover, the height of that little cylinder will be $-L\, dx$, if L is the height of the cylinder constructed above the same orifice and having the same capacity as the proposed vessel; further, this length $-L\, dx$ is that which is traveled in the small time interval dt, and because this small time interval is usually set equal to the distance traveled divided by the velocity, in this case one will have $dt = \dfrac{-L\, dx}{\sqrt{A}}$; let this value be substituted in the equation $-dx = a\, dt$, and there will result $-dx = \dfrac{-aL\, dx}{\sqrt{A}}$, or $a = \dfrac{\sqrt{A}}{L}$. Thence the final equation is this:

$$\ln \frac{1}{x} = \frac{t\sqrt{A}}{L}.$$

If one chooses to express the time by a certain number of seconds, which we shall call n, and if by s is understood the distance which a movable object travels by falling freely from rest within one second, then one will have to set $t = 2n\sqrt{s}$, and thus it will occur that $\ln \dfrac{1}{x} = \dfrac{2n\sqrt{As}}{L}$.

Problem

§35. The motion of denser air flowing out from a vessel through a very small orifice into an infinite, thinner, external air is sought, assuming the same degree of heat in either air.

SOLUTION. Let the initial density of the internal air be D, the density of the external air be δ, the density of the residual internal air after a given time t be x, the height of the homogeneous air (in relation to either the internal or to the external air, for it cannot be different, if each air is furnished at the same temperature, and thus the densities and elasticities decrease in the same ratio) be A. At every point the height of the homogeneous air is sought that has the same pressure or expansion as the external air and the density of which is the same as that of the internal air; at the beginning this height will be $\frac{\delta A}{D}$, and after the time t it will be $\frac{\delta A}{x}$. But it is evident that the velocity of the discharging air will be such everywhere that it corresponds to the difference of the defined heights A and $\frac{\delta A}{x}$; therefore, the velocity of the discharging air after the time t is $\sqrt{A - \frac{\delta A}{x}}$.

Further, the increments of densities $(-dx)$ are proportional to the quantities of discharging air, which have a ratio composed of the velocity $\left(\sqrt{A - \frac{\delta A}{x}}\right)$, the density (x) and the small time interval (dt); thus, therefore, $-dx = a\sqrt{A - \frac{\delta A}{x}}\, x\, dt$, where a is a constant number which, by the method of the preceding paragraph, becomes $\frac{1}{L}$, the significance of this letter as applied there having been retained; and after this value has been substituted, there develops

$$-dx = \frac{dt}{L}\sqrt{Axx - \delta Ax} \quad \text{or} \quad \frac{-dx}{\sqrt{xx - \delta x}} = \frac{dt\sqrt{A}}{L}.$$

After the required integration has been performed this becomes

$$\ln \frac{[\sqrt{x} - \sqrt{x - \delta}] \cdot [\sqrt{D} + \sqrt{D - \delta}]}{[\sqrt{x} + \sqrt{x - \delta}] \cdot [\sqrt{D} - \sqrt{D - \delta}]} = \frac{t\sqrt{A}}{L},$$

or, again setting $t = 2n\sqrt{s}$, as in the preceding paragraph, one will have:

$$\ln \frac{[\sqrt{x} - \sqrt{x - \delta}] \cdot [\sqrt{D} + \sqrt{D - \delta}]}{[\sqrt{x} + \sqrt{x - \delta}] \cdot [\sqrt{D} - \sqrt{D - \delta}]} = \frac{2n \cdot \sqrt{As}}{L}.$$

§36. COROLLARY 1. The entire efflux occurs in a finite time, in which matter this problem differs from the preceding one; for the air ceases to flow out when $x = \delta$, and then

$$n = \frac{L}{2\sqrt{As}} \cdot \ln \frac{\sqrt{D} + \sqrt{D - \delta}}{\sqrt{D} - \sqrt{D - \delta}}.$$

For instance, let A be 26,000 Paris feet, let the proposed vessel contain one cubic foot, and in addition let the orifice have an area of one square line; L will be 20,736; let it be assumed also that at the beginning the internal air was twice as dense as the external; moreover, as is established, $s = 15\frac{1}{2}$ Paris feet. Therefore, it will occur that

$$n = \frac{20736 \cdot \sqrt{3}}{\sqrt{181 \cdot 26000}} \ln \frac{\sqrt{2} + 1}{\sqrt{2} - 1} = 29.2$$

which indicates that either air will be brought to equilibrium in a time slightly longer than twenty-nine seconds, and that after this all efflux will cease. It can also happen, on account of the contraction which the fluid suffers in front of the orifice (see Chapter IV) and to which we paid no attention in the computation, that this time is increased almost in the ratio of 1 to $\sqrt{2}$.

§**37.** COROLLARY 2. If one assumes that the air flows out not immediately through the orifice but through a long pipe, the velocity will therefore not be changed, if only the capacity of the entire pipe is as if infinitely small in proportion to the capacity which the vessel itself has; it seems, moreover, that the density of the air, as long as it is in the pipe, is the same as the density of the air enclosed in the vessel; nevertheless—as I shall demonstrate below—the elasticity of the air in the pipe is not greater than the elasticity of the external air which surrounds the pipe. It follows, hence, that moved air is denser than air at rest, but not more elastic; nevertheless, the difference in densities will also be small; for a wind that moves at about 30 feet per second will exceed in density the neighboring air, of the same temperature and at rest, hardly by a one-thousand-seven-hundredth part.

Problem

§**38.** To define the inflow of air through a very small orifice into a vessel full of rarer air, the same degree of heat again having been assumed everywhere.

SOLUTION. Let the vessel be completely empty at the beginning, and after a time t let the density of the internal air be set equal to x; thus, by remaining on almost the same track which we used in §35, and with the same notation retained, one finds

$$\frac{dx}{\sqrt{\delta - x}} = \frac{dt\sqrt{AD}}{L}, \quad \text{or} \quad t = 2n\sqrt{s} = \frac{2L}{\sqrt{A}} - \frac{2L\sqrt{D-x}}{\sqrt{AD}}.$$

The number of seconds, therefore, in which the entire vessel is filled until equilibrium exists between both airs is expressed by $\dfrac{L}{\sqrt{As}}$; and the time of replenishing is twice that in which it would be filled if the air would flow in constantly at the initial velocity. In the case when the capacity of the vessel is one cubic foot and the orifice equals one square line, the repletion occurs in a time of about thirty-three seconds, unless the repletion is retarded by the contraction of the inflowing air jet.

§39. We have shown various properties of elastic fluids, whether moving or at rest; one remains that is not to be omitted by which elastic fluids differ from nonelastic fluids: namely, that an elastic fluid at rest possesses a *live force*, which cannot, like other moved bodies, elevate itself to a certain height, and indeed we do not consider here the local motion in the former, but that which by its expansion can generate a certain ascent in other heavy bodies. It will be allowed, I hope, to use in the future the phrase *live force contained in a compressed elastic body*, wherein nothing else is understood by this than the *potential ascent* which an elastic body can communicate to other bodies before it will have lost all its elastic force.

It is of merit to note here in advance that just as the descent of a given body through a given height, whichever way it happens, produces constantly the same live force in the body, so also an elastic body or an elastic fluid, after it has been reduced in any manner from a given degree of tension or compression to a given degree, always retains in itself the same live force and can again communicate it to another body by an opposite change.

Let me mention now a little about live forces of this sort contained in a compressed elastic fluid and a few of their measures; this is an argument worthy of attention, since to this are reduced the measures of the forces for driving machines by air or fire or other motive forces of this kind, of which perhaps several new ones could be developed, but not without considerable practical mechanical improvement and perfection.

§40. As we began with air in a vacuum, we shall consider the cylinder *ABCD* (Fig. 62), oriented vertically, with the piston *EF*, which, devoid of all weight, can be moved upward and downward very freely. Let air be enclosed in the space *EBCF*, but assume the entire cylinder to be located in a vacuum. Let the pressure of the air *EBCF* be such that it can sustain the weight p, which will be equal to the pressure of the atmospheric column if that air is natural. Now let another weight P appear; thus it will happen that the lid descends to *GH* and is agitated by reciprocal motions to the points *H* and *F*. In order to define the motion, let us make use of the common hypothesis that the pressures of air, other things being equal, are proportional to the densities.

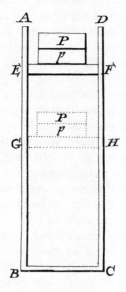

FIGURE 62

And thus $FC = a$, $FH = x$, the velocity of the piston at the position GH is v, and the pressure by which the piston GH is forced to further descent will be $P + p - \dfrac{a}{a-x} p$, and the force which animates the weight lying above the piston must be considered equal to this pressure; therefore, if one divides this force by the mass, one will obtain the accelerative force, which, multiplied by a time increment, or by $\dfrac{dx}{v}$, will give the increment dv of the velocity, and thus

$$dv = \left(P + p - \frac{ap}{a-x}\right) \frac{dx}{v} \bigg/ (P + p),$$

or

$$\tfrac{1}{2}(P + p)vv = (P + p)x - ap \ln \frac{a}{a-x}.$$

But from the descent of the weight $(P + p)$ through the height x, the *potential live force* $(P + p)x$ is generated, and when the piston is in the position GH, the body $(P + p)$ possesses the *actual live force*

$\frac{1}{2}(P + p)vv$, that is, $(P + p)x - ap \ln \frac{a}{a - x}$, which is less than the former by the quantity $ap \ln \frac{a}{a - x}$, and this has gone into the compression of the air.

I say, therefore, that *air occupying the space a cannot be condensed into the space $a - x$ unless a live force is applied which is generated by the descent of the weight p through the height $a \ln \frac{a}{a - x}$, however that compression may have been achieved;* but it can be done in an infinite number of ways. Indeed, I shall now illustrate this rule by one or two examples.

Let the base of the cylinder be of one square foot, the initial height FC two feet; in the space BF let air be contained of the type that is usually the mean on the surface of the earth, which can carry 2240 pounds on the surface EF; let $x = 1$ be assumed, in order to obtain the *live force* by which two cubic feet of natural air can be driven into the space of one cubic foot in a vacuum; and that *live force* will be $2 \cdot 2240 \cdot \ln 2 = 3105$, that is, as large as is generated by the free fall of a body of 3105 pounds through the height of one foot. Therefore, in turn, if one had a cubic foot of air twice as dense as natural air, then a weight of 3105 pounds could be lifted by means of the former to a height of one foot in a vacuum while it assumes the density of natural air.

Further, under the same circumstances let the air be expanded into twice as much space as it was in before, now occupying a height of four feet in the cylinder, and let this again be compressed into the space of one cubic foot, and for this compression a *live force* will be required which is expressed by $4 \cdot 1120 \cdot \ln 4$, which is twice as large as the former. Therefore, if one had in a vacuum a cubic foot of air twice as dense as natural air, then by means of that a weight of 6210 pounds could be elevated to a height of one foot, while it assumes half the density of natural air, or a weight of 9315 pounds while it becomes four times as rare as natural air.

Hence it is a consequence that if air can expand itself into infinite space and it preserves everywhere an elasticity proportional to the density, then a finite quantity of air possesses an infinite live force.

§**41.** However, these things pertain to the estimation of the live force which is contained in air placed in a vacuum; the computation becomes slightly different for the denser air which is located in the atmosphere; here, namely, the maximum degree of expansion cannot be extended beyond equilibrium with the air of the atmosphere; hence it is easy to predict in advance, if one has, for instance, a cubic foot of air twice as dense as natural air, that the live force which can be produced in the atmosphere from this compressed air is far from infinite. Other live forces of this kind could also be determined in this manner.

§42. Let the air *EBCF* be natural and in equilibrium with the external air; also let the pressure of the atmosphere on the piston *EF*, which is certainly in equilibrium with the pressure of the internal air not yet condensed, be indicated by *p*. Let the weight *P* be imposed onto this piston; let the air now be condensed into the space *GBCH*, and let the piston, loaded with the weight *P*, have the velocity *v* at the position *GH*; the remaining notations having been retained, then

$$dv = \left(P + p - \frac{ap}{a-x}\right)\frac{dx}{v}\bigg/P,$$

or

$$Pv\,dv = \left(P - \frac{xp}{a-x}\right)dx,$$

which, integrated, yields

$$\tfrac{1}{2}Pvv = Px + px - ap\ln\frac{a}{a-x}.$$

But now the *live force Px* was generated by the descent of the weight *P* through the height *x*, of which force the portion

$$\tfrac{1}{2}Pvv \text{ or } Px + px - ap\ln\frac{a}{a-x}$$

pertains to the same weight moving at the velocity *v*; therefore, the part of the live force which transferred to the air is $-px + ap\cdot\ln\frac{a}{a-x}$, which is less than the other one defined in §40.

For instance, let there be a cubic foot of air twice as dense as natural air; one will find that the live force which this air yields while it assumes the density of the surrounding natural air is that which is generated by the free fall of a body of 865 pounds through a height of one foot.

In the same sense, a cubic foot of air three times as dense as natural air is found to have a live force such as corresponds to the free fall of a body of 2898 pounds through a height of one foot, which number certainly results if, as in §40, one sets $p = 2240$, $a = 3$, and $x = 2$.

§43. It is evident from this correspondence between the conservation of live forces contained in compressed air and in a body having fallen from a given height that no advantage is to be hoped for from the principle of compressing air for improving the use of machines, and that everywhere the rules shown in the preceding section are valid. But since it happens in many ways that air is compressed not by force, but by nature, or acquires a greater expansion than is natural, there is certainly hope that from natural occurrences of this kind great advances can be devised for driving machines, just as Mr. Amontons once showed a method of driving machines by means of fire. I am convinced that if all the live force which is latent in a cubic foot of coal and is brought out of the latter by combustion were usefully applied for driving a machine, more could thence be gained than from a day's labor of eight or ten men. For the coal not only significantly increases the elasticity of the air while it burns, but it also generates an enormous quantity of new air.

Thus Hales discovered in his *vegetable staticks* that from a half [cubic] inch of coal 180 [cubic] inches of air of the same elasticity as natural air had been generated; therefore, a cubic foot of coal will give air to 360 cubic feet. But if in §41 the live force is sought which can be generated from a cubic foot of air 360 times as dense as natural air, it will be found that the former corresponds to the falling of a weight of 3,938,000 pounds from a height of one foot; and if, furthermore, it is assumed that the elasticity of that air is made four times larger by the heat of the burned coals, then that live force corresponds to the falling of a weight of 15,752,000 pounds from the same height. However, it is difficult to devise a machine suitable for this purpose. There are, furthermore, many natural occurrences which not only heat up compressed air but can also make the surrounding air more elastic by warming it; of this sort are quicklime mixed with fresh water and all fermenting matter; water reduced to vapor by means of fire possesses incredible force; the most ingenious machine so far, which delivers water to a whole town by this principle of motion, is in London, and the illustrious Weidler has described it. But, above all, the astounding effect which can be expected from gunpowder deserves to be considered; indeed, having performed the calculations on some completed experiments, which I shall add below, I learned that the elasticity of ignited gunpowder exceeds more than ten thousand times the elasticity of natural air, and even after everything has been considered carefully, it is likely that the elasticity of the former is incredibly larger; let us assume, moreover, that the elasticity of an expanded blast of ignited gunpowder decreases in the same proportion as the density; under this assumption the live force existing in a cubic foot of gunpowder will be found, if in §42 one sets $a = 10,000$, $x = 9999$, and $p = 2240$, and if one considers [the live force equal to]

$$-px + ap \ln \frac{a}{a-x},$$

which quantity thus becomes equal to 183,913,864. Therefore, in theory a machine is given by means of which one cubic foot of gunpowder can elevate 183,913,864 pounds to a height of one foot, which work, I would believe, not even 100 very strong men can perform within one day's span, whatever machine they may use. It is also probable, as I have said, that the effect of the gunpowder is far greater; but certainly it is not less, since the calculation is based on the height to which an iron shot ejected from a cannon can ascend in a vacuum, in which type of experiment the greatest part of the gunpowder is wasted.

But these things may be better understood if it is noted that the same calculation (which we performed before for demonstrating the effect which arises from condensed air re-establishing itself) appears also for air which, surrounded by natural air, becomes certainly not more dense, but nevertheless more elastic from the increased temperature; so, for instance, every time a cubic foot of ordinary air has acquired twice its expansion by increased temperature, a weight of 865 pounds can be lifted by means of this to a height of one foot, if only a perfect machine is applied.

But the effects of all things shown here depend upon both the increased density and increased temperature of the air.

§**44.** Meanwhile, a live force to be applied for driving machines can be obtained not only from compressed or heated air, but also from rarer or colder air. Indeed, wherever the equilibrium is disturbed, a *live force* exists which can be applied, if a fitting machine can be invented, for lifting loads and driving machinery. But the method of determining the *live force* which can be produced from air of given density and given temperature occupying a given space, the appropriate changes having been made, is the same as that which we have furnished in §42.

§**45.** Again, indeed, let the vertical cylinder $ABCD$ (Fig. 63) with the movable diaphragm EF be considered; assume the air $EBCF$, as in §42, as natural and in equilibrium with the external air; also, let the pressure of air of any sort on EF be called p; next, consider the weight P, which by means of a rope passed through the two pulleys M and N is connected to the diaphragm and pulls the same towards AD; and thus let the diaphragm arrive at GH from the position EF; finally, one assumes again $FC = a$, $FH = x$, and the velocity of the diaphragm in the position GH or of the weight in the position P is equal to v; if with these assumptions §§40 and 42 are consulted, it will now be evident that

$$dv = \left(P + \frac{ap}{a+x} - p\right) \frac{dx}{v} \bigg/ P,$$

or
$$Pv\,dv = \left(P - \frac{px}{a+x}\right)dx,$$
which integrated gives
$$\tfrac{1}{2}Pvv = Px - px + ap \ln \frac{a+x}{a}.$$

But again, by the descent of the weight P through the height x a *live force* Px was produced, during which time, meanwhile, the weight itself, moved with the velocity v, possesses a *live force* of only $\tfrac{1}{2}Pvv$, or $Px - px + ap \ln \frac{a+x}{a}$. Therefore, the *live force* which remains,

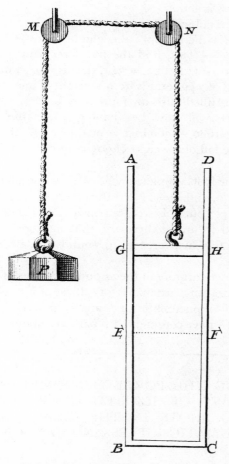

Figure 63

namely $px - ap \ln \dfrac{a+x}{a}$, was transferred into the air, and, with equilibrium restored again between the internal and external air, the *live force* could be transposed arbitrarily to other bodies. Therefore, if one has the space *GBCH* full of air, the density of which is to the density of the external air as *CF* is to *CH*, then the *live force* $px - ap \ln \dfrac{a+x}{a}$ will be in effect.

But whether that live force is contained properly in the external or internal air is just a play on words; it suffices that from the disturbed equilibrium between either air such a live force can be obtained while the restoring process is acting.

For instance, let a cubic foot of air twice as rare as natural air exist, to which hypothesis there apply the conditions $p = 2240$ pounds, $a = \frac{1}{2}$ foot, and $x = \frac{1}{2}$ foot, and the live force which is being discussed will be $1120 - 1120 \ln 2 = 344$, that is, the same as is generated by the free fall of 344 pounds from a height of one foot.

If a cubic foot is filled with air four times as rare as natural, the required live force will now be (having set, namely, $p = 2240$, $a = \frac{1}{4}$, and $x = \frac{3}{4}$) $1680 - 560 \ln 4 = 904$, or as great as that which arises from the free fall of a weight of 904 pounds through a height of one foot.

Finally, if a cubic foot completely void of air is considered, one must set $p = 2240$, $a = 0$, and $x = 1$; and thus the required live force will be $2240(1 - 0 \ln \frac{1}{0})$; but it is well established that $0 \ln \frac{1}{0}$ is infinitely small as compared to unity; therefore, that number is 2240, which indicates that by this live force 2240 pounds can be elevated to a height of one foot.

§46. To the present argument the astounding force of greatly compressed air pertains, but particularly of a blast of ignited gunpowder in the employment of pneumatic rifles and cannons. Let me add to this chapter what I have commented separately about these matters.

CONCERNING THE FORCE OF COMPRESSED AIR AND A BLAST OF IGNITED GUNPOWDER FOR PROJECTING SHOTS IN THE EMPLOYMENT OF PNEUMATIC RIFLES AND CANNONS

I. Let *AG* (Fig. 64) be the length of the barrel in a cannon or rifle located horizontally, and let it be called a; let *AC* denote the length of the space which the compressed air or the cloud of ignited gunpowder occupies at the beginning of the explosion, and let $AC = b$, and the weight of the shot E to be ejected equal to unity; let us further assume that the shot fills out the cavity of the barrel

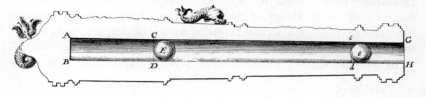

FIGURE 64

exactly and is moved very freely in the latter; the density of the compressed air in the space AD is to the density of natural air just as n is to 1; finally, let the weight of the column of mercury (the base of which is CD and the height of which is the same as in the barometer) be P. Moreover, let us use the hypothesis, whether the shot is propelled by compressed air or by a blast of gunpowder, that the force of that propelling fluid is proportional to the density.

These things having been prepared for the calculation, we shall consider the shot in the position e, by setting $Ac = x$ and the velocity of the shot at that location $= v$; thus the force propelling the shot at the position e will be $\left(\dfrac{nb}{x} - 1\right)P$, which, divided by the mass 1 and multiplied by the element of space dx, gives half the increment of the square of the velocity; hence it occurs that $v\,dv = \left(\dfrac{nb}{x} - 1\right)P\,dx$, or $\tfrac{1}{2}vv = \left(b - x + nb \ln \dfrac{x}{b}\right)P$. If one sets $x = a$, the height is obtained which is due to the velocity at which the shot is exploded; let that height be called α, and therefore

$$\alpha = \left(b - a + nb \ln \frac{a}{b}\right)P.$$

II. For instance, let the length of the barrel in a pneumatic rifle, or a, be 3 Paris feet, the length AC be 4 inches, the air contained in AD be 10 times denser than natural, or $n = 10$, the diameter of the barrel or the shot to be ejected be 3 lines, and its specific gravity be in proportion to mercury as 10 to 17. Therefore, P will be 286, more or less, and hence it is found that $\alpha = 2788$, an indication that the sphere is being ejected at a velocity by which it can ascend in a vacuum to a height of 2788 feet. From the preceding formula is understood that the most vehement thrust of the shot occurs from the same quantity of elastic blast if the length of the barrel is made equal to *nb*. But if attention is paid to the other hindrances which the sphere encounters in its transit through the barrel of the Rifle apart from its inertia and the resistance of the external air, it appears that a much smaller length of the barrel is required for producing the most violent thrust. If the length *nb* is much larger than the length *a*, which is so in stronger thrusts, then one will have, without noticeable error,

$$\alpha = nbP \ln \frac{a}{b}.$$

If the cannon is erected vertically, the calculation becomes somewhat different, but for more violent thrusts the difference cannot be noticed. Therefore, because from now on we shall consider only very vehement thrusts, for the sake of brevity we will set $\alpha = nbP \ln \frac{a}{b}$.

III. Just as we have determined the height due to the velocity at which the shot is exploded in the preceding from the given elastic force of the blast ejecting the sphere, so it is evident in turn that from that observed height the elastic force of the blast can be deduced; thus,

$$n = \alpha \Big/ \left(bP \ln \frac{a}{b} \right).$$

As a consequence, the elastic force of gunpowder can be, if not accurately defined, at least reduced to limits which it will certainly exceed. But one may ask how the height α can be determined in an experiment; to this I answer that it can be accurately enough understood from the time which a shot ejected vertically upwards takes from the instant of explosion until it has fallen to the earth, with the air resistance taken into account in the calculation. Let me transcribe here the experiments reported in the *Commentaries of the Imperial Academy of Science of St. Petersburg*, Book II, pp. 338–39, the calculation of which I performed after the hypotheses had been formed, considering air resistance, that the specific gravities of iron and air are as 7650 is to 1, and that the air in which the shot ascends is of uniform density; the ratio of the specific gravities seems to have been assumed a little higher than was appropriate, but in very high thrusts the error will be compensated by the decrease of the densities of the air towards the higher altitudes.

> The location of the cannon was adjusted with all accuracy to the perpendicular, and in every case it was reset to this position and

fixed; the individual experiments were repeated; the length of the barrel was, moreover, 7.7 English feet, the diameter of the shot was 0.2375 foot; the diameter of the barrel was not measured, nor was the size of the touch-hole; instead, every quantity of gunpowder used was weighed, and with a pendulum the time was defined from the instant of explosion to the instant at which the sphere fell to the earth; the following table shows both what has been observed and what has thence been deduced by calculation:

Quantity of gunpowder,	Time of ascent and descent,	Height of thrust in resistant air from calculation,	Time of ascent in resistant air from calculation,	Time of descent in resistant air from calculation,	Height of thrust in a vacuum from calculation,	Time of ascent and descent in a vacuum from calculation,
Expressed by the number of Holland ounces	Observed in seconds	In English feet	In seconds	In seconds	In English feet	In seconds
I	II	III	IV	V	VI	VII
½	11	486	5.42	5.58	541	11.6
2	34	4550	14.37	19.63	13,694	58
4	45	7819	16.84	28.16	58,750	121

For the same cannon and the same shot, but with the former being shortened by one foot and seven-tenths, such that the residual length of the barrel is precisely 6 English feet, the following table serves, constructed by the same rule.

I	II	III	IV	V	VI	VII
½	8	257	3.95	4.05	274	8.2
2	20.5	1665	9.74	10.76	2404	24.5
4	28	3187	12.5	15.5	6604	40.5
6	32.5	4304	13.9	18.6	11,810	54.3
8	38	5643	15.54	22.46	22,394	74

There are many things which render the success of those experiments doubtful insofar as there is nothing which proves the same elasticity of the blast. I myself would believe the greatest discrepancy to arise from the fact that a very small part of the powder is ignited immediately at the beginning of the explosion, that then a large part is set on fire when the shot is just close to the orifice of the cannon, and that finally the largest part is ejected not yet ignited; perhaps this single reason causes the elastic force of the blast propelling the shot to be a hundred times as large as that which results from the experiment when no account has been taken of this matter; this seems very probable to me from the fact that, with 4 ounces of powder used in a 7.7-foot-long gun, the shot could ascend in a vacuum by this very thrust to a height of 58,750 feet, while for the same quantity of powder and the same cannon, but shortened by 1.7 feet, the thrust would correspond to a height of 6604 feet in a vacuum, which height hardly exceeds the ninth part of the former. From comparison between the two experiments I conclude that the largest quantity of powder was ignited in the longer cannon while the shot was close to the orifice, in fact not further away from the latter than 1.7 feet.

The thrust of the shot is also diminished by the size of the touch-hole as well as by the opening which is left between the sphere and the inner surface of the barrel, through either of which a noticeable, useless part of the blast vanishes; however, not as large a decrease arises thence as that which I had presumed before the calculation had been performed; nevertheless, let me add the calculation in the following, in order that a method be available for stating the very outer limits for the force of gunpowder, which it will certainly exceed in any case.

IV. The one which displays the greatest elasticity of the blast is the third experiment, performed with the gun not yet shortened, which shows that the shot could have risen by the impact received to a height of $\alpha = 58{,}750$ English feet. But the length of the barrel AG was $a = 7.7$; the length AC (as much as I conclude from the area of the barrel and the gravity of the gunpowder) was 0.08. Finally, the value 26.8 is found for P itself (or for the weight of a mercury column, the base of which is a great circle of the shot and the height of which is 30 English inches, in proportion to the weight of the iron shot, designated by unity), the specific gravity between mercury and iron having been taken as 17 to 10; since, according to §3, $n = \alpha \Big/ \left(bP \ln \dfrac{a}{b}\right)$ approximately, there results $n = 6004$. Whence it follows, if the blast of ignited gunpowder has an elasticity proportional to its density, that the maximum elasticity of the former is at least six thousand times as great as the elasticity of ordinary air.

V. But if we now consider the useless part of the blast which escapes through the touch-hole and the aperture left by the shot, we will find a greater elasticity. Since the calculation which is required for solving this question is not a little lengthy and is intricate, I did not hesitate to apply slightly more liberal hypotheses, by which it becomes much easier; although the hypotheses themselves are not true in all rigor, they can nevertheless not produce any noticeable error. *First*, let me assume that either aperture through which the blast can escape is practically infinitely small in proportion to the area of the barrel; with this assumption the velocity at which the blast escapes can be estimated at any instant directly from the pressure alone; but a hypothesis of this kind can be formed without any noticeable error for the entire fluid, even when the openings are not very small at all, as we have deduced the corollary from our theory, and anyone will see quickly that it can be assumed much more easily in a very elastic fluid from the fact that the increment of the *potential ascent* in relation to the internal motion is much less in proportion to the *potential ascent* of a particle springing forth from the orifice in a fluid which is expelled by its own elasticity than that which is ejected by force of gravity; for in the former case the local internal motion is less than in the latter. *Second*, [it can be assumed] that the elastic force of a blast of ignited gunpowder is such that it is not worthwhile to consider the counteracting pressure of the atmosphere; *third*, that the velocity of the shot in the gun, although very large, can nevertheless be considered very small in proportion to the velocity at which the blast escapes through either aperture, because indeed the inertia of that blast cannot be not very small in proportion to the inertia which the shot possesses; by virtue of this hypothesis the blast will escape through either aperture at the same velocity, since otherwise, the velocity in the touch-hole having been set equal to $\sqrt{A}$ and the velocity of the shot equal to v, the velocity of the blast in the aperture left between the shot and the surface of the barrel would have to be called $\sqrt{A} - v$. I now come to the solution.

VI. First it is to be noted that if the elasticities of the blast are considered proportional to the densities, the blast will escape constantly at the same velocity through either aperture, as we have seen in the problem in §34, and that this velocity will be nominally the same as that which is generated by the height of the homogeneous gas, the weight of which can prevent the contained blast from expanding. Hence, the aforementioned velocity will be determined in this manner: let the gravity of the shot be unity, the elasticity or the weight which can keep the blast of powder $ACDB$ just ignited in that state of compression be P, and the weight of the powder used be p; then the weight of the blast of powder just ignited will also be p; and if the length AC is set equal to b, it is evident that the height of the homogeneous gas which has the weight P will be $\frac{P}{p}b$. Therefore, the velocity at which the newly created blast escapes through the touchhole is $\sqrt{\dfrac{P}{p}}\,b$, and it will be ejected at the same velocity during the entire explosion, and this not only through the touch-hole, but also, approximately, through the aperture left between the shot and the barrel.

VII. Now, in addition let the area of the barrel be F, the aperture contained between the shot and the barrel be f, the area of the touch-hole be ϕ, the length of the barrel be a, and the quantity of the blast at the beginning of the explosion be g. Let it then be understood that the shot has come from E to e, and let AC be called x, the residual quantity of blast in the cannon at that instant be z, and the velocity of the shot in that position be v; the remaining notation has already been explained earlier.

Since the elasticity is by hypothesis directly [proportional] to the quantity and inversely to the space, the elasticity of the blast remaining in $AcdB$ will be $\dfrac{zb}{gx}P$; this, indeed, is not all expended for propelling the shot, but only a part of it, which is to the total as $F - f$ is to $[F]$. Therefore, taking dt for an element of time, one has

$$dv = \frac{F-f}{F} \cdot \frac{zb}{gx} P \cdot dt.$$

But by the method shown in §34, where the quantity of air flowing out in a given time element was specifically defined, one finds

$$-dz = \frac{f+\phi}{F} \cdot \frac{z}{x} \cdot \sqrt{\frac{P}{p}}\, b \cdot x\, dt.$$

From a comparison of these two equations, one has

$$-dz = \frac{f+\phi}{F-f} \cdot \frac{g}{b} \cdot \frac{\sqrt{b}}{\sqrt{Pp}} \cdot dv$$

which integrated, with the addition of the proper constant, gives

$$z = g - \frac{f+\phi}{F-f} \cdot \frac{g}{b} \cdot \frac{\sqrt{b}}{\sqrt{Pp}} \cdot v.$$

Now if this value found for z is substituted in the first equation, and simultaneously $\dfrac{dx}{v}$ is entered for dt, then

$$v\, dv = \frac{F-f}{F} \cdot \frac{b}{x} \cdot P \cdot dx - \frac{f+\phi}{F} \cdot \frac{\sqrt{bP}}{x\sqrt{p}}\, v\, dx,$$

or
$$\frac{Fv\,dv\sqrt{p}}{(F-f)\cdot bP\sqrt{p}-(f+\phi)\cdot v\cdot\sqrt{bp}} = \frac{dx}{x},$$

which equation after its proper integration, $x = a$ having been entered, transforms into this:

$$\ln\frac{a}{b} = \left[-F(f+\phi)v\sqrt{p} - F(F-f)\cdot p\sqrt{Pb}\right.$$
$$\left.\cdot\ln\left(1 - \frac{(f+\phi)v}{(F-f)\sqrt{bPp}}\right)\right]\bigg/(f+\phi)^2\cdot\sqrt{Pb}.$$

VIII. Now if that value of v were known from experiment, the value of P itself which denotes the elasticity of the cloud of gunpowder not yet expanded could thence be deduced. In order to illustrate this by an example, let us use the same experiment which we have shown already in Art. IV, in order that it be apparent therefrom what increase of elasticity arises from the escape of the blast. Therefore, the calculation will be performed as follows.

Because we have designated the weight of the shot, which was three pounds, by unity, the four ounces of powder used will have to be expressed by $\frac{1}{12}$; therefore, $p = \frac{1}{12}$. The measurements of the openings which we consider I have not taken; but usually the aperture left by the shot in a gun of this kind constitutes approximately a fifteenth part of the area of the barrel; the area of the touch-hole, I believe, can be neglected altogether here; therefore, I set $F = 15$, $f = 1$, $\phi = 0$. Further, again $a = 7.7$, $b = 0.08$; the height to which the shot could ascend in a vacuum $\frac{1}{2}vv = 58{,}750$, or $v = 343$; therefore, the last equation of the previous article will be this:

$$\ln 96 = \frac{-5251}{\sqrt{P}} + 17.5\cdot\ln\frac{\sqrt{P}}{\sqrt{P}-300},$$

which is satisfied approximately if one takes $\sqrt{P} = 534$, and thence $P = 285{,}156$, which equals the weight of a mercury column of the same area as the barrel on the gun, the height of which is more than 10,000 times as large as the height of the ordinary barometer; moreover, we have found above in Art. IV that the number n (which signified the same) equals 6004. Therefore, we shall now safely affirm (for everywhere the things which we have neglected render a larger force to the powder) that gunpowder possesses an elastic force at least ten thousand times as large as the elastic force of ordinary air. Moreover, it is simultaneously apparent from comparison of the numbers 10,000 and 6004 approximately how much of the force of the powder is lost due to the often-mentioned openings. Indeed, I would have believed this decrement to be greater; but I was confirmed by that calculation in this matter, about which once a man, knowing of such matters, wanted me [to be] more certain, [since] indeed he had observed no noticeable decrement in cannons after the touch-hole was amplified beyond normal by daily use in a siege.

IX. Indeed, in order that from our equation some corollaries can be deduced that are simpler although only approximately true, we shall change the logarithmic quantity into a series. This is, indeed,

$$-\ln\left(1 - \frac{(f+\phi)v}{(F-f)\sqrt{bPp}}\right) = \frac{(f+\phi)v}{(F-f)\sqrt{bPp}}$$
$$+ \frac{(f+\phi)^2vv}{2(F-f)^2 \cdot bPp}$$
$$+ \frac{(f+\phi)^3v^3}{3(F-f)^3 \cdot bPp\sqrt{bPp}} + \cdots.$$

After this value has been substituted in the last equation of Art. VII, there results

$$\ln\frac{a}{b} = \frac{Fvv}{2(F-f) \cdot bP} + \frac{F(f+\phi)v^3}{3(F-f)^2 \cdot bP\sqrt{bPp}} + \cdots.$$

We will notice that here this equation agrees perfectly with the last equation of Art. II if the apertures f and ϕ are set equal to zero; for what is indicated here by $\frac{1}{2}vv$ and nP is indicated there by α and P, the remainder of the notation being identical.

X. This equation will serve in order that it be apparent approximately how much the height of the thrust is decreased by the apertures if these openings are very small. Let α indicate the height which the shot can reach in a vacuum if it is assumed that none of the blast escapes through the openings, and the decrement of that height about to arise from the eruption of the blast through those same apertures will be approximately this:

$$(2\alpha)^{3/2} \cdot (f+\phi)/3F\sqrt{bPp}.$$

Whence in the same gun, with the same quantity of powder used, and with the weight of the shot remaining the same, the decrements of the thrusts will be proportional to the areas of the apertures.

The same decrements follow almost in proportion to the square root of the quantities of powder used, other things equal; because, indeed, the logarithms of large numbers increase at a much lesser rate than the numbers themselves, and because, furthermore, $\alpha = bP \ln \frac{a}{b}$, one could have stated that, the rest being equal, α is proportional to b itself, because P is not affected by b. But the decrement under discussion, other things being equal, is in proportion to the quantity $\alpha^{3/2}/\sqrt{bp}$, or in proportion to the quantity $\frac{b}{\sqrt{p}}$; but p itself, which denotes the weight of the powder used, is proportional to b; therefore, the previously mentioned decrement is approximately in proportion to $\sqrt{b}$, which is the square root of the quantity of powder used. Therefore, from the usual reasoning concerning thrusts, the decrements are much larger in feeble thrusts than in more violent ones, and the experiments described in Art. III seem to confirm this also; I see indeed no other reason why in the first table of experiments the thrust of a shot in a vacuum, with two ounces of powder having been used, should have been more than twenty-six times as high as when half an ounce was used, and why, after the quantity of powder was duplicated to 4 ounces, a thrust only four times as high as that with a quantity of two ounces should result from the calculation.

XI. Whatever other inequalities may appear [by comparison] in either table of the experiments, I derive, as I have said above, to the largest part from the fact that not all powder is ignited, nor is all that which is ignited burned at once at the beginning of the explosion. And I shall certainly not be astonished, since we have investigated this in Experiment 4, Table 1, that the total time of explosion makes up not even the hundredth part of one second. Therefore, since it is certain that the largest part of the powder is ejected not ignited, and since not a small part of the remainder is set on fire more slowly than was assumed in the calculation, and since, furthermore, a noticeable part of the powder is adulterated by vapors and earthen material which does not burn, it follows that the burning parts possess a far greater elasticity than that which was determined from calculation of the experiment in Art. X; perhaps it is ten or a hundred times as large.

But perhaps it is only as large as the experiment has shown, namely, ten thousand times as large as the elasticity of ordinary air; it follows, hence, either that the elastic blast which develops from ignited gunpowder is not ordinary air, or that the elasticities increase in a greater proportion than the densities; indeed, the density of air that arises from powder just ignited cannot be more than a thousand times larger than the density of ordinary air if almost all the powder is composed of compressed air, which I conclude from the specific gravity of the powder with respect to the air.

Meanwhile, the question has been treated for a long time whether the elastic blast which is deduced from bodies is ordinary air or not, which question I will not decide.

If, nevertheless, one assumes that gunpowder is a thousand times as dense as natural air and ten thousand times as elastic, then it will follow from §4 that the air, compressed by an infinite force, cannot be compressed more than 1331 times, and according to the same rule the elasticity of air four times as dense as natural would be to the elasticity of natural air as $4\frac{1}{4}$ is to 1.

But whether experiments performed by others which make the ratio of these elasticities exactly 4 to 1 have been done with sufficient accuracy, and whether the temperature of the air remained the same while it was compressed, I do not know. It is also quite likely that the same blast which lies hidden in the pores of gunpowder is the cause of the elasticity of elastic bodies or of resilient wool; indeed, as long as it is present in the small cavities, the elastic blast is compressed if bodies are driven by some force into an unusual shape, and as long as it is restoring the most capacious form to the small cavities, the body is returning to its original shape and length.

ELEVENTH CHAPTER

Concerning Fluids acting in a Vortex, while also Concerning Those which are Contained in Moving Vessels

§1. From the time at which Kepler and Descartes were employing vortices for explaining various phenomena of nature, many people, reckoning that they were expending their effort wisely, reworked that argument eagerly; but, unless I am wrong, Huygens first penetrated the nature of it correctly in his *Traité de la pesanteur*; let me add certain things which pertain to my purpose, perhaps not sufficiently examined by others.

However, as is customary, one assumes that vortices are reduced to the state of *permanence* or of persistence, so that the fluid, subjected to no change, is moved constantly according to the same law.

§2. Let the cylinder *ABCD* (Figs. 65 and 66), the axis of which is *GH*, be placed vertically and let it be filled to a certain height; let the water be considered as having been formed into a vortex, and let all

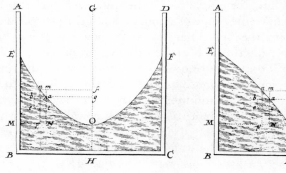

FIGURE 65   FIGURE 66

things be already reduced to the state of permanence. Thus the surface of the water will be depressed toward the axis and elevated toward the sides. We will represent a section through the axis terminated by the surface of the water by the curve *EOF*, and the nature of this curve we will now give from the relation which the velocities have mutually at certain distances from the axis.

Let *ga* and *fn* be drawn infinitely close and horizontal, and let *am* be made vertical. Let $Og = x$, *gf* or $am = dx$, $ga = y$, $mn = dy$. Moreover, it is evident that any volume element whatever located at the surface presses perpendicularly to the surface by its own acting force, composed of the horizontal centrifugal force and the vertical force of gravity, because, if it would press obliquely, there would be nothing which would keep the elemental volume in its place.

Therefore, if the centrifugal force of a volume element located at *a* is expressed by the horizontal *ba* and the force of gravity by the vertical *ca*, and the rectangle *abec* is completed, the diagonal *ae* will be perpendicular to the curve; hence the triangle *eca* is similar to the triangle *amn*, and thus $dx : dy = ec : ca = ba : ca$, or, as the centrifugal force is to the force of gravity at the point *a*.

Moreover, Huygens showed that the centrifugal force of a body driven in a circular course at the speed which it could acquire by free fall through a height of half the radius is equal to its own force of gravity: thus, if accordingly the height corresponding to the gyratory velocity of the volume element is called *V* and the force of gravity *g*, the centrifugal force will be $\frac{2gV}{y}$, from which $dx : dy :: \frac{2gV}{y} : g$, or,

$$dx = \frac{2V\,dy}{y}.$$

§3. If one sets $V = \tfrac{1}{2}y$, *x* will become equal to *y*, and accordingly the line *EO* will be straight, forming a half right angle with the axis *GH*, and the cavity will have the shape of a cone. But if the proportion of the velocities is kept the same, namely, that they are everywhere proportional to the [square] roots of the distances from the axis, and the water is driven around more quickly or more slowly, the angle *EOG* will be more acute, the more quickly [the water] is moved, so that, if the velocity would be infinite, then the water would have to stand perpendicularly to the base, as if it were a wall, and form a cylindrical cavity in the interior, if only there would be a cover at *AD* which would prevent all the water from being ejected.

§4. If it is assumed, a little more generally, that $2V = fy^e$, *dx* will become $fy^{e-1}\,dy$, or, $x = \left(\dfrac{f}{e}\right)y^e$. Hence it follows that the curve will always be concave toward the axis, as in Fig. 65, if e is greater than unity, and convex, as in Fig. 66, if it is less. In the first case the angle EOG is always right, in the other it is always null; only in the case in which $e = 1$ can that angle have any value whatever.

§5. These things can serve for determining to some degree the scale of velocities in an artificially produced vortex: for, if one sees that the surface is concave, one will judge correctly that the velocities increase in a greater ratio than the distances from the axis increase; if it is convex, one will deduce the contrary. If the curve does not seem to be of the parabolic type, this will be proof that the velocities cannot be compared with some force determined from the distances. The greater the observed line EM terminated by the horizontal OM, the greater the absolute velocity of the particles, or the letter f.

§6. But I think that in a homogeneous fluid a vortex cannot remain in the proper state through some notable time if the centrifugal forces of equal portions increase from the axis toward the periphery: for if this were so, since there is nothing which would sufficiently repress the centrifugal force of the portions nearer the axis, thus it would occur that those nearer portions would perpetually recede from the axis, and those more remote would be propelled toward it, and equilibrium or the state of permanence could never be attained under this condition. Hence it appears that this quantity $\frac{2gV}{y}$ (which certainly expresses the centrifugal force of equal portions in homogeneous fluids) either increases together with y or at least does not decrease, and thus if we go again to the special hypothesis formed previously ($2V = fy^e$), e cannot be less than unity. Therefore, in all vortices discussed here that have been reduced to a state of permanence, the surface will never be convex, as in Fig. 66, but always either concave, as in Fig. 65, or conic; and because e is either greater than unity or equal to the same, it cannot occur otherwise than that the velocities increase in proportion either equal to or greater than the roots of the distances from the axis. When I consider these things accordingly, I do not understand in what way Newton could assume that two vortices of a fluid everywhere homogeneous are reduced to a state of perpetual persistence in which in the one *the periodic times of the portions are as their distances from the axis of the cylinder, but in the other as the squares of the distances from the center of the sphere.* For in the first of these vortices the velocities would be equal everywhere, and in the other they would clearly decrease from the axis to the periphery.

It is more probable in the majority of vortices which have already attained the state of absolute persistence that the periodic times of the individual portions of either a homogeneous or heterogeneous fluid will be the same as if the entire cylinder were solid, moreover that the portions which are specifically heavier will be nearer to the circumference. In this case v becomes proportional to y itself and V proportional to the square of the same, and the curve EOF will be an Appolonian parabola, the vertex of which is at O and the axis of which is OG.

I presume especially that things will be approximately so if the vortex is generated by the rotation of a cylindrical vessel about the axis HG, or even by uniform agitation with a stick near the sides of a vessel, the phenomena of which sort of vortices Mr. Saulmon showed in the *Commentaries of the Royal Academy of Science of Paris for the year 1716*.

§7. The pressures which the different portions of the cylinder $ABCD$ sustain from the fluid are proportional to the heights of the vertical columns corresponding to the same portions; indeed, it is not required that we add to this weight the impulse of the fluid developing from the centrifugal force, because that impulse has already obtained its effect in elevating the water. And if the vessel were not cylindrical but of some irregular shape, it would be permissible to assume a cylinder, the axis of which coincides with the axis of rotation, full of fluid such that the point O in the proposed vessel as well as in the fictitious cylinder is located in the same place; for thus, at any point in the cylinder the pressure will be as great as it is at the same point in the proposed vessel. It appears from this very thing that the surfaces of the vortices can be defined from a principle other than that which we used above: indeed, after the horizontal line OM has been drawn, and the vertical Na [has been drawn] with pn infinitely close to it, it follows that the height Na or Og is proportional to the centrifugal force of all the particles which are at ON, and that the difference of the two neighboring heights, namely, am and gf, is proportional to the centrifugal force of the particle Np. Thence the final equation which we presented in §2 is again derived, namely, $dx = \dfrac{2V\,dy}{y}$.

§8. Now let us see what must happen to bodies floating on a vortex; but in order that the matter may be made the more clear and simple, in place of the body we will consider a small globule of the same specific gravity as the eddying fluid.

Such a globule united with the fluid is driven immediately by two forces, the one tangential, drawing its origin from the impetus of the fluid, the other centripetal, which develops from the centrifugal force of the fluid. These forces maintain a constant ratio to each other, namely, the square of the *respective* velocity of the fluid, whether the body is at rest or is carried in a circular motion.

Moreover, it deserves to be noted by those who adhere to Cartesian principles in explaining the phenomena of gravity that the tangential force is incomparably greater than the centripetal force: indeed, the former is to the latter as the distance of the body from the axis of the vortex is to eight-thirds of the diameter of the globe; the proof can be seen in the *Commentaries of the Imperial Academy of Science of St. Petersburg*, Book II, pp. 318 and 319.

§9. Although I know that many things have been alleged by various people in order that they might show that a delicate material driven very suddenly into a vortex can indeed dislodge bodies toward the axis, on the other hand it does not follow thence that at the same time those bodies are transferred by the vortex; nevertheless, I was not able to remove this doubt after I learned that the tangential force is almost infinitely greater than the centripetal force. Perhaps this difficulty is obviated no better if we state that there are two vortices, contrary and of equal strength, about the same axis. For it seems that most phenomena of nature cannot be conciliated with the hypothesis of vortices unless we assume that two or more vortices can cross over one another very freely in any direction whatever: for the common gravitation alone of all the celestial bodies toward one another, which cannot be doubted, shows well enough either that one should bid farewell to the hypothesis of vortices, or that the very free crossing of several vortices in all directions should be concluded. Therefore, if two vortices of equal strength were assumed to be contrary and about the same axis, then the contrary impetuses would destroy the tangential forces of each vortex; but at the same time each vortex would join in depressing a body toward the common axis.

§10. Another difficulty occurs in that the gravity of bodies cannot be sought from the effect of two contrary vortices moving about the same axis. For thus the bodies would not gravitate toward a common point or quasi point, but toward the axis, and they would glide toward the same in a perpendicular motion, which conflicts with the vertical descent of bodies and the roundness or quasi roundness of the earth and of celestial bodies.

This other difficulty is also overcome if two axes are assumed perpendicular to each other or approximately so, about each of which two contrary vortices of equal strength are driven. For the force composed of all the vortices can be understood to be so constituted that a body moves approximately toward the point at which both axes intersect one another; nevertheless, the earth would always be compressed somewhat toward the plane crossing through both axes. But certainly it will be possible to contend with this inconvenience, if only it is an inconvenience, by increasing the number of vortices very much: for if the vortices are considered as almost infinitely many, they all can cross over themselves with the same facility as rays of light, which do not impede each other at all.

I wanted to add these things here for the sake of those who are pleased by vortices, in order that they may see whether this motion can be conceived more easily than that which Huygens assumed: for the phenomena of nature can be explained equally by each. I showed this idea a little more accurately in the dissertation which, having been rewarded with a prize in the year 1734, the Royal Academy of Science of Paris chose to print.

§11. Because no one can doubt that all planets *gravitate* toward the sun and satellites toward their own planets according to Newton's thinking, and that the cause of this gravity is connected with that by which terrestrial bodies tend toward the center of the earth, the hypothesis of vortices will have to be extended to the entire system of the world if it is applied for explaining the gravity of terrestrial bodies. Thus, indeed, planets floating in a fine material would be moved in a resistant medium, and, gradually losing some of their own motion, they should have to drift toward the center of the sun in the manner of a spiral: but since this is not apparent from the most ancient observations, the hypothesis of vortices postulates that an eddying fluid is assumed, rare and delicate beyond all measure, that is moving at a velocity which the human mind can barely comprehend: for, the rarer the fluid, the quicker one must assume the motion to be. Perhaps the perpetuity of the motions will be explained more by a certain reciprocal communication of motion, which is such that a celestial body which has just propelled certain particles is propelled in turn by them with a similar force.

§12. I now come to the remaining properties of gravitating bodies which follow from the hypothesis of vortices. Let us consider that a body which transmits no particles of fluid through its pores is resting in an eddying fluid; thus the body tends toward the center of the vortex, and its centripetal force will be precisely equal to the centrifugal force of the eddying fluid which is located within the same volume at the same distance from the center. Therefore, any bodies whatever located in the same position in the vortex have the same centripetal force if they have the same volume, even if the quantities of material in any one body are unequal in any way whatever, and if bodies of this sort can be moved freely toward the center of the vortex, they will be carried at unequal velocities, reciprocally proportional, of course, to the square roots of the quantities of material, if the measured distances are equal.

§13. Those things which were mentioned in the preceding paragraph are easily applied to the gravity of bodies, if only the origin of gravity is the centrifugal force of some delicate material driven very quickly in a vortex. But because experience shows that all terrestrial bodies descend at the same velocity in a vacuum, and all bodies suspended by an equal thread make tautochronous vibrations, we thence conclude that *ultimate heavy particles*, through which, indeed, heavy fluid cannot penetrate, are of equal specific density in all terrestrial bodies, that is, they contain equal quantities of *solid matter* in equal volumes, and this no less in the heavy particles which compose gold than in those which compose feathers. But lest these things be understood other than as I wish, I will have to explain what I mean by *ultimate heavy particles* and by the *solid matter* innate to them.

§14. Therefore, *heavy particles* are those, having been appropriately named such, which are impenetrable to a fine eddying material: indeed, particles of this sort act the same way as bodies placed in a vortex, which we discussed in §12; nevertheless, although they may be impenetrable to the fine material just mentioned, I would not believe that they are perfectly solid, which Huygens seems to have presumed in his treatise *De Gravitate*, that is, such that their entire space is filled with material, without pores or inter-flowing fluid; I think, rather, that these *heavy particles* do have their own pores again, and in them there is some far more delicate fluid which traverses the heavy particles with the same freedom with which a heavy fluid flows through observable bodies: but the remainder, which coheres to itself in the *heavy particles*, I call the *solid material* pertaining to the same particles.

§15. From these [considerations] it is clear that the different specific gravities of bodies should be sought by no means from the different density of the *heavy particles*, but from the fact that these particles can be unequal in number, or even in magnitude, in different bodies within the same volume, such that in the more compact bodies, or those of greater specific gravity, the *heavy particles* either are positioned with less interstices or are greater in volume.

But even if the *heavy particles* should have different specific densities in different bodies, on that account the bodies would not have different specific gravities, the remaining things having been set equal; moreover, such bodies, having fallen from above, should descend toward the center of the earth at different velocities from one another.

Therefore, it should be possible that bodies of equal specific gravity should descend in the vacuum so commonly mentioned at an unequal velocity no less than we see bodies of different specific gravity descending at equal velocity. Moreover, in bodies of this sort the laws of motions would be greatly different from what they are nowadays when masses are estimated from weights alone.

§16. Finally, because all terrestrial bodies, as far as is known from experience, have their *heavy particles* of equal specific density, as was indicated in §13, I may indeed be easily induced to believe that the same thing occurs in all planets considered separately. But it is altogether probable to me that the planets compared to each other have their individual *heavy particles* of different specific density, because I do not see any reason why these particles should be similar in all planets. But in any planet whatever, its centrifugal force or its attempt to recede from the sun depends upon the density of its *heavy particles*. Therefore, it is not yet allowable to infer that *the centrifugal forces of the planets are in an inverse square ratio of their distances from the sun from the fact that the periodic times follow a ratio of the three-halves power of the distances*: for such a conclusion supposes the same density of the *heavy particles* on all planets.

§17. The centrifugal forces of planets are certainly equal to the contrary forces by which they are drawn toward the sun. But because, as I mentioned in the paragraph above, it is not yet certain in what ratio with respect to the distances from the sun the centrifugal forces of the planets are changed, therefore it is not permissible to say anything definite about their forces of gravity toward the sun. And indeed there are many things in the hypothesis of vortices which constitute and determine the forces of gravity at different distances. For when the force of gravity is equal to the centrifugal force of a fine material which cannot penetrate the heavy particles of a body, it follows that the force of gravity is greater, the greater is the quantity of fine material to which transit is denied; but because we know that a body, impenetrable to one fluid, often offers the freest transfer to another finer fluid, it can occur, if only we assume that the eddying material is unequally fine at different distances from the center of the vortex, that one and the same planet, at unequal distances from the sun, is driven unequally toward the sun, which same thing can apply more easily in different planets because it happens that the structure of the heavy particles can be different.

In addition to these things are the different density of the eddying material, the velocity, and the distance from the center, which [all] contribute to forming the force of gravity. But if they are taken into

FLUIDS IN VORTICES OR IN MOVING VESSELS

consideration, it will certainly appear that the forces of gravity can decrease with increasing distances from the center of forces. Nevertheless, on account of the fact that the centrifugal forces of equal volumes of eddying material do not decrease equally, I consider that the latter [situation] cannot occur on account of the reasoning shown in §6.

But let these things suffice which we have discussed generally and incidentally concerning the nature of vortices and their application to the Phenomena of gravity: the intention was not to commend the hypothesis of vortices, but only to draw certain conclusions from it without which I should believe this hypothesis cannot subsist.

I come now to the other part of the chapter in which we will consider briefly the state of fluids which are contained within moving vessels; the subject is very fertile and variable in infinite ways. But we will treat a few matters, or examples, to which many others can be referred.

§18. If water is contained in a perforated vessel and the same vessel falls freely, it is self-evident that no water will flow out during the fall of the vessel, because, certainly, the upper particles do not gravitate toward the lower. If the vessel indeed descends in an accelerated motion, but slower than that by which bodies are accelerated naturally in a vacuum, the water will flow out, but at a lesser velocity than if the vessel were at rest; the contrary will be the case if the vessel is drawn upward by an accelerated motion. Finally, if the vessel is borne horizontally by an accelerated motion (for now we will not attend to the remaining directions), it can happen that the velocity of the water flowing out is either greater or less than the ordinary velocity in relation to the position of the orifice. But the velocities of the water will thus be determined.

§19. For example, let the cylinder $ACDB$ (Fig. 67) be full of water right up to AB, the base CD of which has a very small orifice at E through which water flows, while in the meantime the entire vessel is drawn upward by a descending weight P by means of a string running over two pulleys H and G. Finally it is assumed that as much water is constantly supplied from above as flows out through the orifice E; but let the weight of the cylinder and the water contained in it be indicated by p. Thus it appears that any volume element of water standing in the vessel is stimulated to ascend by a force which is in proportion to the natural force of gravity as $\frac{P-p}{P+p}$ is to 1. But because the reaction of a volume element against the base is equal to the force by which any volume element is stimulated to ascend, it exerts another pressure on the base, in addition to the natural pressure, which will be expressed by $\frac{P-p}{P+p}$. But both pressures taken together will be to the natural pressure alone as $\frac{2P}{P+p}$ is to 1, so that

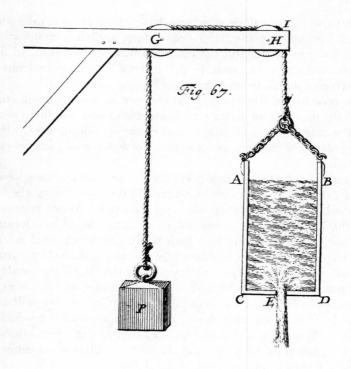

FIGURE 67

the base is pressed not at all otherwise by the water lying above than if the cylinder were at rest and the height of the water were $\frac{2P}{P+p} \cdot AC$, and from this itself it follows that the height due to the velocity of the water flowing out uniformly is $\frac{2P}{P+p} \cdot AC$.

Therefore, if $P = 0$, no water will flow out with the vessel falling in a naturally accelerated motion; if $P = p$, the water will flow out at the ordinary velocity, because then the vessel is at rest; and if $P = \infty$, the velocity of the water flowing out will be to the ordinary velocity as $\sqrt{2}$ is to 1.

§20. Now one seeks what must happen to a fluid which is contained in a vessel to which a uniformly accelerated horizontal motion is imparted. But it is very easy to see from this alone that now the inertia of the particles is horizontal or opposite to the direction in which the vessel is moved, while that of their gravity is vertical. But each remains constantly the same.

Therefore, after the fluid arrives at the state of persistence or *permanence*, its surface will be plane but inclined toward the direction of motion. Moreover, the angle of inclination will be determined as follows.

Let there be a cylindrical vessel $ACDL$ (Fig. 68), positioned vertically, which is moved in a uniformly accelerated motion over the

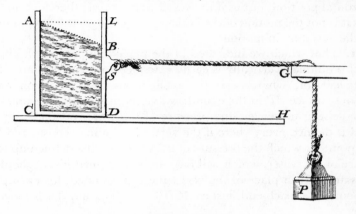

FIGURE 68

horizontal plane CDH by means of the weight P attached to the vessel at S with the help of the pulley G, and let the weight of the vessel and the water contained in it be to the weight P as p is to P; let the natural gravitation be unity; the force of any volume element in the direction GS, with respect to its own gravitation, will be $\dfrac{P}{P+p}$. Therefore, if AB is in the same plane as SG and as the surface of the water, and if AL is drawn, it is evident that the action of the natural gravity will be to the reaction arising from the weight P as BL is to AL, or as 1 is to $\dfrac{P}{P+p}$; and with the entire sine designated as 1, the sine of the angle LAB will be

$$\frac{P}{\sqrt{2PP + 2Pp + pp}}.$$

Hence it is understood also that the base *CD* experiences a greater pressure at *C* than at *D* from the water lying above, and this in proportion to the heights *AC* and *BD*; and, if the same base is perforated by a very small orifice, that the water will be ejected at a velocity which corresponds to the height of the vertical column lying above. Thus, indeed, this will be after everything has reached the state of *permanence*; if the weight *P* is variable, the surface *AB* will never remain in the same position; moreover, the velocity at which the vessel is moved in individual locations depends on that weight. Therefore, if the total weight is removed after the vessel already has acquired motion, the vessel will continue to be moved at its own velocity, but the surface of the water will lose its slope and again be composed in a horizontal position, just as if the vessel were at rest; therefore, in these cases it is not the motion of the vessel which changes the state of fluids, but the variation in motion.

§**21.** That which we indicated in the preceding paragraph about a vessel positioned vertically is easily extended to a vessel of any shape whatever: for, whatever is the inclination to the horizon of the aqueous surface *AB* in the cylindrical vessel, it will be the same in all other vessels; moreover, the pressure of the water on the walls of a vessel is defined everywhere if the vertical column is considered from that point for which the pressure of the water is to be defined up to the surface of the water, which will have to be imagined if that should be necessary. If in place of the vessel there is assumed, for example, a pipe curved at each end, just as *ACDL* (Fig. 69), and this is moved in

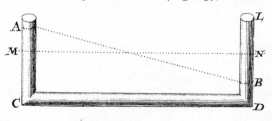

FIGURE 69

the direction *CD*, then each surface *M* and *N* will change position to *A* and *B*, until the straight line *AB* obtains the required inclination defined previously; also, it can occur that part of the water flows out through *A* before equilibrium is present; if the leg *DL* is directed downward as in Fig. 70, the water will remain as if suspended; indeed, in each case the inclination of the line *AB* will be the same, the remaining things being equal.

However, in Fig. 69 the line *MA* will be greater, the longer is the horizontal leg *CD*, such that the smallest accelerations or retardations can be observed; this often can be useful for other things such as measuring accelerations of ships and the pressures which rowers exert in individual immersions of the oars; however, in these cases the entire

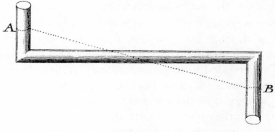

FIGURE 70

motion of the fluid which is developed in individual repetitions should be investigated, because the state of persistence or *permanence* cannot be assumed.

Because of this same reasoning, it is not yet allowable to determine wholly from the preceding what must happen when vessels containing a fluid are pierced.

However, the rules of percussion can be deduced from the ordinary laws of pressure, since indeed a percussion is nothing other than an immense pressure lasting for a very short time.

§22. For example, let the cylindrical pipe *ABCD* (Fig. 71), placed horizontally, be full of water, and let the sphere *P* impinge on the

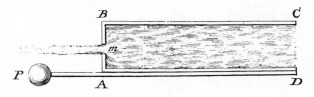

FIGURE 71

extension *AP* of the pipe; then the water will suddenly press the base *BA* violently toward *P*; in order to understand this pressure properly, we will assume first that the pipe has no weight; thus it appears from the equality between action and reaction that during the impulse of the sphere the base is not impelled differently by the water than it would be impelled in the opposite direction by the sphere if the latter impinged directly against the base. But if the weights of the water and the pipe are assumed to be in proportion as p is to π, the impulse of the water against the base will be to the residual impulse as $p + \pi$ is to p; for the impulse is distributed equally over all the material of the water as well as the pipe, and only the fluid reacts on the base.

But now let us assume a very small orifice m in the base BA; through this, nevertheless, water is considered to flow very freely; thus we understand that a particle of water will be ejected through the small orifice m during the impulse; however, the quantity of that water cannot be determined, for it depends upon the rigidity of the material AP receiving the impulse: indeed, if that material is very rigid, a greater pressure is to be substituted for the impetus, but lasting for less time; for example, let the same impetus be considered for two different cases: moreover, in one let the pressure be quadrupled, in the other let the duration of the pressure be quadrupled, which can happen when the material is more rigid in the former case than in the latter; thus, approximately double the quantity will flow out in the impulse of the lesser pressure and greater duration than in the other case. In this way the rigidities of materials can be explored: but they can be found as well from sound.

TWELFTH CHAPTER

Which shows the Statics of Moving Fluids, which I call Hydraulico-Statics

§1. Among those who gave measurements of the pressure of fluids existing within vessels, few have gone beyond the common rules of Hydrostatics which we showed in Chapter II; nevertheless, there are many other rules which pertain to the appropriately named Hydrostatics, such as whenever a centrifugal force or the force of inertia is united with the action of gravity, each of which we discussed in the preceding chapter; dead forces of this type can be devised and combined in infinitely many other ways. But these are not the things which seem to me to be most desirable, since it is not difficult to give general rules for this procedure. I desire, rather, [to treat] the statics of fluids which are moved within vessels in a progressive motion, such as of water flowing through conduits to leaping fountains: indeed, this is of multiple use, and it has not been treated by anyone, or, if some people can be said to have made mention of it, it was not at all properly explained by them; indeed, those who have spoken about the pressure of water flowing through aqueducts and the strength required of the latter for sustaining that pressure did not hand down any laws other than those for extended fluids with no motion.

§2. It is singular in this *hydraulico-statics* that the pressure of water cannot be defined unless the motion has been known correctly, which is the reason that this doctrine escaped notice for so long; indeed, up to now Authors were hardly anxious to investigate the motion of water, and they estimated velocities almost everywhere from the height of the water alone; however, although the motion often tends so quickly toward this velocity that the accelerations clearly cannot be distinguished by observation, and all the motion seems to be generated in an instant; nevertheless, it is of interest to understand these accelerations correctly, because otherwise the pressures of the flowing water often cannot be defined, and on that account I estimated that it is a matter of greatest moment to consider those changes, however *instantaneous*, from the beginning of motion up to a given limit with all care, and to confirm them by experiments, which I did at different places in this treatise, but especially in Chapter III.

§3. If the motion could be defined everywhere, it would be easy to develop the most general statics in moving fluids: indeed, if one assumes an orifice which is infinitely small in that very place at which the pressure of the water is desired, one will seek to learn first at what velocity the water would erupt through that tiny orifice and to what height that velocity would be due; moreover, one understands that the pressure which is sought is proportional to this very height.

From this principle the pressure is to be sought which the horizontal plate LQ in Fig. 43 sustains if it has not been perforated. Indeed, since it has been shown by us in the second corollary of §31, Chapter VIII, that, if the orifice H is infinitely small in proportion to the orifices M and N, and the ratio of these orifices M and N is indicated by α and γ, then the height due to the velocity of the water erupting through H will be $\dfrac{\alpha\alpha(LB) - \gamma\gamma(NQ)}{\alpha\alpha + \gamma\gamma}$, we will thence judge that the pressure of the water against the nonperforated plate LQ is proportional to this very height. We gave the same proof in another way in §19 of the cited chapter. Hence it follows that it can occur that the section LQ experiences no pressure, however great the height of the water above it may be, as for example when $\gamma = \alpha\sqrt{LB/NQ}$; indeed, the pressure can even be changed into suction.

§4. Similarly, the pressure of the water against the section LQ is obtained if, for instance, the latter is perforated by an orifice of finite size in proportion to the two remaining [orifices]. For if the section is perforated by an infinitely small orifice with respect to that which exists at H, the water cannot but erupt at a common velocity through either one. And since this velocity is known (from §30, Chapter VIII) for the orifice H, the velocity is also obtained at which the water must erupt through the tiny orifice which we conceive, and thus we know the pressure of the water. For example, let the orifices M, H, and N be equal to one another, and also let the height BL have a ratio to the height LQ as 10 is to 3, and the pressure against the plate LQ will be one-tenth of what it is with the orifices H and N closed off.

Finally, if one should desire the pressure of the water in another location, he will simply add the height by which the section LQ exceeds that point to the height of the thrust through the orifice H. The same method serves for determining water pressures in the rest

STATICS OF MOVING FLUIDS 291

of the vessels which we treated in Chapter VIII. But all these matters differ from those which pertain to the motion of fluids through conduits, because the water, on account of the infinite size of the vessels assumed by us, is as if at rest in cavities, and nevertheless it exerts a far different pressure from what is otherwise customary. Moreover, in conduits the water changes its pressure more, the greater the velocity at which it flows through, and it exerts almost all its customary pressure if that velocity is very small.

This is so whenever the velocities of fluids can be determined by the methods presented by us just above. But the matter must be handled by a singular method when the water flows through conduits, and I comprehend this doctrine especially under the title of *hydraulico-statics*. Here, not so much can the pressure be defined from the velocity as, reciprocally, the velocity from the pressure, if a small orifice is made in the walls of the conduit. And in the present chapter I decided to treat especially that *hydraulico-statics*, the application of which is very broad.

PROBLEM

§5. The very wide vessel *ACEB* (Fig. 72), with the cylindrical and horizontal pipe *ED* attached, is to be kept constantly full of water;

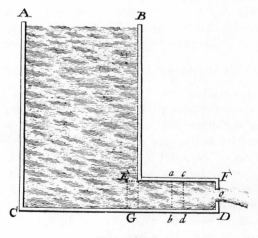

FIGURE 72

and at the extremity of the pipe let there be the orifice *o* emitting water at a uniform velocity; the pressure of the water against the walls of the pipe *ED* is sought.

SOLUTION. Let the height of the aqueous surface AB above the orifice o be a; the velocity of the water flowing out at o, if one excludes the first instants of flow, will have to be considered uniform and equal to $\sqrt{a}$, because we assume the vessel to be kept constantly full; and, with the ratio of the areas of the pipe and its orifice assumed equal to $\frac{n}{1}$, the velocity of the water in the pipe will be $\frac{\sqrt{a}}{n}$. But if the entire base FD were missing, the ultimate velocity of the water in the pipe itself would be $\sqrt{a}$, which is greater than $\frac{\sqrt{a}}{n}$. Therefore, the water in the pipe tends to greater motion, but its pressure is impeded by the added base FD. By this pressure and repressure the water is compressed, which very compression is confined by the walls of the pipe, and hence these sustain a like pressure. Thus it appears that the pressure of the walls is proportional to the acceleration, or the increment of velocity which the water would receive if the entire obstacle to motion would vanish in an instant so that [the water] might be ejected immediately into the air.

Therefore, the problem is now changed into this: if during the flow of water through o the pipe ED were broken at cd at an instant, one seeks the magnitude of the acceleration the volume element $acbd$ would thence be about to obtain; indeed, the particle ac taken at the walls of the pipe will sense that much pressure from the water flowing through. To this end the vessel $ABEcdC$ is to be considered, and with regard to it the acceleration of an aqueous particle close to efflux is to be found, if this would have the velocity $\frac{\sqrt{a}}{n}$. We handled that matter very generally in §3, Chapter V. Nevertheless, because the calculation is short in this particular case, we will here again subject the motion in the shortened vessel $ABEcdC$ to evaluation.

Let the velocity in the pipe Ed, which [velocity] is now to be considered as variable, be v; let the area of the pipe, as before, be n, the length $Ec = c$; let the length of the aqueous particle ac, infinitely small and about to flow out, be indicated by dx. There will be an equal volume element at E entering the pipe at the same instant that the other, $acdb$, is ejected; moreover, while the volume element at E, the mass of which is $n\,dx$, enters the pipe, it acquires the velocity v and the *live force* $nvv\,dx$, which entire *live force* was generated anew; indeed, the volume element at E, not yet having entered the pipe, had no motion on account of the infinite size of the vessel AE; to this *live force*, $nvv\,dx$, is to be added the increment of *live force* which the water at Eb receives while the volume element ad flows out, namely,

$2ncv\,dv$; the sum is due to the *actual descent* of the volume element $n\,dx$ through the height BE or a; therefore, one obtains $nvv\,dx + 2ncv\,dv = na\,dx$, or $\dfrac{v\,dv}{dx} = \dfrac{a - vv}{2c}$.

Moreover, in all motion the increment of velocity dv is proportional to the pressure multiplied by the differential time, which here is $\dfrac{dx}{v}$; therefore, in our case the pressure which the volume element ad experiences is proportional to the quantity $\dfrac{v\,dv}{dx}$, that is, to the quantity $\dfrac{a - vv}{2c}$.

But at that instant at which the pipe is broken, $v = \dfrac{\sqrt{a}}{n}$, or $vv = \dfrac{a}{nn}$; therefore, this value is to be substituted in the expression $\dfrac{a - vv}{2c}$, which thus is transformed into $\dfrac{nn - 1}{2nnc}\,a$. And this is the quantity to which the pressure of the water against the portion ac of the pipe is proportional, whatever area the pipe may have, or by whatever orifice its base may be perforated. Therefore, if in a particular case the pressure of the water would be known, it would be understood at the same time in all remaining [cases]: but, indeed, we have this [pressure] when the orifice is infinitely small or n is infinitely large with respect to unity: for then it is evident from itself that the water exerts its entire pressure, which conforms to the total height a, and this pressure we will designate by a; but when n is infinite, unity vanishes with respect to the number nn, and the quantity to which the pressure is proportional becomes $\dfrac{a}{2c}$. Therefore, if we wish to know in general how great the pressure is when n is any number whatever, the following analogy must be used. If the pressure a conforms to the quantity $\dfrac{a}{2c}$, what then will be the pressure for the quantity $\dfrac{nn - 1}{2nnc}\,a$? And thus the desired pressure is found equal to $\dfrac{nn - 1}{nn}\,a$. Q.E.I.

§6. COROLLARY 1. Because the letter c vanishes from the calculation, it follows that all portions of the pipe, those which are nearer to the vessel AG as well as those which are more remote, are pressed equally by the water flowing through, and certainly less than the elements of the base CG, and the difference is the greater, the larger is the orifice o; and, further, the walls of the pipe do not sustain any pressure if in the latter the entire barrier FD is missing, so that the water flows out from a full orifice.

§7. COROLLARY 2. If the pipe is perforated somewhere by a very small orifice that is necessarily in some ratio to the orifice o, the water will spring forth at the velocity by which it could ascend to the height $\dfrac{nna - a}{nn}$ if only no foreign hindrances were interfering. Indeed, this will be the height of the thrust in Fig. 73, or $ln = \dfrac{nna - a}{nn}$. But

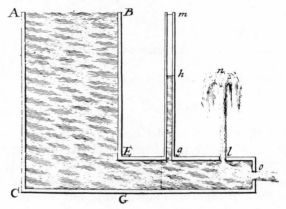

FIGURE 73

if the small tube gm is attached, vertical or even inclined in some way, connecting with the horizontal pipe, but so, nevertheless, that the extremity of the inserted tube does not project into the cavity of the horizontal pipe lest the water flowing past strike against that extremity, the vertical height gh of the water standing in the inserted tube will also be equal to $\dfrac{nna - a}{nn}$; and it is not necessary in this latter case that the tube gm be very narrow.

SCHOLIUM

§8. Therefore, this theory can be confirmed very easily by experiment, and this will be of more importance because up to this time no one has defined equilibria of this sort, the use of which is very widely evident, because by the same method the pressure of water flowing through conduits can be obtained very generally for aqueducts in-

STATICS OF MOVING FLUIDS   295

clined in any way whatever, curved, of varied area, and at any velocity of water whatever; then, as well, because not only this [theory] of pressures, but the entire theory of motions besides, which we gave above, is confirmed by experiments of this sort, because they prove that the accelerations of the water were defined correctly by us. But one must take care in the experiment that the horizontal pipe is very smooth on the interior, perfectly cylindrical and horizontal, and that it is wide enough so that no noticeable decrement of motion can arise from the adhesion of the water to the walls of the pipe; let the vessel itself be very wide and be kept full continuously. Also one must observe how great is the characteristic of elevating standing water in the glass tube gm, which characteristic pertains to all capillary or rather narrow tubes; for this elevation is to be subtracted from the height gh; or, rather, a pipe of equal thickness is to be assumed with the orifice o blocked off, the point m is to be noted, and then, with the water allowed to flow, the point h is also to be noted; moreover, according to the theory the descent will be $mh = \frac{1}{nn} a = \frac{1}{nn} (EB)$.

Finally, one must pay attention as well to the stream of water flowing out at o, for its contraction also causes the water in the horizontal pipe to flow through at a velocity less than $\frac{\sqrt{a}}{n}$. I treated that contraction and the method of preventing it in Chapter IV. But although it can happen with these inconveniences that no noticeable error remains in the experiment, nevertheless, if we wish to apply greater accuracy, the quantity of water flowing out in a given time will have to be discovered by experiment, which [quantity], compared with the area of the pipe, will give very correctly the velocity of the water flowing within the pipe, which in the calculation we have set equal to $\frac{\sqrt{a}}{n}$. But if in the experiment it will be found to be less, such, for example, as is due to the height b, then the pressure of the water flowing by will be approximately $a - b$.

§9. COROLLARY 3. If the orifice at o is blocked off at first by a finger, and afterwards the water is allowed to flow, the pressure a at the first moment of flow is changed into the pressure $\frac{nna - a}{nn}$, but that change of pressures does not occur in an instant; if, indeed, one is to speak accurately, it occurs at last after an infinite time, because, as we saw in Chapter V, the entire velocity of the water, which was assumed by us in the calculation to correspond to the whole height a, is never present exactly; nevertheless, it tends toward this velocity with an incredible acceleration immediately after the first drops have been ejected, so that it seems to have acquired the total [velocity], as much as can be judged by observation, without any noticeable delay, unless the aqueducts are very long, for then the accelerations of the water can be discerned clearly by eye, an example of which I gave in §13, Chapter V. Therefore, in those conduits bearing water to a leaping fountain from a reservoir located very far away, if the pressures are investigated at some point by experiment in the manner that I mentioned above, it is found that the pressure is diminished quickly indeed, nevertheless not in an instant, and it will be possible to distinguish the differences of the pressures.

But in order to define the force of the water generally, one must assume for v that velocity which the water has at that same place and that same instant at which the force is desired, and if this velocity is known to conform to the height b, the force of the water will be $a - b$. Hence, since those things which were offered in Chapter V have agreed with the present, it will be possible to define what the pressure will be at any moment.

For these [statements] it is not difficult to anticipate the laws of this *hydraulico-statics* if both the shape of the vessel and the velocity of the water flowing through the conduits are assumed at will as anything whatever. Indeed, the pressure of the water will always be $a - b$, where by a is understood the height due to the velocity at which water will flow out of an abrupt conduit and vessel kept constantly full after an infinite time, and by b the height due to the velocity at which the water actually flows through. It is clearly amazing that this very simple rule, which nature affects, could remain unknown up to this time. Therefore, I will now show it more expressly.

Problem

§**10.** To find the pressure of water flowing at any uniform velocity whatever through a conduit arbitrarily formed and inclined.

SOLUTION. Let there be a conduit ACD (Fig. 74), through the orifice o of which water is considered to flow at a uniform velocity due to the vertical height oS; let the line SN be drawn, and let the infinitely wide vessel $NMQP$ be assumed full of water right up to NP, from which the conduit draws its water perpetually and equally; I assume these things accordingly in order that a cause be present, or a uniform propelling force, which propels the water at a given velocity or maintains an equal flow of water. And without this hypothesis our problem would be indeterminate, because the same velocity in the same conduit pertaining to any instant can be generated in infinitely many ways, and therefore, in order that a measure of the cause propelling the water be obtained, uniformity must be assumed in the motion of the water.

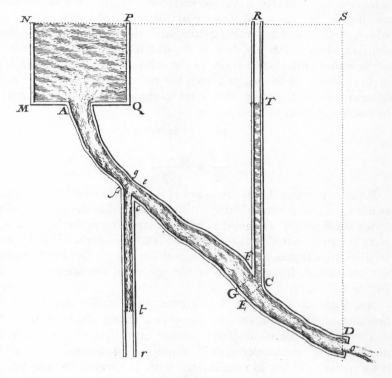

FIGURE 74

Now the pressure of the water is to be defined at CF (or cf); and to this end we will consider again that the conduit is broken at the section CE (or ce) perpendicular to the conduit, and we will examine what acceleration or retardation the volume element $CEGF$ (or $cegf$) will receive after the first instant of rupture; for this reason we have to define generally the *instantaneous* motion through the shortened vessel $NMECAQP$ (or $NMceAQP$). Therefore, let the velocity of the infinitely small volume element $CEGF$ (or $cegf$) at that very point of cutting off be v, and let its mass be dx; the *live force* of the water moving in the shortened vessel will be proportional to the quantity vv; hence we will set it equal to αvv, understanding by the letter α some constant quantity which depends upon the areas of the suddenly broken conduit; however, its precise determination is not required here. Let it be noted that the *live force* of the water in the fictitious vessel $NMQP$ is neglected on account of its infinite area; nevertheless, even if it would not be of infinite area, no variation would hence arise in the calculation. Now we have the increment of *live force* of the water moving in the shortened vessel equal to $2\alpha v \, dv$, to which if there is added the *live force* generated at the same time in the ejected volume element, there arises $2\alpha v \, dv + vv \, dx$, which is the total increment of *live force* due to the *actual descent* of the volume element dx through the vertical height of the water above the point C (or c), which we will designate by a; therefore, that total increment of *live force* is hence to be made equal to $a \, dx$, such that

$$2\alpha v \, dv + vv \, dx = a \, dx$$

or

$$\frac{v \, dv}{dx} = \frac{a - vv}{2\alpha}.$$

If the remaining things occur as in §5, and the velocity v is assumed as if it were due to the height b, it will be found that the pressure of the water at CF (or cf) is as great as in water standing at the height $a - b$. Here it can be noted that the height b is to the height oS, if there are no alien hindrances to the motion and the stream flowing out at o is not contracted, in proportion as the square of the orifice o and the section CE (or ce).

§11. Corollary. When b is greater than a, the quantity $a - b$ becomes negative, and thus the pressure is changed into suction, that is, the walls of the conduit are pressed inward; moreover, then the situation is to be considered as if, in place of the aqueous column CT lying above and set in equilibrium with the water flowing by, the aqueous column ct were attached, the tendency of which to descend is prevented by the attraction of the water flowing by, just as if, for example, the area ce of the conduit were equal to the orifice o, whereupon $b = oS$, not considering the *accidental* hindrances to the motion; hence, if the tube cr descends from the conduit, and if it is full of water from its origin c right to the point t placed on a level with the orifice o, the water ct will remain suspended without motion; but if the point t is placed below o, the water will descend through the tube cr, and it will flow perpetually at r, and, nevertheless, as anyone can now estimate after this theory has been considered, the velocity of the water flowing out at r will be that which is due to the height of NP above r, and even if all hindrances are removed, this velocity will correspond rather to the height tr, if only the tube is very narrow in proportion to the conduit. If the point t is placed higher than the point o, the water will ascend on its own, and after it will all have entered the conduit, air will be drawn through the tube, and soon the aqueous stream flowing out at o will be disturbed by the admixed air, its clearness and solidity having been spoiled. Therefore, one sees when the pressure will be positive and when negative: indeed, the pressure is the greater in a tube, the larger it is [in area] and the lower it is placed. Certainly in theory the height $b = \frac{1}{nn}(oS)$, if $\frac{1}{n}$ denotes the ratio between the area of the orifice and of that section of the pipe for which the pressure is to be defined. But when hindrances diminish the motion notably, it will be agreed upon in estimating pressures rather that the velocity of the water, as it actually is, be found by experiment and the height required for that velocity be substituted for b; similarly, the pressure will be estimated more accurately if for a not the height of the aqueous surface NP above the place of efflux is substituted but rather the height of the velocity at which the water actually flows out from the conduit broken in the same place. Nevertheless, these corrections are not always important. But I will now illustrate that general theory by certain examples.

§12. EXAMPLE 1. Let there be a vessel $ABFG$ (Fig. 75), from the middle of the base of which the pipe DE descends, having the shape of a truncated cone diverging toward the lower regions. Let water be supplied perpetually at AG, so that the vessel is thus kept full.

Moreover, let the height of the aqueous surface above the orifice E be a, and above D (which is the point at which the pressure of the water is desired) be c, the area of the orifice at E be m, and the area of the horizontal section at D be n. The pressure of the water at D will be $c - \frac{mm}{nn}a$, which quantity is negative by virtue of the hypotheses, so that the walls of the conduit are pressed inward by an aqueous column of height $\frac{mm}{nn}a - c$.

Therefore, if the curved pipe DLN is understood to be inserted in the other pipe DE, the water flowing past D will be in equilibrium with the water DLN when the height of D above N is $\frac{mm}{nn}a - c$. If this height is less, the water will ascend on its own, and it will not stop ascending as long as the orifice N is submerged in water, so that thus water can be elevated from a lower place to a higher without any external force, if it flows in at AG in sufficient quantity. But, indeed, when the vertical height of D above N is greater than $\frac{mm}{nn}a - c$, the water will ascend in the leg LN until it will be equal to the other.

Finally, it must be recalled here that in passing I indicated that experience shows that water is indeed far from flowing out through pipes diverging from the vessel to which they are attached at its total velocity which it should obtain by virtue of the theory; I indicate reasons for this in §26, Chapter III.

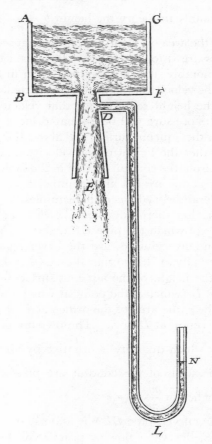

FIGURE 75

Hence it occurs that the height of D above N is somewhat less than that which should be found by theory. The error may be corrected if in place of $\dfrac{mm}{nn} a$ the height of the velocity is used which the water has at D; this height is obtained from an experiment performed on the quantity of water flowing out in a given time.

§13. EXAMPLE 2. If to a similar vessel a vertical pipe is attached, which is represented in Fig. 76 by *CE*, in which the areas everywhere

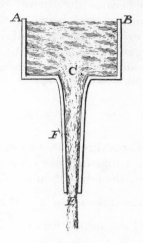

FIGURE 76

have an inverse ratio to the square roots of the heights of the water lying above, that pipe is not affected by the water flowing through, and it does not sustain either pressure or suction anywhere.

Hence it follows that the natural shape of a vertical aqueous filament, as long as it is continuous, is the same as that of the pipe *CFE*, which both reasoning and experience confirm; moreover, the filament will be attenuated more quickly, the less the height of the aqueous surface above the orifice *C*, or the slower the water flows out; it appears that the aqueous filament is of this nature in order that the same quantity of water may flow across the individual sections and that the velocity is not changed anywhere, wherever the filament is cut off, which same property occurs as well in the pipe *CFE*, so that these things agree with each other very well.

§14. EXAMPLE 3. Let water discharge from a reservoir through a conduit in the base of which there is an orifice through which water springs forth vertically just as in a leaping fountain; I say that the pressure of the water at individual points in the conduit is everywhere equal if its areas are respectively as $\sqrt{\dfrac{a}{x-b}}$, where *a* expresses the height of the water in the reservoir above the orifice of efflux, *x* the height of the same water above a point chosen at will in the conduit, and *b* an arbitrary constant height, and then the pressure of the flowing water everywhere will be to the pressure of standing water as

b is to a. But because, with the remaining things being equal, wider conduits are less resistant to rupture than narrower, and this indeed in ratio of the radii, or because the attempt of water in rupturing a conduit, with the remaining things being equal, follows a square root ratio of the areas, it is evident that the conduit will be subject to the same danger of rupture at any location if the area (y) in proportion to the orifice ejecting water (unity) everywhere follows the law of the following equation:

$$\left(x - \frac{a}{yy}\right)\sqrt{y} = b, \qquad \text{or} \qquad xxy^4 - bby^3 - 2axyy + aa = 0.$$

In a conduit of equal area throughout its entire length, the pressing force of the water for rupturing the conduit will be everywhere proportional to the strength of the conduit, if the thickness of the walls of the conduit follows the ratio of $x - \dfrac{a}{mm}$, the area of the conduit in proportion to [that of] the orifice (unity) having been understood by m.

§15. EXAMPLE 4. It can happen that the height of the aqueous surface with respect to the place at which the pressure is to be investigated is negative, such as in curved siphons drawing water from one vessel to another placed lower. Then the pressure becomes negative on two accounts, namely, equal to $-a - b$, with a denoting the height of the point above the surface of the water and b the height due to the velocity of the water at that point.

Truly these things will suffice, as I believe, for correctly understanding the statics of moving fluids. Now I come to certain other phenomena, the solution of which depends on those rules which we have just presented.

§16. In Chapter III, §25, I made mention of the cohesion of water flowing through pipes; however, to define the true measures of that cohesion everywhere is a matter which cannot be explained without that previously mentioned *hydraulico-statics*: for it does not suffice to have considered the vertical heights above the orifice of efflux, as it is commonly thought, but it is necessary to know also the velocities conforming to the water, and these are understood from the areas. But in order that the general law may appear at once in defining the force of cohesion or the inclination with which fluids tend toward mutual separation, I say that that force of cohesion is equal to the force by which the walls of a conduit are pressed inward, which we defined in §10. This proposition does not seem to me to need any other proof; for just as the compression of water, or the force by which its portions are pressed mutually against one another, is equal to the standing aqueous column lying above, so in turn the tendency to separate the fluids is to be reckoned equal to the [effect of] the attached standing vertical aqueous column which is in equilibrium with the water flowing by. In place of examples we will accept the same things which we used above for indicating the negative pressures of water.

I. In Fig. 75, explained in §12, if in the tube DLN the height of D above N is such that the water standing in it is in equilibrium with the water flowing past at D, then, in order that the water not be torn apart at that place, the force of cohesion at D must be as great as that which the weight of an aqueous column of similar base and vertical height DN has. Hence that which I mentioned in §25, Chapter III, is understood: *that the length of a pipe can be increased so that finally the water stops being continuous in the pipe, but rather it is divided into columns, and this happens in cylindrical pipes when they descend beyond 32 feet; moreover, in diverging pipes a lesser descent is required, so that, for example, if the lower orifice were twice as large as the upper orifice attached to the reservoir, pipes could not descend below eight feet without the danger of the dissolution of the water being present.* However, in these examples considered theoretically, the water is assumed to flow at its full velocity without diminution of motion.

II. From the same reasoning it is evident that if pipes converge toward the lower regions, then they admit a descent greater than 32 feet; and finally, in the case of Fig. 76, explained in §13, the pipe can be continued without end, as also in infinitely many other ways.

III. But if the height of the aqueous surface in a reservoir is negative with respect to the proposed point, as occurs when water is to be carried across a mountain, never, no matter how the problem is attacked, can the height exceed 32 feet, which is evident from §15. For even if the water is to flow through at an infinitely small velocity, a force of cohesion is already required which is equal to the entire aqueous column, and a greater force is required if it flows through at an appreciable velocity. Hence I consider the remedies employed by some Writers as useless: certainly I know that without other artifice water often remains suspended beyond a height of 32 feet, and Mercury beyond 30 inches; but this effect is uncertain and not consistent. Certain people also affirm that the flow of water through curved siphons occurs in a vacuum; but whether the vacuum is such that not even a sixtieth part of the air is left in the receptacle, and whether the height of the pipe exceeds by more than half a foot the surface of the water to be drawn, I do not know. Thus, therefore, I wish that those things which I mentioned about the subsequent dissolution of the water should not be considered other than hypothetically spoken. It will suffice that I have determined accurately by what force the water is urged to mutual separation.

§17. Further, there are other phenomena of nature, the true explanation of which depends on that *hydraulico-statics* theory: for example, that smoke ascending through a chimney draws air after itself with great impetus through an orifice made in the chimney; that the wind blowing from a rather narrow place into a more open one loses some of its elasticity, just as it is gathered from this that opened windows are closed by air attempting exit from a room on account of its greater elasticity; and others of this sort, which individual cases it is not permitted to study.

The pressures of moving fluids can be varied, indeed, in infinite ways; nevertheless, I believe that all can be reduced to our principles; we have examined two forms of that theory: I deduced the first from the known motion which the fluid will have, if at the point where the pressures are to be determined the vessel is perforated with an infinitely small orifice; the other I deduced, as they say, *a priori* from our general theory; often they both pertain at the same time, as one requires the help of the other, and then another estimation of pressures arises which I will indicate by a single example.

§18. Let us consider, in the vessel which Fig. 72 shows, that the horizontal pipe has, not only at its extremity but also at its insertion EG, a section in a vertical place perforated in the middle, the remaining positions indicated in §5 being maintained; the walls of the pipe ED will endure a different pressure from the water flowing through than if there were no section EG added, and certainly a lesser one, although [the water] flows through at a lesser velocity. In order that this pressure be accurately defined, the path to be followed is the same as cited in §5: namely, first of all the velocity is to be sought at which the water flows in the pipe ED after it has already been made uniform. Then one should inquire as well into the value of $\dfrac{v\,dv}{dx}$ if the pipe is assumed to be broken off somewhere.

But how this can be found is a matter which pertains especially to Chapter VIII, with the precautions of §14, Chapter VII, having been heeded at the same time. In Chapter VIII the motion of fluids flowing through many orifices is shown generally, and in §14, Chapter VII, it is demonstrated in particular how the *potential ascent* which is generated in volume elements is to be estimated when these [elements] flow through the orifice, not into practically standing water but into water carried by a motion which cannot be neglected.

STATICS OF MOVING FLUIDS   305

If one proceeds properly along these indicated paths, he will discover that the velocity with which the water flows uniformly through the pipe ED conforms to the height $\dfrac{mmppa}{mmnn + nnpp - mmpp}$, where by m, p, and n are indicated, respectively, the areas of the orifices made in the sections EG and FD, and as well of the pipe ED; moreover, by a one understands the height of the water above the horizontally positioned pipe ED.

Further, if one assumes that the pipe is broken off at cd and that the volume element ad is being moved at the velocity v, or that the height due to this velocity is vv, and if at the same time one indicates the length Ec by c and the very small length ac by dx, one will encounter the following equation:

$$2cv\,dv + \frac{nn}{mm} vv\,dx = a\,dx$$

or

$$\frac{v\,dv}{dx} = \frac{mma - nnvv}{2mmc}.$$

Now for vv let the value just indicated, $\dfrac{mmppa}{mmnn + nnpp - mmpp}$, be substituted, and there will be

$$\frac{v\,dv}{dx} = \frac{mmnn - mmpp}{2c(mmnn + nnpp - mmpp)}\,a,$$

to which the sought pressure is proportional. But if the area of the final orifice, indicated by p, is as if infinitely small, the pressure becomes a. Generally, therefore, the pressure sought, by virtue of §5, is equal to

$$\frac{mmnn - mmpp}{mmnn + nnpp - mmpp}\,a.$$

§19. If the area n of the pipe is as if infinite in proportion to the areas of the orifices in the sections, the pressure becomes $\dfrac{mma}{mm + pp}$: and so great also is the height to which the water flowing out at o can ascend by its own velocity; therefore, this conforms with §4, Chapter VIII, because the shape of the vessel, or [its being] of infinite area everywhere, does not cause the velocity of the water springing forth to differ.

When there is no plate at F, it happens that $p = n$, and the entire pressure vanishes. This deserves to be noted because it shows the reason why in diverging pipes the suction is not as great as it should be according to the hypothesis in which all the *live force* is assumed to be conserved. Indeed, in the present case we took into consideration a *live force* which is continually diminished. And so also the sides of the pipe experience no pressure when the section which is at *EG* has an orifice infinitely smaller than that which exists at *FD*. Finally, it is worth noting also that, although fluids being moved through conduits constructed without any cross-sectional plates generally effect a pressure which corresponds to the difference of the heights due to those velocities at which the fluid flows after an infinite time through a cut off conduit and at which it flows actually through the uninterrupted conduit, in the present case this law is nevertheless least valid, to which I wish that those would pay attention who want to show the general proposition of §10 synthetically by our observed *hydraulico-statics* theory. For perhaps there will be some to whom this matter will seem so intrinsically obvious that it hardly need be proven; but particular laws of this type which occur in *hydraulico-statics* show that those, if there are any, deceive themselves by a certain false resemblance to the truth.

§20. It will be to the point to undertake experiments also concerning these things which were mentioned in §18, for [determining] the velocity of the water flowing out at *o* as well as the pressure; for hence in addition to the laws of pressures that theory of accelerations will also be confirmed which obtains when a certain portion of the *live force* is continually used up uselessly, which problem we treated especially in Chapter VIII. Moreover, in undertaking an experiment, as much as it can be done, let those hindrances be avoided of which we have already often made mention.

§21. Let me inject here a question which certainly does not pertain to the statics of fluids, but to the hydraulics or motion of fluids, but which cannot be solved without those previously given *hydraulico-statics* rules. In Fig. 72 (here I no longer consider any plate at *EG*) one seeks, if the pipe is perforated by an orifice at *ac* having a finite ratio to the area of the pipe as well as to the area of the orifice *o*, and if the motion of the water has already been made uniform, one seeks, I say, at what velocity the water will erupt through each aperture.

At this time let the height *BE* again be *a*, the area of the pipe be *n*, the area of the orifice at *o* be *p*, the area of the orifice *ac* be *m*, and the velocity of the water flowing out through *o* be *v*. The velocity of the water which flows across the orifice *ac* will be $\frac{p}{n}v$. Therefore, at that same place it exerts a pressure on the walls of the pipe which is

$a - \frac{ppvv}{nn}$ (according to §5), and on that account I assume that the height which can generate the velocity at which the water springs forth through the orifice ac will also be approximately as great, but that this velocity itself is $\sqrt{a - \frac{ppvv}{nn}}$. With this having been established, the velocities at the orifices o and ac will be as v is to $\sqrt{a - \frac{ppvv}{nn}}$; and thus any volume element whatever entering the pipe at GE, when it arrives at the region of the first orifice, is separated into two portions, one of which flows out through ac, the other through o; and these portions are, respectively, proportional to the velocities at which the efflux occurs on either hand multiplied by the areas of the orifices. Therefore, if the mass of the entire volume element GE is called g, the portion of it flowing out through ac will be equal to

$$gm\sqrt{a - \frac{ppvv}{nn}} \bigg/ \left[pv + m\sqrt{a - \frac{ppvv}{nn}}\right],$$

and the other portion flowing out through o equals

$$gpv \bigg/ \left[pv + m\sqrt{a - \frac{ppvv}{nn}}\right].$$

If these portions are multiplied, *respectively*, by the squares of their velocities, their *live forces* will be obtained, the sum of which is to be equated to $g \cdot a$, that is, to the *actual descent* of the volume element g through the height a. Thus, if it is reduced, the following equation is obtained:

$$n^3vv - n^3a = mpv\sqrt{nna - ppvv}$$

or

$$vv = \frac{2n^6 + mmnnpp + nnmp\sqrt{4n^4 + mmpp - 4nnpp}}{2n^6 + 2mmp^4} a,$$

and this quantity expresses the height for the velocity of the water flowing out at o, by which knowledge also is obtained the similar height for the other orifice ac, which indeed is $a - \frac{ppvv}{nn}$.

§22. If $p = n$, it happens that $vv = a$; therefore, the water then springs forth at the total customary velocity through the orifice o, and nothing flows out through the other orifice ac. Further, in either orifice the velocity corresponds to the entire height a, if p is as if infinitely small. But if m is infinitely small, it certainly occurs that $vv = a$, but the height of the velocity pertaining to the small orifice ac is $a - \frac{pp}{nn}a$, as was already indicated in §7. If $m = p$, it occurs that

$$vv = \frac{n^4 a}{n^4 - nnpp + p^4}; \quad \text{and} \quad a - \frac{ppvv}{nn} = \frac{(nn - pp)^2 a}{n^4 - nnpp + p^4}.$$

Finally, it can be observed that the water is always ejected through the orifice o at a greater velocity than that which corresponds to the height a, which certainly occurs because the water at Ed makes somewhat of an impetus against the water at dF.

Meanwhile, although all these Corollaries agree splendidly with the nature of the argument, nevertheless, the solution of that problem cannot be considered other than approximately true.

HYDRAULICO-STATIC EXPERIMENTS FOR CHAPTER XII

Pertaining to §§3 and 4

The pressures which have been shown in the aforementioned paragraphs can be confirmed by a simple experiment, if the vessel which Fig. 43 shows and which is described in §30, Chapter VIII, is carefully prepared, and if in its cross section LQ a glass tube is inserted vertically, either end of which is open; thus it will be observed, with the orifices H and N blocked off and the entire system filled with water, that the water in the glass tube ascends to the level AB, or it exceeds it according to the nature of capillary tubes. Then also, if the finger is removed from the orifice N, it will be observed that the water in the glass tube descends, and after the measurements have been taken, it will be found, unless I am mistaken, that the residual height of the water in the glass tube (after the height due to the effect of capillary tubes has been subtracted) is $\frac{\alpha\alpha(LB) - \gamma\gamma(NQ)}{\alpha\alpha + \gamma\gamma}$, just as it was mentioned in §3, where the denominations of these letters are explained.

Further, if from each orifice H and N a finger is removed, then the residual height of the water in the glass tube will be just that which is indicated in §4. Similarly, a glass tube can be inserted in the section QN, and this then bent [upward], so that it can be learned whether the pressures at the section QN have also been defined correctly.

But the experiments which pertain to the pressures of water carried through pipes I myself undertook in the presence of our Society, and they are described in Vol. IV of the [St. Petersburg] *Commentaries*, p. 194. Therefore, I will present those things here as they are described there.

I used a wooden box, the width of which was one foot, the length three feet, the height 14 inches. I filled this with water, and I implanted horizontally in its final portion a cylindrical tube accurately made from iron. But that iron pipe was made as follows: namely, it had a length *AB* (Fig. 77) of 4 English inches, 2 lines, a

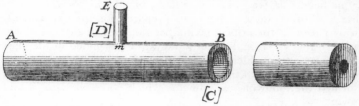

FIGURE 77   FIGURE 78

diameter *BC* of 7 lines; the pipe was perforated in the middle by a small orifice *m*, and at the same place the tube *DE*, likewise of iron, having six lines in length and one and one-half lines in diameter, was welded so that the small orifice *m* would lie in the middle of the base. A little later I attached to this small tube a glass tube of equal area, as it appears in Fig. 79, which shows the method of the whole experiment. Further, I took care that three covers be made, [each] attached to the iron pipe and perforated by an orifice of different size; such a cover is represented in Fig. 78.

With all these things brought together in that way, which Fig. 79 shows, and having insured that the water did not flow through openings other than the aperture at *BC*, I blocked off the orifice at *BC*, and then I observed, in the vertically placed glass tube, the

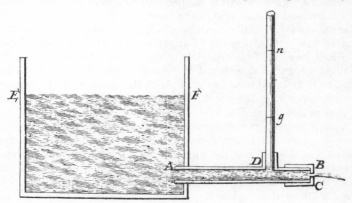

FIGURE 79

point *n* to which the water ascended, and this I marked by wrapping a silk thread around it; but first I had determined the capillary effect of that glass tube, and this I had found to be five lines, so that, with the tube vertically immersed in the water, the difference between both surfaces of water was five lines; accordingly, the point *n* was elevated above the surface *EF* by the same number of lines, and hence in the calculation any height *Dn*, *Dg*, is to be considered diminished by five lines.

In the individual experiments the box was kept full of water so that the height *AF* was 9 inches, 7 lines, but the height *Dn* was 10 inches. With all these things thus prepared for the experiment, then by reason of the orifice having been opened at *BC*, efflux was granted to the water, and directly the water descended in the glass tube, as from *n* to *g*, which point *g* I marked again with another silk thread wrapped around the tube beforehand. And thus at last we performed the following experiments which correspond to §5 and following.

EXPERIMENT 1. When the diameter of the orifice in the cover *BC* was $2\frac{1}{5}$ lines, the descent *ng* was a little greater than one line, so that no difference could be observed between the theory and the result of the experiment.

EXPERIMENT 2. With another cover applied, in which the diameter of the orifice was $3\frac{2}{5}$ lines, or a little greater, the observed descent *ng* was six lines and two-thirds, again clearly as the theory indicates.

EXPERIMENT 3. With the third cover applied, in which the diameter of the orifice was 5 lines, or somewhat less, we observed a descent *ng* of 28 lines. According to the theory it should have been about 29 lines, and, indeed, the orifice was seen to have not quite five lines in diameter. The very small difference is to be attributed to the hindrances which the water experiences in flow through the tube which are greater than in the preceding experiments on account of the increased motion within the tube.

EXPERIMENT 4. Finally, with no cover attached, we allowed the water to flow out through the full orifice, and then almost all the water had gone out from the glass tube; nevertheless, some portion remained which we discovered to be eight lines high. But five of them are to be attributed to the effect of the capillary tube, the remaining three are due to the hindrances which the water encounters in flow from *D* to *B*.

Thus, therefore, the experiments agree with theory correctly. Moreover, hence it is not difficult to see in advance that it can happen that the walls of the pipe not only are not pressed toward the exterior, but also that they are compressed inward toward the axis of the pipe (see §11). Moreover, I was shown this by the following additional experiment.

EXPERIMENT 5. In place of the cylindrical pipe *AB* I applied a conical one; the external orifice of which was greater than the internal orifice, and at the same time I made use of a curved glass tube such as Fig. 80 shows. And, while before flow the water stood at *n* in the glass tube, the water descended in the same tube right to *g* when water flowed through the conic pipe; and the point *g* was

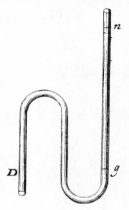

FIGURE 80

below *D*, [serving as] proof that the conical pipe was under pressure during flow. But in these cases there are significant hindrances to the motion which make the velocities of the water at the external orifice much less than those which correspond to the height of the water; and for this reason the height of the point *D* above *g* was not as great as it would have been otherwise, although there was some [height]. I obtained the same but altogether more notable effect in another way (see §12). This other experiment I performed in the following year in the presence of the Academicians, the Most Serene Prince Emanuel of Portugal being present.

EXPERIMENT 6. In Fig. 81 *ACFB* represents a cylinder, in the base of which was implanted the conic pipe *DGHE*; and the latter had a small tube at the side at *l* which was joined by the extremity of the curved glass tube *lmn*; the height *CA* was 3 inches 10 lines; *El*, 4 lines; *lH*, 2 inches $9\frac{1}{2}$ lines; the area of the conic pipe at *l* was to the area of the orifice *GH* as 10 is to 16; *ln* was 5 inches 6 lines, and its orifice *n* was submerged in water in the small vessel *M*.

With a finger placed over the orifice *GH* and the vessel filled, the water trickled through the glass tube *lmn* into the vessel *M*; but with the finger removed and the water now flowing out through *GH*, the water ascended of its own in a reciprocal motion from the small vessel *M* through the tube *nml*, and together with the remainder flowed out through *GH*, during which time the entire small vessel *M* would have been emptied. But water was supplied continuously from above, so that the vessel was kept full. If a portion of the orifice *GH* was blocked off by a finger, it was easy to cause the water in the glass tube *lmn* to move up or down at will.

If anyone should wish also to find out by experiment whether the theory agrees with the problem of §18, he will not have organized his work badly, since indeed he will thus illustrate not only this new *hydraulico-statics* of ours, but also the theory of Chapter VIII, equally new and treated by no one, by a splendid example, and this very easily.

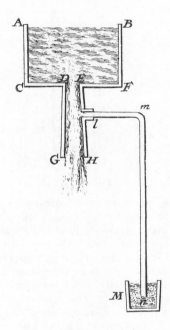

FIGURE 81

After these things now have been collected in writing, I myself undertook the experiments of which I just made mention. I used the same mechanism for this that I just described and which is represented in Fig. 79; but in addition, as the nature of the matter requires, I placed another cover on the pipe at A; and the height of water AF was 8 London inches; the diameter of the iron pipe AC was again 7 lines. Also I used the same covers as before. But in every experiment I observed the descent which the surface n made when the finger was removed from the cover BC; moreover, at the same time, after the measurement of the vertical height of the orifice C above the floor had been taken, I observed the distance of that vertical line from the place at which the aqueous stream struck. This distance I shall call the *amplitude of the thrust*; but this vertical height was 19 inches in the individual experiments. With these things thus prepared I performed the following experiments.

EXPERIMENT 7. When the diameter of the orifice of the interior cover was $2\frac{1}{5}$ lines and the diameter of the orifice of the exterior cover $3\frac{2}{5}$ lines, the descent *ng* was a little less than 7 inches, and the *amplitude of the thrust* was 9 inches. However, in the theory shown in §18, the descent *ng* is indicated as 6 inches 10 lines and the *amplitude of the thrust* as $9\frac{1}{2}$ inches.

EXPERIMENT 8. Next the diameter of the internal orifice was 5 lines and the diameter of the other orifice $3\frac{2}{5}$ lines; the descent *ng* was almost 17 lines and the *amplitude of the thrust* 24 inches. In theory *ng* is $17\frac{3}{4}$ lines and the *amplitude of the thrust* 23 inches.

EXPERIMENT 9. Further, when the diameter of the internal orifice was $3\frac{2}{5}$ lines and the diameter of the exterior orifice 5 lines, the descent *ng* was almost the same as in Experiment 7, namely, about 7 inches. But the *amplitude of the thrust* was greater, that is, 11 inches. In theory *ng* is 6 inches 11 lines and the *amplitude of the thrust* almost 11 inches.

EXPERIMENT 10. Finally, with the diameter of the interior orifice being $3\frac{2}{5}$ lines and the diameter of the exterior orifice $2\frac{1}{5}$ lines, the descent *ng* was about one inch and the *amplitude of the thrust* 23 inches. In theory *ng* is 14 lines and the *amplitude of the thrust* is $22\frac{1}{2}$ inches.

Actually all these experiments agree splendidly with the theory; perhaps a greater agreement would have resulted if it would have been possible to obtain the measurements of the orifices with greater accuracy; nevertheless, no one, as I believe, is displeased by those minimal differences in numbers. Moreover, they arise for the most part from the compression of the water at *AC* which is produced while the volume elements entering the conduit through the interior orifice lose part of their motion; hence the amplitude of the thrust is slightly greater and the descent *ng* is less in theory than in the experiments; I did not wish to add the measure of this matter, although it would have been within my power, lest the calculation become more intricate.

THIRTEENTH CHAPTER

Concerning the Reaction of Fluids flowing out of Vessels and the Impetus of the Same, after They have Flowed out, on the Planes against which They Strike

§1. Water, while it is being ejected from a vessel, acts in the same way against the vessel out of which it is flowing as a shot against the cannon or rifle from which it is expelled: it certainly repels the vessel. And this, indeed, Newton already noted in *Principia Mathematica Philosophiae Naturalis*, first edition, p. 332, and from this he correctly deduces the ascent of mortar shells which are filled with gunpowder properly mixed with charcoal. For after the material has been ignited, it projects the mortar shells upward while it expires slowly through the orifice.

But neither did the cited author (since it was not in accordance with his purpose) handle the argument generally enough for the importance of the matter, nor did he give the true measurement of it. Finally, in the two later editions he ignored it altogether. However, he considered that *that force of repulsion is equal to the weight of an aqueous cylinder the base of which is the orifice transmitting the water and the height of which is equal to the height of the aqueous surface above the orifice*. Indeed, this quantity is deduced correctly from the opinion which Newton favored at that time about the velocity of the water flowing out of a vessel, when he stated that the water can ascend to one half the height of the surface by its own velocity.

But just as now the falsity of the latter proposition is unknown to no one any longer, so also the defect of the other anyone hence easily gathers, although at first glance it seems true enough.

§2. At first we will consider the matter in the very simple case in which, certainly, we assume the water to flow horizontally out of a vessel of infinite area. Moreover, I have demonstrated that the total force of repulsion is not present immediately at the beginning of flow unless insofar as the total velocity itself is present in the water flowing out, such that if the vessel is not of infinite area, the force of repulsion together with the velocity of the water flowing out increases little by little, or even decreases, according to the nature of the circumstances. However, at first let us disregard these instantaneous changes by assuming that the flow from an infinite vessel becomes constant. And thus the force of repulsion is best defined if whatever force is required for producing the motion is sought. Indeed, to this end one has to look not only for the velocity of the water flowing out, but also for the quantity of it; but the quantity depends partly on the magnitude of the orifice and partly on the contraction of the stream, which latter is variable; indeed, we saw in Chapter IV that it can be entirely avoided. If some contraction exists, nevertheless, the section of the most greatly contracted or attenuated stream is to be considered instead of the orifice, and then I say that *the force of repulsion will be equal to the weight of an aqueous cylinder the base of which is the orifice transmitting the water* (that is, the section of the most greatly contracted horizontal stream) *and the height of which is equal to double the height of the aqueous surface above the orifice, or, more accurately, to double the height appropriate to the velocity of the water flowing out.* Therefore, if there is no contraction of the stream, just as there is none when water flows out through a short pipe, the repulsion will be twice or almost twice as great as that which was defined by Newton.

§3. In order that we may show this proposition, a certain Mechanical principle will have to be considered here, for which I have often found use in solving other questions. The principle is this:

If a body has acquired the same velocity from rest through direct motive pressures, variable in any way whatever, and if the individual pressures are multiplied by their proper differential times, the sum of all the products will always be the same; that is, if the pressure is p and the differential time is dt, then $\int p\, dt$ will be constant. I showed this matter more clearly in the *Commentaries of the Imperial Academy of Science of St. Petersburg*, Book I, p. 132.

§4. Let us assume now a cylinder of practically infinite area from which water flows out horizontally at uniform velocity; let us disregard the influence which gravity exerts on the particles after they have flowed out, so that the individual ones continue to be moved horizontally and uniformly. But the particles are accelerated and they experience pressure as long as the maximum value of the velocity is not yet present, and they obtain this value when they have arrived at the place of the greatest contraction of the stream. For this reason I said that the section of the stream formed at that place is to be considered instead of the orifice of efflux. Let the area of that section be 1, and let the water there have a velocity which is due to the height A. Let it be assumed that the cylinder of water has flowed out which has 1 for its base and L for its length. If the time is expressed by a length divided by a velocity, the velocity appropriate to the height will have to be expressed by $\sqrt{2A}$, and the time of flow by $\dfrac{L}{\sqrt{2A}}$. With these things set forth in advance, we will investigate the motive pressure which can impart the velocity $\sqrt{2A}$ to the cylinder L in the time $\dfrac{L}{\sqrt{2A}}$. Let the pressure be p, and let it be considered, for the sake of a shorter calculation, to have acted during the time t and to have given the velocity v to the cylinder. Then $dv = \dfrac{p\,dt}{L}$ and $v = \dfrac{pt}{L}$, whence $p = \dfrac{Lv}{t}$. Now let $\sqrt{2A}$ be substituted for v and $\dfrac{L}{\sqrt{2A}}$ for t, and thus $p = (L\sqrt{2A}) / \left(\dfrac{L}{\sqrt{2A}}\right) = 2A$. Therefore, the pressure constantly exciting the water to efflux is equal to the weight of the aqueous cylinder of which the base is the above-defined orifice transmitting water, the height of which is equal to double the height appropriate to the velocity of the water flowing out; and also just as great is the reaction which repels the vessel. Q.E.D.

§5. The proof is the same if the water flows out not through an orifice but through a horizontal cylindrical pipe at a constant velocity, or even through a pipe of size varying in any way. This latter can also be proven directly if the pressure required in the individual particles is expressed correctly so that these [particles] receive the required increments or decrements of velocities.

§6. The height which we called A differs very little indeed in experiments from the height of the water above the orifice of efflux, especially if the water flows out from a very large vessel through a simple orifice which is not very small. But the orifice of efflux more often differs notably from the minimum section of the stream, which we consider as the orifice transmitting the water; the quantity of water flowing out in a given time, compared with its velocity, indicates this in experiments.

Hence it occurs that our proposition of §3, after it has been challenged by experiment, ordinarily does not differ much from the proposition of Newton shown in §1. But if everything is carefully avoided which can produce a contraction of the stream and which can diminish the velocity, the repelling force according to our theory will become almost double that which was defined by Newton, and accordingly, such a value is also confirmed by experiments.

But in order that we may bring the matter clearly to light and treat it now rather generally, we will handle it so that we determine the repelling force from the beginning of flow, while the velocities are being changed continuously; and, indeed, our first theory does not have meaning other than when the velocity remains unchanged. In order that we may be more intelligent in handling this slightly more intricate question, it will help here to set forth certain rather general things in advance.

§7. *Momentum* is the product of the velocity and the mass. If the velocities are unequal, the *absolute momentum* will result if the individual particles are multiplied by their own velocities, respectively, and the sum of the products is taken. The *momentum* is generated by the motive pressures acting for a given time, and the effect is to be considered equal to the cause. Therefore, the sum of the motive pressures multiplied by their proper differential times is to be evaluated from the momentum generated. And because any motive pressure reacts on the vessel from which the water flows out, the total repelling force for any instant whatever will be equal to the new momentum divided by the differential time in which it is generated. With these things having been set forth, I proceed to the question itself.

§8. Therefore, let the vessel *ACDB* (Fig. 82) be of infinite size, and let the tube *EHID*, the areas of which are assumed unequal in some way, be fastened horizontally to it. Let the area of the orifice *HI* be 1, and the length of the tube be *m*. The velocity at *HI*, variable

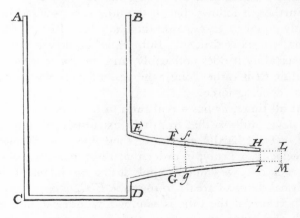

FIGURE 82

in any way, equals $\sqrt{2v}$, or that which is due to the height v. I say at first that the absolute momentum of the water contained in the tube will be equal to $m\sqrt{2v}$, that is, as if the tube were cylindrical and equal in area to the orifice *HI*, because, of course, the velocity of any section *FGgf* is reciprocally proportional to the mass [*sic.*]

Now, indeed, let us consider that in a given infinitely small differential time there flows out through the orifice *HI* the small column *HLMI*, the length *HL* or *IM* of which we consider equal to α. The mass of this column will be α, and it will have the momentum $\alpha\sqrt{2v}$, but in the same time the mass of water contained in the tube acquired the momentum $\dfrac{m\,dv}{\sqrt{2v}}$ (for it had $m\sqrt{2v}$). Therefore, the absolute momentum generated in the given differential time is $\alpha\sqrt{2v} + \dfrac{m\,dv}{\sqrt{2v}}$; but if this is divided by the same differential time (which is to be expressed by $\dfrac{\alpha}{\sqrt{2v}}$), as we saw in §7 the required pressure repelling the vessel will result, which, therefore, if it is called p, will be

$$p = \left(\alpha\sqrt{2v} + \frac{m\,dv}{\sqrt{2v}}\right)\bigg/\frac{\alpha}{\sqrt{2v}}, \quad \text{or} \quad p = 2v + \frac{m\,dv}{\alpha}.$$

(α) It appears from this that the last definition of the question depends on the ratio which exists between dv and α; this, in fact, we defined generally in Chapter III: however, no attention was paid to the hindrances which are due to this case. Therefore, the shape of the tube also contributes something here.

(β) Further, it follows that, if the flow is considered uniform, p is constantly equal to $2v$, because then $dv = 0$. In fact, this conforms with what we showed in §5. But, while the flow is being increased (which certainly it does noticeably, and this for long enough time if the conduit *EI* is rather long), the vessel experiences a continuously different repelling force.

(γ) At all times, dv has a real ratio to α. Therefore, the repelling force is never null, so that from the first instant of flow the vessel is repelled, even if hardly any water then flows out on account of its trifling velocity. Truly, in order that the general use of our rule be clear to everyone, we will now apply it to a special case by attributing a cylindrical shape of area 1 to the tube *EHID*.

§9. Therefore, if the tube is assumed cylindrical, entirely open at *HI*, with the other assumptions and designations having been retained, the *live force* of the water contained in the tube will be mv; the increment of this is $m\,dv$, to which is to be added the *live force* of the little column $HLMI$, or αv, and their sum is to be made equal to the product of the height of the surface AB of the water above the orifice HI, which we will call a, and the differential mass α. Therefore, $m\,dv + \alpha v = \alpha a$, and from this there results $\dfrac{dv}{\alpha} = \dfrac{a-v}{m}$. But with that value substituted in the equation of the above paragraph, there results $p = a + v$ from which I deduce the following conclusions:

(α) The length of the tube contributes nothing to the repelling force which the vessel sustains if the velocity is assumed to be the same, because the letter m vanished from the calculation. However, this length (just as we showed in the above more than well enough) causes the velocities to assume faster or slower increments, for the longer the tube the more slowly the water will be accelerated, and vice versa, so that it acquires in an instant from rest its maximum rate of speed if the length of the tube is null. But if this same tube is of infinite length, the water can acquire a noticeable degree of speed only after an infinite time.

(β) Therefore, it can occur, when the height of the water has not been changed, the expenditure of water being howsoever small, that the repelling force is notable and lasts arbitrarily long. And, indeed, this can be obtained in a double manner, either by prolonging the tube or by closing off the orifice rather often before the water has attained a notable velocity. However, the former method assumes a free flow of the water through the tube; indeed, when the flow of water has been retarded by external hindrances, never to be avoided in overly long tubes, the repelling force is also diminished.

(γ) Let me be allowed here to mention in a few words a certain proposition from *Principia Mathematica Philosophiae Naturalis*, 2nd edition, of Newton. After he had changed his thinking shown in the first edition of the cited work about the velocity of water flowing out of a vessel, and after he had recognized in the second edition that, if it is ejected vertically upward, it ascends to the full height of the surface of the water, the Author presented the following words in the second book, proposition 36, corollary 2: *The force by which the entire motion of the water flowing out can be generated is equal to the weight of a small cylindrical column of water, the base of which is the orifice EF (see Newton's figure), and the height of which is 2GI or 2CK.* That thinking was once opposed by me and by some, and again confirmed by others. But now, after I have thought about this theory of moving water, it seems to me that the dispute is to be settled thus: when the water has arrived at a uniform motion, which, certainly, is Newton's hypothesis, then that force is defined correctly by the height $2GI$, but at the beginning of flow, when the velocity is still null, the force corresponds to the simple height GI, and soon, with the velocity increasing, the force animating the water to efflux increases simultaneously, and finally it rises to that magnitude which Newton assigned. Now these things are obvious to anyone, because the force generating the motion of the water about which Newton speaks cannot but be equal to the repelling force, which we saw to be equal to $a + v$. Also the Illustrious Ricatti, with whom I had a discussion concerning this argument, when asked *whence that force corresponding to twice the height of the water could arise, whereas it is apparently manifest that, with the orifice blocked off, the volume element adjacent to the latter is pressed by the force corresponding to the simple height*, answered that *one must distinguish the state of rest from the state of motion*.

§10. If the tube attached to the vessel is not cylindrical, the calculation will have to be performed thus:

Let the area of the conduit at FG or fg be y, the distance of the section $FGgf$ from the orifice ED be x, and let the other designations be retained. The *live force* of the water contained in the tube will be $v \int \frac{dx}{y}$, and its increment will be $dv \int \frac{dx}{y}$, to which, as it was done in the preceding paragraph, is added the *live force* of the small column $HLMI$, or αv, whereupon $dv \int \frac{dx}{y} + \alpha v = \alpha a$, from which it thus appears that

$$\frac{dv}{\alpha} = (a - v) \bigg/ \int \frac{dx}{y}.$$

With this value having been substituted in the equation of §8, there results

$$p = 2v + m(a - v) \bigg/ \int \frac{dx}{y}.$$

Therefore, since in the uniform flow of water $v = a$, it follows that $p = 2a$. In addition, as long as the flow of water is accelerated, the motion of the water in the vessel $ACDB$ near the orifice DE, which we have disregarded in all this work, is not to be neglected here. But that motion cannot be determined correctly, and, therefore, the expression which I gave for the repelling force does not apply accurately if the water has not yet been understood to flow uniformly; but when the water flows steadily, the expression prevails very accurately.

§11. After we have thus proven for the uniform efflux of water that the repelling force is always equal to the weight of an aqueous cylinder

REACTION OF FLUIDS FLOWING OUT OF VESSELS 321

constructed above the orifice and rising up to double the height of the water, it is pleasing to show it also indirectly by *reduction to the absurd*, so that also those not knowing the rules of mechanics may perceive the truth of this somewhat paradoxical proposition.

To this end we will consider water flowing vertically down from a cylinder, disregarding the hindrances taking something from the velocity of the water and [disregarding] the contraction of the stream, which can be avoided. The vertical pipe which is seen in Fig. 76 corresponds to the orifice, and all the things behave as stated in Chapter XII, §13: the water has constant flow; the walls of the vessel and conduit are understood to be free from gravity; the height of the cylinder is assumed equal to a, and the height of the small tube equal to b; the height $cF = x$; the area at E equals 1. The area at F will be $\frac{\sqrt{a+b}}{\sqrt{a+x}}$, and at C it will be $\frac{\sqrt{a+b}}{\sqrt{a}}$. Finally, the area of the cylinder is set equal to M. After these things have been assumed, we will seek the weight of all the water $ABCE$. We will express the weight of the water ABC by Ma, and thus the weight of the water CE will be $2a + 2b - 2\sqrt{aa + ab}$; therefore, the weight of all the water $ABCE$ will be $Ma + 2a + 2b - 2\sqrt{aa + ab}$. Therefore, under the assumption that the water is at rest in the vessel and the tube, the force required for suspending the water is $Ma + 2a + 2b - 2\sqrt{aa + ab}$.

But now we will investigate a similar force when the water flows out through E at its full velocity (by which certainly it can ascend to the height $a + b$). But this will be obtained if the repelling force is subtracted from the former force. If, therefore, this repelling force is assumed, as we stated, equal to $2a + 2b$, the force suspending the water during flow will be $Ma - 2\sqrt{aa + ab}$.

But, indeed, assume that the pipe CE is not present, and through our same rules the suspending force while the water is discharging through the orifice C will again be $Ma - 2\sqrt{aa + ab}$, indeed because the weight of the water ABC is Ma and because the area of the orifice C is $\frac{\sqrt{a+b}}{\sqrt{a}}$, which, multiplied by double the height a, gives $2\sqrt{aa + ab}$. Therefore, our estimate of the repelling forces shows that the suspending force during the efflux of the water is the same whether or not there is a small tube present, and the tube may have any length whatever as long as it has the shape described in §13, Chapter XII; and the necessity of this agreement and identity appears also without calculation from the very nature of the matter, because the tube thus formed makes no change in the water flowing through, since the stream of water assumes of its own accord the same shape that the tube has, as long as the water coheres. But if we estimate the repelling force differently, we will generally never obtain that agreement between suspending forces. Thus, for example, if according to common sense we say that the repelling force is equal to the weight of the often mentioned simple cylinder, then, while the water is assumed to flow out of the vessel ABC through the conduit CE, it will be $a + b$; and if this force is subtracted from the weight of the entire water $ABCE$, or $Ma + 2a + 2b - 2\sqrt{aa + ab}$, there remains $Ma + a + b - 2\sqrt{aa + ab}$, which is the force required for suspending the system $ABCE$ while the water flows. Moreover, we saw that this force must be the same if the conduit CE is absent: but then the suspending force is $Ma - \sqrt{aa + ab}$, because the weight of the water ABC is Ma and the repelling force by hypothesis is the simple cylinder erected above the orifice C to the height a. Therefore, according to this hypothesis it should always occur that

$$Ma + a + b - 2\sqrt{aa + ab} = Ma - \sqrt{aa + ab},$$

or

$$a + b = \sqrt{aa + ab},$$

which is absurd. A similar absurdity can be shown if the stream is considered to ascend vertically upward, and here in vain it would be stipulated for confirming common sense that the stream flowing out of CE cannot be assumed continuous unless some viscosity of the water is assumed at the same time (for otherwise the stream will be broken off in little drops directly in front of the orifice) and that the viscosity changes the state of the situation; for, certainly, neither are the velocities of the water changed by the mutual cohesion of the water at CE nor do the sides of the conduit CE sense any pressure, just as I demonstrated in Chapter XII, §13, so that I may pass in silence over the fact that the cohesion of the water does not arise from viscosity but from some other magnetic property or from mutual attraction, by virtue of which the center of gravity of no system can acquire either a greater or a lesser velocity. But, clearly, this exception for vertically ascending streams [taken] by opponents has no significance when the water remains there continuously, even if the water has no viscosity or mutual attraction.

But I could confirm our thinking in infinite other ways and by particular examples if I wished to pause here for a longer time. Thus, for example, in Fig. 29, described in Chapter V, §4, if the height

$NS = 1$, the orifice $LM = 1$, and the orifice $RS = 2$, then $PB = \frac{1}{3}$; the repelling force which develops from the flow of water through RS is equal to $2 \cdot \frac{2}{3} = \frac{4}{3}$, and I can show that the repelling force which results from the efflux of water from the simple cylinder RN through LM is also $\frac{4}{3}$, and thus that the total repelling force is $\frac{8}{3}$, which constitutes precisely double the aqueous cylinder above the orifice LM standing to the height $NS + PB$. Moreover, such an agreement in no way appears in other theories erroneously conceived, so that there can be no further doubt concerning ours, except by those utterly unskilled in these matters. But if I wished to prove that what I said —that the repelling force of the water flowing out of the simple cylinder RN through LM is $\frac{4}{3}$—it is required that the repelling force be defined when water flows from a non-infinite vessel at some given nonvaried velocity. But lest I become too involved in this matter, I leave this to be accomplished by others, and it should now no longer be a great task. I proceed to other things.

§12. The proofs which we gave up to now are valid only for straight tubes in which certainly the motive force of any of the volume elements and the repelling force arising from it are in accordance with all the others, and they have a common direction. But when the tubes attached to the vessel through which the water flows are curved, another method of proof is to be employed. In order that we omit nothing further in this new argument, we will also show this case. And it will not happen that one regrets the work, since from this will appear the true laws of pressures which nature follows, not only in these cases but also in many others.

§13. And so let us consider a tube attached to an infinite vessel, certainly of uniform area, but curved according to any curvature AS whatever (Fig. 83), so that A is the point of insertion and S the point of efflux. Let the tangents at A and S be drawn, namely AR and SB, and let AB be perpendicular to SB. The velocity of the water flowing through the tube will be uniform and that which is due to the height A. The area of the tube everywhere is 1. *I say that the total repelling force taken in the direction SB will again be $2A$*, and this alone will be present.

For the sake of the proof, the infinitely close lines nq and ep are drawn perpendicular to SB, and nm is drawn parallel to the same SB; let $Sq = x$, $qp = dx$, $qn = y$, and $em = dy$. The radius of the curve at en will be $\dfrac{-ds\,dy}{d\,dx}$, the elements en, which I will call ds, having been considered as constants. Moreover, the little column of water intercepted between e and n has a centrifugal force to be determined thus:

the gravity of the column is ds (because its base is 1 and its height is ds) and if the radius of the curve were $2A$, there would result, by the theorem of Huygens, the centrifugal force of the particle equal to its

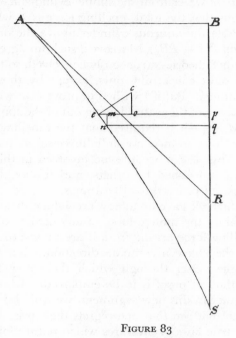

FIGURE 83

gravity, and the centrifugal forces are, other things equal, in reciprocal proportion to the radii. Therefore, the centrifugal force of the little column is $\dfrac{-2A\,ddx}{dy}$; this centrifugal force is expressed by ec perpendicular to the curve, and co is drawn parallel to BS itself. The force ec is resolved into oc and eo. There will be (on account of the similarity of the triangles eoc and nme) the force

$$oc = \frac{-2A\,ddx}{ds},$$

the force

$$eo = \frac{-2A\,dx\,ddx}{dy\,ds} = \frac{2A\,ddy}{ds}$$

(on account of ds being constant).

But the elementary force oc alone acts in the direction SB, while the other eo is to be neglected with respect to this direction. Let the integral of the elementary force oc be taken with such a constant that the integral vanishes together with the abscissa. This integral is $2A - \frac{2A\,dx}{ds}$, because at S it occurs that $dx = ds$. Now, in order that the force be obtained in the direction of the tangent SB for the entire tube, $\frac{RB}{RA}$ is to be substituted for $\frac{dx}{ds}$; therefore, the entire force following the tangent SB is $2A - \frac{2A(RB)}{RA}$. This certainly develops from the centrifugal force of any volume element whatever, but another force remains to be considered: namely, when the water flows continuously from the infinitely large vessel into the tube at a uniform velocity corresponding to the height A, the vessel is repelled along the direction RA by a force $2A$ (by §4); if this is resolved into a tangential along SB and a perpendicular along BA, the prior quantity $\frac{2A\,(RB)}{RA}$ alone will have to be considered. And because it has a direction common with the force $2A - \frac{2A(RB)}{RA}$ developing from the centrifugal force and just defined, it will have to be added to the same, and thus the sum $2A - \frac{2A(RB)}{RA} + \frac{2A(RB)}{RA}$, or $2A$, expresses the repelling force in the direction of SB.

In order to show further that the vessel is repelled in no other direction, we will return to the elementary force eo, which we saw to be $\frac{2A\,ddy}{ds}$, the integral of which is $\frac{2A(AB)}{AR}$, which is cancelled precisely by the force $2A$ repelling the vessel in the direction RA after the latter has been resolved in the proper manner. Q.E.D.

§14. This simplicity of the most general theorem, by which certainly the repelling force in the direction contrary to the uniformly flowing water is indicated constantly by $2A$, can be the *argumentum ad hominem*, as it is called, for its excellence, against those who either do not understand our reasoning or who do not desire to examine it with sufficient attention. If, in truth, one states that the repelling force of the water flowing into the tube in the direction AR from the infinite vessel is A, one sees that the system is repelled in the direction SB by a force which is $2A - \frac{A(RB)}{RA}$, which is absurd, as even the formula itself seems to indicate to me. And in this opinion the force in the direction perpendicular to the former would not be zero. For the vessel should be pressed back in the direction BA by the force

$\dfrac{A(AB)}{AR}$, which again to me is absurd, and the falsity of this I learned from experiment in the case in which ARS was a right angle and $AB = AR$.

Many other theorems in favor of this argument, taken in the full extension which it can have, could be elicited and demonstrated for the flow of water not yet uniform through a tube irregular in any way whatever, if only at the same time attention is paid to what was pointed out in §8. But because there is not space [enough] to go through the individual ones, I progress to examining another force equal to the prior but in the opposite direction, that indeed which a stream flowing out exerts on a plane when it impinges perpendicularly on it.

§15. Concerning the impetus of an aqueous stream impinging on a plane, many have written and performed very many experiments. I also contributed something to this matter in the *Commentaries of the Imperial Academy of Sciences of St. Petersburg*, Book II. Experiments are conspicuous in the works of Mariotte in his *traité du mouvement des eaux*, in the *History of the Academy of Science* [Paris] contributed by Mr. Duhamel, p. 48, and elsewhere. Indeed, they do not all agree very closely; nevertheless, most of them seem to indicate at first glance that the pressure of the aqueous stream flowing uniformly is equal to the weight of an aqueous cylinder, the base of which is the orifice through which the water flows, and the height of which is equal to the height of the water above the orifice. To this thinking the majority, in fact all, adhered and do adhere up to this time, because it agrees wonderfully with other experiments also, especially those which are customarily performed on spheres moved in a resisting medium. Therefore, I myself followed the same [thinking] in the cited *St. Petersburg Commentaries*, although many made my mind uncertain, and I indeed did not hesitate in this work itself, which I have at hand, to make use of that as an example in Chapter IX, §31 and §32. However, after I had thought over the matter more attentively, had applied new principles, and at the same time had undertaken other experiments of a new type, at last I saw clearly that that common opinion about the impetus of a stream of water had to be changed in the same manner as that of Newton about the repelling force, so that in place of the orifice the section of the contracted stream should be considered, and in place of the height of the water twice the height corresponding to the actual velocity of the water should be applied. For I have proof that the force of repulsion shown in §2 is entirely equal to the impetus of the stream if all of it strikes the plane perpendicularly; it follows, hence, that the impetus of the stream is greater, the smaller the concentration of the stream and, with the latter simply vanishing and the water erupting at the same time at the full velocity which it can have in theory, then the impetus is twice as great as is commonly stated. Indeed, because the velocity always lacks something and the stream is seldom not contracted to almost one half, it is a fact that most experiments have seemed to support the simple height in the Cylinder in estimating that impetus. Moreover, I would wish that it be noted properly that I discuss here only solitary streams which the planes receive entirely, but not fluids surrounding bodies and making an impetus on the same, such as Winds or rivers. Indeed, I say that these two types of impetus which authors have confused up to this time are to be distinguished properly from one another on account of reasons to be explained briefly below.

§16. With respect to the aqueous stream I think as follows: I assume that water flows out horizontally at a uniform velocity from the infinitely wide vertical cylinder *ABM* (Fig. 84) through the lateral ori-

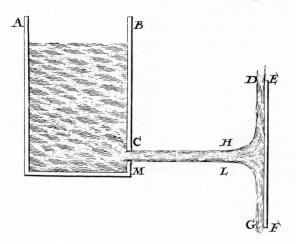

FIGURE 84

fice *CM*, and that the stream impinges perpendicularly against the plate *EF*; thus I see easily, since the following particles hinder the prior ones so that they cannot rebound, that the individual particles will be deflected to the sides, and this in a motion parallel, or almost so, to the plate *EF* (if only the latter is large enough so that the entire stream, however dispersed, is intercepted). And because all things are in a *state of permanence*, it is permissible to assume that the plate *EF* is fixed to the vessel and that the stream is surrounded by the lateral surfaces *CHDGLM*, so that the water can be assumed to flow out from the vessel *ABCHDEFGLM* through the circular opening *DEGF*. If this were so, we have shown in §13 that the volume elements flowing out at *DE* would certainly produce a repelling force in the direction of *EF*; but at the same time it appears that the repelling force at *GF* is opposite to the former, so that here no attention has to be paid to this class of repelling forces. But as far as the direction perpendicular to the layer *EF* or to the cylinder *BC* is concerned, we showed at the end of the same §13 that in this direction clearly no repulsion occurs. Therefore, the plate *EF* is propelled just as much as the cylinder is repelled. And this is what I wished to show. And hence it follows now that *the total pressure of the aqueous stream which strikes the plate is the same as the weight of the aqueous cylinder which has as base the cross section of the stream (after the latter has reached a uniform area) and as height twice that required for the velocity of the water (after this has similarly been made uniform).*

§**17.** I do not doubt that there will be many to whom this wholly new proposition seems suspect and contrary to experiments. Indeed, I would wish those to consider that the experiments performed up to now by no means correspond accurately to the common rule, and in most cases our Rule differs little from the common, although in theory they are greatly different; then also I want those to have been informed beforehand that I have undertaken other experiments which individually confirm my thinking exactly, and clearly reject the old one! At the end of the Chapter I will review the experiments performed by me. Perhaps also the method of proof which I used will seem insufficiently accurate to some, but I have another direct proof which is supported by a new Mechanical property once observed by me, and which I will communicate here, both because anyone can deduce the said proof very easily, and also because he can apply the same to other uses. And thus it is presented.

If a body is moved at a uniform velocity but changes its direction continuously by any cause whatever acting in any way whatever until it has acquired a direction perpendicular to the first, and if the individual pressures deflecting the body are resolved into two groups, the one parallel to the first direction, the other perpendicular to it, and, finally, if the individual parallel pressures are multiplied by their proper times, I say that the sum of the products will be constantly the same and indeed equal to that which can generate the entire motion from rest or absorb the entire generated motion.

In this dynamical relation, if we use it in our present problem, the plate *EF* is to be considered, which by its reaction on the water changes the direction of the latter until it has become perpendicular to the first. Therefore, the proposition of the preceding paragraph, with the help of this relation, will be proven in the same manner which we used in §4 for determining the repelling force with the help of the principle shown in §3. Therefore, that idea, which we must understand about the impetus of water, seems true; however, it assumes that the individual particles of water rebound to the sides in the direction of the plate, by which pattern I observed the water not always to recede; indeed, I even saw that some particles, though few, spring backwards; however, the latter produce a greater pressure than those which are deflected to the sides. And from this itself I am convinced firmly that if an aqueous stream with a great impetus impinges obliquely against a plane, for example at an angle of thirty degrees, a pressure will thence arise which is more than half of that which arises from the same stream impinging directly, while according to the ordinary rules it should exert exactly half the force; the reason for this matter is that in an oblique impulse more particles can rebound than in a direct one, in fact almost all [can rebound] if the velocity should be great.

However, if all particles are assumed to rebound so that the angle of incidence is equal to the angle of reflection, then each impulse will have to be considered the same. The best method of estimating the pressures of the water here is that in which the reasoning is supported *a posteriori*.

§18. It follows further from the previously discussed well-known relationship that the same effect arises from the pressures whether the plate deflects the water to the sides or a cause is assumed which absorbs all the motion which the aqueous particles having flowed out of the cylinder have acquired. Hence it is understood what would happen if the orifice *CM* (Fig. 85), through which water flows out of the cylinder *ABM*, were submerged in other water standing in the vessel *PQFE*. Certainly the cylinder *ABM* would be repelled against *PQ* within the vessel *PQFE* if the latter were not connected to the cylinder; but if the vessels were fixed to one another, the system would undergo no prevailing pressure; for as great a pressure as there is against *PQ* from the water flowing out, also as great an opposite pressure develops against *EF* from the continual destruction of the motion which the particles having flowed out of the cylinder have acquired.

§19. I spoke about the pressure of a stream which, even if expanded, is intercepted completely by a plate. I come to the other type of impetus of water which indeed plates submerged in fluid on all sides sustain; however, I consider that this cannot be defined *absolutely*, because the individual particles impinging on the plate are deflected differently. But if the deviation of any particle whatever is assumed to be known, the solution of this question will no longer be difficult after the theorem which we used in §17 has been changed a little, and

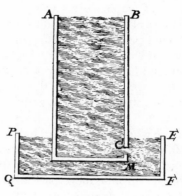

FIGURE 85

this, given more generally, is as follows: *if the angle of the change of direction in a moving body is not right, but less than right, then the sum of the products (which was the subject of the discussion before) will also be less in proportion as the sine of the changed direction is to the entire sine.*

Therefore, for any particle whatever it should be investigated how much it is forced to change its direction of motion by the obstacle or by the plate placed across its path. But this sort of definition can hardly be shown accurately in theory; experience does not prove the theorems customarily brought forth in this matter: such as that the force of a stream impinging directly against a disk is twice as great as the force of the same stream against a sphere of the same diameter, and others which are similar. However, the fact that the quantity of pressure for a sphere which is given customarily by authors agrees accurately enough with experiments made by Newton and others and recounted in *Principia Mathematica Philosophiae Naturalis* is, I consider, after thinking over everything well, to be attributed to a fortuitous case.

In the *Commentaries of the Imperial Academy of Science of St. Petersburg*, Book II and following, I gave theorems which were developed for motion in resisting media, considered theoretically, and also several physical observations. Therefore, I do not wish to repeat those here, although they pertain to our purpose; there is not space to delay any longer on these hydrodynamical meditations. Hence, I hasten to the end. I pursued this new theory about the reaction and the impetus of fluids, which upsets the opinion accepted by all authors up to now in a

REACTION OF FLUIDS FLOWING OUT OF VESSELS 331

matter of great importance, in a singular Dissertation which should be inserted in its own time in the *Commentaries of the Imperial Academy of Science of St. Petersburg*, and I confirmed the same with indubitable experiments. I come now to another argument, not at all unworthy of the attention of Geometers.

§20. It entered my mind at one time that these things which I had pondered about the repelling force of fluids while they are ejected, which I exposed here for the most part, can be applied usefully to instituting a new method of navigation. For I do not see what would hinder very large ships from being moved without sails and oars by this method: the water is elevated continually to a height and then flows out through orifices in the lowest part of the ship, it being arranged so that the direction of the water flowing out faces towards the stern. But lest someone at the very outset laugh at this opinion as being too absurd, it will be to our purpose to investigate this argument more accurately and to submit it to calculation, for it can be useful, and it is very fertile for many geometric investigations.

Let me begin with this, for which then it will appear under what circumstances the maximum success should be expected from that new navigation.

§21. Therefore, it is to be noted that a ship is retarded continuously by water drawn in on account of the inertia of the latter when the same velocity is communicated to it at which the ship is borne, and while it is communicated, the ship is forced backwards by the reaction of the water, but at the same time it is pressed forward by the efflux of the same. That meeting of the contrary actions places limits to the force propelling ships to be obtained from a given absolute potential. For, if the prior action were not present (which, to tell the truth, I did not consider for a long time), *a force, however great, for propelling ships could be obtained by the work, however little, of men*, which I thus demonstrate.

In Chapter IX (see especially §26) I showed that the work of men expended in elevating water, which I designate by the term *absolute potential*, is to be estimated from the product of the quantity of water multiplied by the height of elevation, so that, for instance, by the same labor, according to all measures, both four cubic feet can be elevated to a height of sixteen feet, and sixteen cubic feet to a height of four feet. Now, I say further that a uniform pressure is present, propelling ships forward, as long as the fluids flow out at an equal velocity; this pressure is to be estimated from the quantity of water flowing out and from the root of the height of the water placed in the vessel above the orifice; let the quantity of water flowing out in a given time be Q and its height be A, then the magnitude of the orifice spewing forth the water will have to be considered proportional to the quantity $\frac{Q}{\sqrt{A}}$ for that time; but the repelling force which in this case promotes the ship is indeed equal to the magnitude of the orifice multiplied by twice the height of the water (by §4), that is, equal to the quantity $\frac{Q}{\sqrt{A}} 2A$, or $2Q\sqrt{A}$. From a comparison of both propositions it follows that the labor of men engaged in elevating water for thence obtaining the force for propelling ships is as QA is to $2Q\sqrt{A}$, or as $\sqrt{A}$ is to some constant quantity. Therefore, the less the height to which the water is elevated, the greater a force propelling vessels is obtained from the same labor, *so that by work of men, however little, an arbitrarily great force for propelling ships can be obtained.* But also the inertia of the water which is taken in (about which we spoke at the beginning of this paragraph), retarding the ships, obtains a greater proportion to the force propelling the ships, the less the height A is taken, to which proper attention is to be paid here.

§22. It is clear from the preceding paragraph that the height to which the water is to be elevated is of the class of those which somewhere have a maximum. But in order that the height most beneficial to our purpose be determined, other questions present themselves to us for being examined first.

Problem

Let a ship be assumed to progress at the uniform velocity which is generated by a free fall through the height B, and let it be assumed that water flows continuously into the ship, such as in the form of rain, and certainly at that quantity which a cylinder constantly full to a height A would supply through an orifice of magnitude M, with all alien hindrances removed. There is sought how much resistance the ship experiences from that perpetual and uniform inflow of water and its inertia.

SOLUTION. Let any time t whatever be assumed; if this is established from the distance which the fluid flowing in travels at its own velocity divided by the same velocity, then the velocity is to be expressed by $\sqrt{2A}$, and the quantity of water flowing in during the time t will be equal to the cylinder constructed above the base M of length $t\sqrt{2A}$. But that quantity receives in the time t, while it is discharged from the ship, the velocity due to the height B, to be expressed by $\sqrt{2B}$;

REACTION OF FLUIDS FLOWING OUT OF VESSELS   333

therefore, the uniform force is to be sought which can in the time t communicate the velocity $\sqrt{2B}$ to the aqueous cylinder $Mt\sqrt{2A}$, and that force will, on account of the reaction which acts on the ship, have to be considered equal to the resistance sought. Let the previously mentioned force be p, and let it be assumed to have given the velocity v to the aqueous cylinder $Mt\sqrt{2A}$ in the time θ, and there will result $dv = \dfrac{p\, d\theta}{Mt\sqrt{2A}}$, and $v = \dfrac{p\,\theta}{Mt\sqrt{2A}}$. Now let $\sqrt{2B}$ be substituted for v, and t for θ, and it will occur that $\sqrt{2B} = \dfrac{p}{M\sqrt{2A}}$, or $p = 2M\sqrt{AB}$.

Therefore, the resistance sought is equal to the weight of an aqueous cylinder of which the base should be equal to the orifice M and of which the length should be equal to double the mean proportional between the heights A and B.

Problem

§23. Let there be a cylinder in the ship of height A above the surface of the sea, through the orifice of which, placed at the same surface, of area M, water flows out toward the stern without any impediment, and let the cylinder be kept constantly full of water. Determine the force propelling the ship continuously.

Solution

The force propelling the ship is equal to the reaction of the water while it flows out, or to the repelling force diminished by the force defined in the preceding paragraph developing from the intertia of the water which is continuously drawn in. The repelling force is equal, through §4 of this chapter, to $2MA$, and this advances the ship; the other force, which retards the ship, is, through the preceding paragraph, $2M\sqrt{AB}$. Therefore, the absolute force advancing the ship is $2MA - 2M\sqrt{AB}$.

§24. COROLLARY. If the ship has no velocity, the force urging it will be $2MA$; but if the ship is moved at the same velocity at which the water flows out in the opposite direction, it occurs that $B = A$, and then the ship will be propelled by no force. If, then, the ship were moved very freely across the sea, it would nevertheless not acquire from the action of the water which is taken in continuously and flows out below a velocity greater than that at which the water flows out, not because the water flowing out from a uniformly moved vessel repels the vessel with a lesser force than from an unmoved vessel, but because then the inertia of the water produces a resistance equal to the repelling force.

Problem

§25. For a given potential of the laborers who elevate the water and a given height to which the water is elevated, find the size of the orifice of efflux and the repelling force.

SOLUTION. Let the potential be such that by it a number N cubic feet of water can be elevated in one second to a height of one foot, which potential a number of laborers to be designated by $5/4 N$ can develop according to the second experiment inserted following Chapter IX. Let the height to which the water is continually raised be equal to A, expressed in feet, and let the area of the orifice in square feet be equal to M. The number of cubic feet of water which the laborers can elevate to the height A in a single second by the given potential will be equal to $\dfrac{N}{A}$ (through §22, Chapter IX). Therefore, the orifice will have to be constructed of an area so that in one second that number of cubic feet of water can flow out through it if the water flows very freely. But let us assume instead of seconds the time which a body takes while it falls freely through the height A. This time is to be expressed here as $\frac{1}{4}\sqrt{A}$ (it having been assumed for the sake of a more simple calculation that a body falling freely from rest travels 16 feet in one second), and in this time the number of cubic feet of water to be designated by $\dfrac{N}{A} \cdot \dfrac{1}{4} \sqrt{A}$ or $\dfrac{N}{4\sqrt{A}}$, must flow out. But actually $2MA$ flows out, that is, the aqueous cylinder of which the base is M and the length is double the height A; therefore, $\dfrac{N}{4\sqrt{A}}$ is equal to $2MA$, whence the area of the orifice is

$$M = \frac{N}{8A\sqrt{A}}.$$

However, the repelling force becomes equal to $2MA$ or $\dfrac{N}{4\sqrt{A}}$.

Scholium

§26. In any ship, water is to be elevated to a different height, so that, by the same force which is expended in drawing in the water,

REACTION OF FLUIDS FLOWING OUT OF VESSELS

the maximum force advancing the ship is obtained, and two things are required for defining that most useful height for a certain number of laborers. *First*, it must be known what velocity the proposed ship acquires from a given potential: with regard to this postulate we assume that the ship receives a velocity which would be generated by free fall through the height C from a pressure which is equal to the weight of one cubic foot of water, or about 72 pounds; and since from now on we will always express all measures in feet, the weight of one cubic foot of water will have to be expressed by unity. *Second*, the relation between the velocities of the ship and forces propelling the ship is to be assumed as known; here it is commonly stated that velocities are in proportion to the square roots of the propelling forces; however, experiments do not confirm this hypothesis exactly for slow motions; meanwhile, nevertheless, we consider this [hypothesis] preferable to all the remaining. If someone wishes to explore the matter under another hypothesis, he can perform the calculation by the same method which we will now use.

Problem

§27. To find the height most useful to our purpose to which the water is to be elevated continuously, namely, such that for the same potential applied for elevating the water the force advancing the ship becomes a maximum.

SOLUTION. Let all designations applied in the previous argument be retained. First of all, one is to find the velocity of the ship or the height required for this velocity, which we called B. But, because the velocities of the ship are assumed proportional to the square roots of the forces propelling the ship, the heights of the velocities will be proportional to the forces themselves. Therefore, the following analogy will have to be established.

Just as the weight of one cubic foot of water is to the height C (see §26), so the pressure driving the ship or $2MA - 2M\sqrt{AB}$ (see §23) is to the height corresponding to the velocity of the ship, which therefore will be $2MC(A - \sqrt{AB})$. But this height we called B; therefore,

$$B = 2MC(A - \sqrt{AB}).$$

Hence the pressure driving the ship becomes equal to $\dfrac{B}{C}$, and therefore proportional to the height B, because C is a constant quantity; therefore, both the pressure advancing the ship and the height corresponding to the velocity of the ship become maximum at the same time. If, therefore, for the present purpose the quantity $2MA - 2M\sqrt{AB}$, which expresses the pressure propelling the ship, is differentiated, one can set $dB = 0$. But before the differentiation is performed, it is appropriate to substitute for M its value from §25, and then the pressure advancing the ship becomes $\dfrac{N}{4\sqrt{A}} - \dfrac{N\sqrt{B}}{4 \cdot A}$, in which the letter N is a constant, but the letters B and A are variables. Now let the differential of this be taken, and by making $dB = 0$, one sees that the former becomes equal to 0, and thus it will be found that $A = 4B$.

Therefore, the force advancing a ship is greatest when the height to which the water is elevated is four times the height appropriate to the velocity of the ship.

Let $A = 4B$ be substituted in the equation $B = 2MC(A - \sqrt{AB})$ found above, and it will be seen that $M = \dfrac{1}{4C}$, and because (through §25) $M = \dfrac{N}{8A\sqrt{A}}$, there then results

$$A = (\tfrac{1}{2}NC)^{2/3}, \quad \text{and} \quad B = \tfrac{1}{4}(\tfrac{1}{2}NC)^{2/3}.$$

§28. Corollary. If, according to the precept of the preceding paragraph, the area $\dfrac{1}{4C}$ is attributed to the orifice through which water flows out of the conduit from below towards the stern, that is, one which is to an area of one square foot as a measure of one foot is to four times the height appropriate to the velocity of the ship animated by a force of 72 pounds, it will then occur that the ship is moved at half the velocity at which the water flows out, and the repelling force of the water flowing out will be

$$2MA = \dfrac{1}{2C}(\tfrac{1}{2}NC)^{2/3}.$$

But the force advancing the ship will be half of this, so that half the effect is lost by the inertia of that water which is continuously drawn in.

Scholium

§29. After we have thus demonstrated how that method of navigation is to be undertaken most usefully and with the greatest success,

I believe that now this matter has to be illustrated by an example which, I should think, does not agree poorly with the very nature of the matter, in order that it be shown at the same time how the occurrence might take place, more or less.

Let us consider a trireme, commonly called a *galley*, with 260 rowers; let us assume that this galley, drawn by the weight of one cubic foot of water, or 72 pounds, completes a length of two feet in one second, the generating height of which velocity, indicated by C, is $\frac{1}{16}$, under the assumption that a heavy body falling freely from rest travels 16 feet in the first second. Because, further, 260 laborers are furnished, any one of which according to the second experiment pertaining to Chapter IX, can elevate four-fifths of a cubic foot to a height of one foot in one second, there will be $N = (4/5) \cdot 260 = 208$. Therefore, let the orifice through which the water flows out be of an area of 4 square feet; the laborers will be able to maintain the water elevated in the conduit above the orifice at a height of approximately $3\frac{1}{2}$ feet, which is indicated by the letter A, and if one takes the fourth part of this height, one will have B = $\frac{7}{8}$ foot, so that the ship will progress by that navigation at the velocity that a weight acquires by free fall through a height of $\frac{7}{8}$ foot; thus, therefore, the ship completes a length of $7\frac{1}{2}$ feet in any one second and 27,000 feet in one hour, that is, more than two Gallic miles; such a great velocity of a ship can indeed barely, or not even barely, be obtained by rowing.

But now let me apply the calculation to another hypothesis, which I trust those understanding nautical matters will not reject completely, for it agrees with many observations which I myself made at sea: let me suppose that the sails of a trireme, expanded perpendicularly to the keel, have a surface of 1600 square feet, and that a wind which travels through a distance of 18 feet in one second strikes them, impinging directly, but that the ship thus travels a distance of 6 feet in one second in the same direction. So the wind strikes the sails at a relative velocity of 12 feet; I estimate the force of that wind equal to the weight of $\frac{9 \cdot 1600}{850}$ cubic feet of water, or almost 17 cubic feet of water.

If these things are so, it follows that a ship can be propelled by the elevating of water by 260 laborers at that velocity at which it travels through a length of $6\frac{1}{2}$ feet in one second.

An estimate not very different from this follows from those things which Mr. Chazelles has in the *Commentaries of the Royal Academy of Science of Paris for 1702*, p. 98, Paris edition. But in order that they can be applied properly to our purpose, it will have to be noted that, in rowing, the force propelling the trireme is not to be estimated from the pressure of the rowers against the oars, but from the pressure which the extremities of the oars submerged in the water exert against the water. In order that we may define this [pressure] approximately, these things will have to be observed first: 260 rowers were furnished, rowing with all their might; in the first minute 24 strokes (*palades* in French) of the oars were made; the entire agitation of the oars occurs in three motions, which I assumed to be of the same duration, and of which only one advances the trireme. In this way the trireme was carried forward at a velocity by which it traveled a length of $7\frac{1}{5}$ feet in any one second. The part of the oar inside the ship was 6 feet, and outside the ship 12 feet. But the surfaces of all the oars (*les pales* in French) which are impelled against the water, gathered into one, make 130 square feet according to Mr. Chazelles. He noted further that the internal extremity of the oar describes a distance of 6 feet in any one agitation, and because any one agitation is completed in a time of $\frac{60}{24}$ of one second, and at the same time consists of three motions, which I assume tautochronous, it appears that any retraction of the oar occurs in the time of $\frac{20}{24}$, or $\frac{5}{6}$ of one second, and in this time the internal extremity of the oar completes a length of 6 feet. Further, on account of the length of the surface of the oars which is impelled against the water, not the entire [surface] is to be considered at a distance of 12 feet. Therefore, I will assume that to be at a distance of 10 feet, just as if the part of the oar beyond the ship projected a length of 10 feet. The extremity of this part describes 10 feet in the time of $\frac{5}{6}$ of one second; but because the trireme itself has a velocity by which it travels 6 feet in the same time, it is to be understood that the extremities of the oars are impelled against the water at a relative velocity which describes 4 feet in a time of $\frac{5}{6}$ of a second. Therefore, the force propelling the trireme is equal to the force which the water would exert against a surface of 130 square feet if it were to strike against it at that velocity by which it travels 4 feet in $\frac{5}{6}$ of one second. I find this force, according to common estimation, to be more or less equal to the weight of 40 cubic feet of water; however, that force is not applied continuously, but only during that time in which the oars are drawn back: therefore, two thirds of that force is to be removed, so that the force which propels the trireme continuously is to be considered finally as equal to the weight of $13\frac{1}{3}$ cubic feet of water.

It follows thence, if the velocities of the ship are assumed to be in proportion to the square roots of the propelling forces, that this same trireme, if driven by the weight of one cubic foot of water, would have had a velocity by which it could travel approximately two feet in any one second. This hypothesis is the same as that which we applied in the first place, so that it again follows therefrom that the trireme will acquire from that navigation a velocity by which it can travel $7\frac{1}{2}$ feet in any one second, which velocity is a little greater than that which was given to the trireme by the very strong rowing of 260 oarsmen.

With these things having been well considered, I am at a loss as to which kind of navigation is to be preferred, rowing or elevation of water, and I should believe that the success of either is almost equal, and I dare to affirm for certain that if a ship is advanced less by the elevation of water, the defect will be slight; but perhaps it will be advanced more. Meanwhile, I do not doubt but that this new idea of navigation appears to be groundless and ridiculous to those ignorant of these things. But I feel otherwise, and I should wish that attention be paid further to the following:

First. That water can be elevated easily in every kind of ship where oars clearly cannot be provided, so that by that new navigation even very heavy warships such as are used in naval battles can be driven as it pleases in the absence of any wind.

Second. That thus in theory an example occurs in which the motive or propelling forces are given, which can be called intrinsic. Ingenious minds will be incited by this example to devising other principles for this kind of motion, to perfecting them further, and to applying them to use in navigation.

Third. That in many ways the work of men in elevating water can be assisted other than by the use of oars; there are indeed natural things, extraordinary and furnished with almost incredible value as compared to their moderate expense, by which the same can be produced as by the work of men; the use of these things can serve especially in establishing short passages during serene and tranquil periods. I wrote in Chapter X, §40 and following, concerning the innate value of natural things of that sort and concerning the effects to be obtained therefrom and their measures, but especially I would wish that attention be paid to §43, by which all to whom productive genius was given by nature for devising machines should be incited to attempt the perfection of that matter.

Fourth. That several other purely mechanical devices can be applied similar to that which was given in §27, by help of which, certainly, the effect of the same work in advancing ships increases not a little; but it is not permitted now to treat all things according to the true nature of the matter.

EXPERIMENTS WHICH PERTAIN TO CHAPTER XIII

In order that it be possible to understand the repelling force correctly by experiment, a vessel can be furnished which has the form of a parallelopiped, and the weight can be taken empty as well as full of water, and afterwards the ratio can be investigated between the area of the vessel and the area of the orifice which must be in the side of the vessel, just as also the ratio between the heights of the water above the orifice and above the base. Then it will be possible to deduce the ratio between the weight of the vessel full of water and that of the aqueous cylinder lying vertically above the orifice. Further, from the observed amplitude of the thrust, the velocity of the water will be obtained; from this, if at the same time one applies in addition the quantity of water flowing out in a given time, also to be observed, one deduces the area of the contracted stream, which one would be able to compare with the area of the orifice.

After all these things have been investigated, let a vessel be suspended from a very long thread, with care having been taken at the same time so that it cannot have any motion other than that which is opposite to the direction of the water flowing out. Then at last let efflux be granted to the water, and it will be observed that the thread forsakes the vertical position, and from the angle of declination the repelling force will be ascertained, and one will be able to compare this with the measurements which we indicated.

EXPERIMENT 1. At one time I myself did all the things that I just indicated, and it was seen that our rule of §2 is properly confirmed; nevertheless, I was not able then to perform the experiment with sufficient accuracy with respect to time, nor did I repeat it later.

EXPERIMENT 2. At another time I tried the matter differently: namely, I placed a vessel full of water, of which I had taken all required measurements, at the stern of a small boat; the boat was floating on the water in a tub. Then, with the water flowing out of the vessel (however, so that it did not strike against the boat), the boat progressed in the opposite direction; I determined the velocity of the boat very accurately from the space traveled through in a given time. Then I inquired as to what elemental weight should be appended to the boat [by means of string and a pulley] in order that, stimulated by that weight, it would acquire the same velocity. Then, after the comparison of that weight had been made with the weight of an aqueous cylinder of given diameter, I saw that our theory was confirmed very accurately.

REACTION OF FLUIDS FLOWING OUT OF VESSELS 341

EXPERIMENT 3. With the water from the vessel (superimposed on the boat) flowing out into the boat, the latter remains wholly unmoved. This indicates that the impetus of the aqueous stream is equal to the repelling force, as I showed in §16 and §17. Then, as well, if the aqueous stream was impinging directly on a plane affixed to the boat, the latter similarly stayed unmoved, which again proves the equality of impetus and repelling force; but if the stream was striking obliquely against the plane, the boat obtained a certain, but slower, motion.

Finally, if the water flowing out of the boat was intercepted so that the orifice was submerged in the water standing within the boat, similarly the boat remained without motion, as proof that the same pressure arises from the stream, whether it happens that all its motion is confined or that it is declined at a right angle, just as was demonstrated in §18. I confirmed very precisely by several other methods the equality between the repelling force and the force of the aqueous stream striking perpendicularly against a plane. Moreover, I verified this force conforming to our theory and contrary to the opinion common to everyone until now by an experiment overcoming all objections; I conducted this in my home in the presence of Mr. Emanuel Koenig, my paternal uncle Nicolaus Bernoulli, and my Father, with such great confidence that, after all measurements had been taken, I predicted with all precision how great the pressure of the aqueous stream would be, although the experiment had never been performed by me before. I communicated all these things found out through new mechanics principles to the Academy of Science of

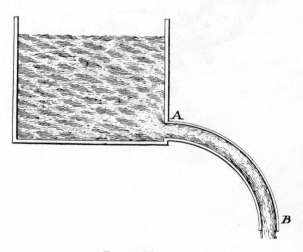

FIGURE 86

St. Petersburg, in the *Commentaries* of which they should be inserted at some time.

EXPERIMENT 4. In order that I might also show the falsity of the accepted rule concerning both the repelling force and the impetus of the water, I furnished the vessel which Fig. 86 shows, connected to the curved conduit AB of uniform area, of which the direction at A was horizontal, at B vertical. I saw clearly that the vessel was not repelled horizontally; therefore, by §14, the rule is false which adheres to the simple cylinder defined in that place.

FINIS

HYDRAULICS
BY
JOHANN BERNOULLI

JOHANNIS
BERNOULLI
HYDRAULICA

Nunc primum detecta ac demonstrata directe ex fundamentis pure mechanicis.

ANNO 1732.

Photographic reproduction of the title page for the first part of Johann Bernoulli's treatise, Hydraulica, *as published in the* Memoirs of the Imperial Academy of Science in St. Petersburg. *It translates: "Johann Bernoulli's* Hydraulics, *now first discovered and directly shown from purely mechanical foundations. 1732." The date is false, as Dr. Rouse explains in his Preface to this volume, page x.*

LEONHARD EULER
Most Sagacious Mathematician

TO THE AUTHOR

Previously indeed I highly praised Your Theory of flowing water because of the true and genuine Method which You alone, Most Excellent Sir, first revealed for fully treating Problems of this type. But now, after I had examined another portion of Your Studies, I was thoroughly astounded by the very fluent application of Your principles to the solution of the most intricate Problems, because of which most useful and also most profound finding Your very distinguished Name will forever be revered among future generations. But You also so distinctly and plainly explained the most obscure and most abstruse question about the pressure which the sides of vessels experience as a result of water flowing through them that there remains nothing more to be desired concerning this rather troublesome matter. Although indeed no one has undertaken this matter except Your very renowned Son, who, however, defined pressure in a rather indirect manner only so far as the entire motion has already acquired the steady state, nevertheless, after the genuine method had been brought to light, You at once determined most accurately the pressure in every state of water, because of which Your most praiseworthy discovery I congratulate You from my heart, Most Excellent Sir, and for this communication I give You the greatest of thanks.

CONTENTS

	Page
PREFACE	351

FIRST PART

Treating the motion of Water through Vessels and Cylindrical Conduits which are Composed of Several Cylindrical Pipes attached to one Another in Succession — 356

SECOND PART

Containing the Direct and Universal Method for Solving all Hydraulics Problems whatsoever which can be Formed and Proposed Concerning Water Flowing through Conduits of any Shape — 391

HYDRAULICS PROBLEM — 446

PREFACE

Hydrostatics, which deals with water standing in vessels which are closed below the water surface, has its laws demonstrated and its principles deduced from reason, whence the performance and the phenomena are explained clearly and distinctly such that concerning this Science there is scarcely any more to be desired. The situation is different in *Hydraulics*, where not so much is done concerning the gravitation of water and its pressure, but where, besides this, both the motion which is thence produced if water can flow out of a given aperture, or is forced to go from one pipe to another of different size, and any other effects to be regarded which attend this motion must be determined demonstratively. Surely this Science, commonly called *Hydraulics*, is extremely difficult and up to this time has not been subjected to the laws and rules of mechanics. Whatever material Authors have written on this matter, they rely either on experience alone or on theories that are wholly uncertain, having insufficient foundation.

In the book *Hydrodynamics* which my Son published not long ago,* he undertook that subject under luckier auspices, but he relied upon an indirect foundation, namely the conservation of live forces, which is most certainly true and was proven by me as well, but is still not accepted by all Philosophers. It was I who first presented this hypothesis in the Dynamics of solids (after Huygens used a similar principle to determine the center of oscillation), and from that hypothesis I firmly exhibited the same solution for a water-course which is given by the ordinary principles of dynamics accepted by all Geometers;† this clearly general conformity of the solutions elicited by either procedure should by itself be sufficient for overcoming the obstinacy of Skeptics. Thus far no one has given a direct method by which, *a priori* and only through the principles of Dynamics, one can

---

* Daniel Bernoulli, *Hydrodynamica, sive de viribus & motibus fluidorum Commentarii*. Strassburg, 1738.

† See Nos. CXXXV, CXXXVI, CXL [*Opera Omnia*, Johann Bernoulli, Lausanne et Geneve, 1743].

investigate the nature of the motion of water issuing forth from vessels through orifices or flowing through conduits of nonuniform size.

Having wondered from what source there is so much more difficulty in successfully applying the principles of dynamics to fluids than to solids, finally, turning the matter over more carefully in my mind, I found the true origin of the difficulty; I discovered it to consist of the fact that a certain part of the pressing forces important in forming the *throat* (so called by me, not considered by others) was neglected, and moreover regarded as if of no importance, for no other reason than that the throat is composed of a very small, or even an infinitely small, quantity of fluid, such as occurs whenever fluid passes from a wider place to a narrower, or *vice versa*, from a narrower to a wider. In the prior case, the throat is formed before the transition, in the other, after the transition.

On the other hand, I will demonstrate that in the forming of the throat, however small a size it may have, a pressing force is required, nevertheless, which is not negligible and by no means infinitely small but finite and determinate, and so far not at all to be disregarded but wholly worthy of being taken into account. Now, that force required for the latter effect, which amazingly enough can be observed, plainly does not depend upon the length of the throat, which can be understood to be greater or lesser as long as it is considered extremely small; it always consumes the same portion of the pressing forces in its formation, if all other conditions are unchanged.

What the throat may be and in what manner it may be formed will be understood from the very discussion of the matter, and at the same time it will be evident that the formation of the throat is accomplished without noticeable loss of live forces with respect to the amount which is present in the whole aqueous mass. Hence the reason is apparent wherefore, safely and without error, the Theory of live forces can be applied in Hydraulics, even if those who use this theory pay no attention to the throat, provided that they are not ignorant of the existence of the throat and that they see that it detracts nothing from the conservation of live forces; for otherwise they cannot contend that they themselves have arrived at the truth of the matter wholly and scientifically.

I shall treat this investigation in two parts. In the first I will consider the phenomena of flowing water and efflux from cylindrical or prismatic vessels, be they either simple or composed of several [sections], such as conduits composed of various pipes of different size or of cylindrical pipes joined as syphons. In the other part I shall examine completely all perforated vessels, whatever may be their shape, whether regular or irregular, and the conduits and pipes attached to them.

In order to have a clearer understanding of things, I am setting forth the following Definitions and Lemmata, the validity of which is manifest from Dynamics as well as Hydrostatics.

I. A uniform *accelerative* force [i.e., force per unit mass, or *acceleration*] is that which impresses a given velocity on a given body in a given time.

II. A *motive* force is that which, when it acts on a body at rest, excites it into motion, or which can cause a body already moving to accelerate, decelerate, or change its direction.

III. Motive forces are in proportion to the products of masses and accelerative forces. Thus, for example, in order to move twice the mass with the accelerative force tripled, or, which is the same thing, in order to move three times the mass with the accelerative force doubled, a six-fold motive force is required.

IV. The motive force divided by the mass gives the accelerative force, but divided by the latter gives the mass.

V. The absolute gravity g, or the cause of gravity, whatever it may be, is an accelerative force which, when it causes a prescribed mass m of a body to move, produces in it a motive force gm. However, in our thinking it will be permissible to separate it from the body and thus to consider it in the same way as if it were acting externally upon the body. We therefore consider that that same body, free from gravity, will be accelerated by an external motive force gm according to the same law by which it is accelerated naturally. However, it is convenient to call that same force gm, inasmuch as it exists beyond the body, an *immaterial* motive force; therefore, if that force, translated in another manner, acts on another mass M, the latter will be accelerated by an accelerative force gm/M.

VI. An immaterial and invariable motive force, acting without impediment on a body, accelerates it in the same manner whether it be at rest at this point or already in motion. Since this force always follows the body, there is no relative motion between them, and thus a motive force acts on a body in motion in the same way as if it were completely at rest. This is the reason why heavy bodies, while descending, are continuously and uniformly accelerated in accordance with time, it having been supposed, certainly, that the intensity of the accelerative force is not changed during the action, that is, neither augmented nor diminished, just as in fact the force of gravity continually maintains the same intensity on a descending heavy body from the beginning of the descent.

VII. The *intensity* of an invariable motive force is the measure according to which, on a body to be moved, there is produced a greater or lesser accelerative force; thus gravity, on a body falling vertically, has a greater intensity than it has on the same body sliding on an inclined plane. In the first case, to be sure, a greater accelerative force is produced than in the other; in either, however, gravity is invariable.

VIII. A *variable motive force* is one of which the intensity is changed while acting. Thus, for instance, the force of a stretched elastic has a greater intensity, and as a consequence impresses on the body to be propelled a greater accelerative force at the beginning than during the progression of the relaxation. From the above, these Rules result:

$$\begin{aligned}
\text{Let the space traveled by a body} &= x, \\
\text{the mass of the propelled body} &= m, \\
\text{the motive force within the limit of the region traveled} &= p, \\
\text{the velocity acquired} &= v, \\
\text{the time through } x &= t.
\end{aligned}$$

Hence $dt = \dfrac{dx}{v}$; there will be $\dfrac{p\,dt}{m}$ or $\dfrac{p\,dx}{mv} = dv$, and therefore $\int p\,dx = \tfrac{1}{2}mvv$, which is very well known.

IX. The lower portions of the water contained in any vessel are pressed upon by the aqueous mass lying above in accordance with the depth alone, whatever shape the vessel may have. That is, if in one's thinking the aqueous mass be divided into horizontal strata of infinitely small thickness, every one of these strata is pressed the same amount as if an aqueous cylinder of water were lying over it having the same altitude as that which corresponds to the depth of the stratum itself in the vessel.

X. Hence the following is concluded directly: if the areas of the strata, each having the same infinitely small thickness, are m, m', m'' m''', etc., and their corresponding weight elements are also in proportion as m, m', m'', m''', etc., their own gravitations can be imagined as separable from the strata, so that their substance remains alone without weight. But the same pressure will arise in the individual strata as if they had remained in their natural state if, in place of the gravitations which have been removed, just as many others are substituted, which together press the uppermost surface of the water; [this is done,] certainly, by observing the following analogy at any instant: as the area of an arbitrary stratum is to the area of the uppermost surface, so the proper gravitation of the stratum is to the gravitation to be substituted.

XI. I call that mental substitution *Translation*. In order that I might explain myself, let some stratum from the lower regions have an area m, let its gravitation, or its own weight element, be π, and let the area of the uppermost surface be h. The *translated gravitation* to the uppermost surface will be $\frac{h}{m}\pi$, which together with all the remaining translated in this way, constitutes the total immaterial motive force by which all the water in the vessel is pressed downward, in the same way as it happens naturally.

Admonition

It is now appropriate to say in advance that through all this treatment of the motion of flowing water I avoid consideration of the foreign and accidental hindrances which can alter the motion determined through the rules. Such hindrances are the imperfect fluidity of the water, also its adhesion, the friction at the walls of vessels, excessive slenderness of pipes, narrowness of orifices or apertures, tenacity of fluid particles on account of which they do not very easily separate from one another, and others existing of this sort to which I do not attend.

I would also wish it to be noted that it is not of absolute necessity that the strata of water always be considered in a horizontal position. They are more conveniently assumed perpendicular to the direction of the motion of the water. Thus, for example, when water from a larger vessel flows out into a narrower horizontal pipe through an orifice or aperture lying in a vertical plane perpendicular to the wall of the pipe, the water contained in the pipe is considered to be divided very appropriately into vertical strata parallel to the plane of the aperture, and more so in addition because by its very nature it tends toward this sort of orientation. We see, for instance, that a column of water in some pipe not greatly exceeding two lines in diameter [1 Paris line = 0.226 centimeter] has both its extreme surfaces disposed in positions perpendicular to the sides of the pipe, whether the pipe itself is oblique to the horizontal or altogether horizontal. The line joining the centers of gravity of the strata, whether it be straight as in rectilinear pipes or curved as in curvilinear pipes, will be called the *center line*, or simply the *centric*. Of course, individual strata which have their substance concentrated at their own centers are understood to have that motion which the [actual] strata themselves have.

FIRST PART

Treating the Motion of Water through Vessels and Cylindrical Conduits which are Composed of Several Cylindrical Pipes Attached to one Another in Succession

SECTION I

First, let the conduit *ABCFDE* be given (Fig. 1), composed of two cylindrical pipes of different size, *AGDE* and *GBCF*, of which the former has a base *GD* open at the orifice *GF* through which it connects to the narrower pipe *BF*. Now let the whole conduit *BE* be full of a homogeneous liquid of no weight of its own but driven from a section at the orifice *AE* by a given motive force [per unit density] = p, which, by pressing equally, is expanded through the whole surface *AE* of the liquid. The law of acceleration is sought according to which the liquid flows through the conduit. Moreover, I consider the conduit to be always full of liquid, which occurs by understanding that a new supply of liquid is provided freely from another connected source, flowing into the pipe *GE* at any instant at the same velocity, for the purpose of replacing that which is flowing out through the other orifice *GF* into the pipe *GC* and from there is escaping into the air through the opening *BC*.

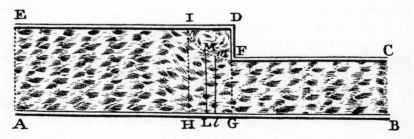

FIGURE 1

Section II

From Hydrostatics I assumed that the immaterial motive force p by which the surface of the liquid AE is being pressed is extended instantaneously to the surface GF of the liquid contained in the pipe BF, and this whether the liquid is standing still in the entire conduit or is flowing, as long as it remains full.

Section III

While the liquid goes from one pipe to the other, in any case the velocity will be changed in a manner reciprocal to the areas; on the other hand, no change is sudden, but successive and gradual, proceeding through all possible intermediate values from the lesser to the greater, or from the greater to the lesser.

Section IV

Hence when the liquid flows with a parallel motion in the direction from AE toward GD so that at any moment the same velocity pertains to the individual portions of the liquid, before the portions near to GF itself arrive at the orifice GF, it is necessary that they begin to be accelerated, at least through the small distance HG, and that they continue accelerating until, at the entrance GF itself, they will have acquired the velocity of the liquid flowing through the tube BF in a motion uniformly parallel and common to the individual particles.

Section V

And, accordingly, there is formed along the indefinitely small length HG something like a throat, $IFGH$, contracting from the wide into the narrow, through which the liquid must pass, the acceleration being continuous but nevertheless augmented gradually, with a rather small portion of the liquid (which fills the small space IFD) remaining at perpetual rest.

Section VI

Let the curve IMF defining the throat be of any nature whatever, for it is not necessary to assume it of some prescribed shape. Directly, indeed, I will show: that there is always the same motive force uniquely required for this purpose; that the liquid be driven through the throat, whatever length HG it may have, as long as it be infinitely small; and that the line IMF which connects the extremities I and F may be of any nature whatever.

358   HYDRAULICS, PART I

Section VII

Let no one consider that that motive force (which pushes a small, in fact infinitely small, portion of liquid through the throat) must be and is always very small and thus can be disregarded. Indeed, the motive force is by all means one of finite quantity, because, although the quantity of material moving is infinitely small, on the other hand the accelerative force must be infinitely large compared to the former, in order that certainly in the infinitely short time in which the liquid passes through the small space HG, a finite change in velocity can nevertheless be created, since that which had been the velocity at H is to that which now prevails at G as GF is to HI.

Section VIII

The neglect of this motive force as if of little import has been the reason why no one up to this day could have given from statical and purely mechanical principles the laws of liquids flowing through non-uniform conduits. But those who undertook to determine those laws exactly returned, by my example indeed, to the principle of live forces, the application of which to this problem and to others in solids as well as in fluids they perhaps never would have considered if they had not followed me, who by all means first showed how to derive these laws from the conservation of live forces. But I myself, being dissatisfied since this method was indirect and also founded on a theory of those forces which is still not universally accepted, did not hesitate to search for a direct method which would be supported solely by dynamical principles denied by no one. Finally, after a rather long meditation, I achieved my aim in the year 1729, when I saw the crux of the whole matter to lie in the contemplation of the throat, previously considered by no one. And so now I am undertaking to share my discoveries, already explained privately to certain friends, with the public as well. Since the generation of the throat has now been indicated, it is pleasing to pursue this task as far as I can with [any] clarity.

Section IX

Let there be considered [Fig. 1] the abscissa $HL = x$, the ordinate $LM = y$, and an element of the former $Ll = dx$. Let the area AE or HI of the pipe HE be called h, the area BC or GF of the pipe GC be called m, and the velocity of the liquid in the pipe GC be v. Accord-

MOTION OF WATER THROUGH VESSELS

ingly, the velocity of the liquid in the pipe HE will be $\frac{m}{h}v$, as the velocities are reciprocally proportional to the areas. By the same reasoning, the velocity of the liquid $LMml$ at any place in the throat will be $\frac{m}{y}v$, which may be set equal to u. Now therefore let there be an accelerative force γ which excites the section Lm of the liquid. From the nature of the acceleration, it follows that $\gamma\, dx = u\, du$, and therefore $yy\, dx = yu\, du$. That is, the motive force by which the liquid section $LMml$ is excited is equal to $yu\, du$. But this motive force, according to §II, is generated by a single motive force existing in the pipe HE and distributed over the entire area AE; in order that this may be explained, $yu\, du$ must be made into $hu\, du$ in proportion as LM is to HI, or as y is to h. The particular motive force in pipe HE (translated certainly from $yu\, du$ itself) will be $hu\, du$, which can produce the motive force $yu\, du$ in the section $LMml$ of the throat; and by integrating through the whole throat, one has $\frac{1}{2}h\left(vv - \frac{mm}{hh}vv\right)$, or $\frac{hh - mm}{2h}vv$, which designates the motive force in the pipe HE required uniquely for creating the acceleration in the throat necessary to change the lesser velocity to the greater, which must be done in order that the liquid may flow into the narrower pipe GC.

COROLLARY 1. Hence it is evident that the nature of the curve IMF, as well as the length of the throat HG, does not enter into the determination of the motive force for generating the motion in the throat. Thus, if the areas of the end sections HI and GF are given as h and m and the velocity as v, one always has the motive force in the pipe HE, generating the motion in the throat, equal to $\frac{hh - mm}{2h}vv$.

COROLLARY 2. If now with a continuous flow of liquid the velocity v in the pipe BF remains always constant, it is manifest that the other velocity—in the pipe HE—remains constant as well, and accordingly that the motive force, or the pressure p, adds nothing more to accelerating the motion in either pipe; thus it is clear that the entire force p is uniquely applied to forming the throat and maintaining it in its proper state. Consequently, $p = \frac{hh - mm}{2h}vv$.

COROLLARY 3. Let us consider the pipe HE or GE to be vertically erect in the manner of a cylindrical vessel and to be connected to the horizontal pipe GC [Fig. 2]. Also consider the force p to be the very weight of the column of liquid contained in GE, so that (with g having been taken to designate the natural accelerative force of heavy bodies, and also HA or $GA = a$) one has $p = gah$ equal to the weight [per unit density] of the liquid contained in GE, from which $gha = \frac{hh - mm}{2h} vv$. But in order that v be determined by the vertical altitude z through which some freely falling weight has acquired the velocity v, $g\,dz$ must equal $v\,dv$, from which $gz = \frac{1}{2}vv$. Therefore by substituting gz for $\frac{1}{2}vv$, we will have $gha = \frac{hh - mm}{h} gz$, from which emerges $z = \frac{hh}{hh - mm} a$, which gives the following hydraulics Theorem.

Section X

THEOREM. *Let the cylindrical vessel AGFE (Fig. 2) be vertically erect and furnished at the base with a horizontal cylindrical pipe FB open at either*

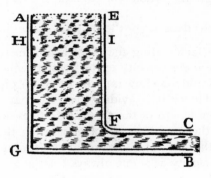

FIGURE 2

end. Likewise let both the vessel and the pipe be continuously full of water, so that, of course, as much water as flows out through the opening BC is continuously supplied through AE at the same velocity that the water has in the vessel. I say that the velocity of the water which flows out (if it starts from rest) converges very rapidly to that which is acquired by a weight falling freely through a height $\frac{hh}{hh - mm} a$.

The truth of this is evident from Corollary 3 preceding.

COROLLARY 1. Whence if the opening BC be very small with respect to the area AE of the vessel, such that m can be neglected with respect to h, there will result $z = a$, that is, the velocity of the water flowing from the pipe will be equal to that which a weight, having fallen freely from a height *EF*, acquires. This is a very well known Theorem, but up to now not shown from dynamic principles, especially if the attached pipe *BF* were present, since previously the Theorem was believed correct only for a small orifice placed at *F*.

COROLLARY 2. The greater is the opening *BC* with respect to the area *AE* of the vessel, the greater becomes the maximum efflux velocity of the water. Thus for *m* large, the value of the fraction $\frac{hh}{hh - mm}$ is increased, until for *m* increasing to *h* the maximum velocity is infinite; and also from here it is evident that this is true because in that case both the vessel and the pipe are of the same size and they form one continuous bent pipe. And the force of the weight of the water in the portion *AF*, always full, continuously accelerates the entire aqueous mass such that finally its velocity, generated in an infinite time, becomes also infinite itself. Now, with the length of the pipe *FC* being called *b*, the mass [per unit density] of all the water in the bent pipe *AGC* will be $ha + hb$; and this will not be accelerated in any way other than as is some solid body which is animated by an accelerative force equal to $\frac{gha}{ha + hb} = \frac{ga}{a + b}$, and certainly such a body by falling for an infinite time would acquire an infinite velocity.

COROLLARY 3. Surely if *m* were greater than *h*, that is, if the horizontal pipe were larger than the vertical vessel, the maximum velocity would certainly never be attained even in an infinite time, since $\frac{hh}{hh - mm}$ would be negative, with the result that, during flow to eternity (if the flow lasts into eternity), the acceleration of the water flowing out will not cease to be augmented. In this case, for instance, an inverse throat will be formed in the pipe, looking back at the orifice *BC*, which, as will be evident from the following, is of the nature that it increases the motive force rather than decreases it, during which time it disappears, so to speak, toward the pressure being produced in the rear, whereby the water can descend rather freely in the vessel.

SCHOLIUM

So far we have considered the vessel and the pipe constantly full of water and the water flowing out at its maximum and therefore constant or uniform velocity, so that no additional motive force is required for accelerating the water either through the vessel or through the pipe, but that the total motive force *p* is used for controlling the throat which is formed in front of the entrance from the wider region to the narrower. Now we shall consider the velocity of the flow of water as if it were increasing, starting from rest, such that its own equally special portion of the motive force p is required for developing the acceleration in the vessel as well as in the pipe. First we shall examine the case in which a constantly full vessel is connected to a pipe.

Section XI

Let x be the length of the region through which the water passes in the pipe from rest. Then $\frac{m}{h}x$ will be the length through which it passes in the same time in the vessel. Thus, similarly, with the existing velocity in the pipe equal to v, the velocity in the vessel will be $\frac{m}{h}v$. From this the accelerative force in the pipe equals $\frac{v\,dv}{dx}$, and this multiplied by the mass of the water mb will give the motive force $\frac{mbv\,dv}{dx}$, which, translated to the vessel (by §2), will give the equivalent $\frac{hbv\,dv}{dx}$, from which, certainly, that in the pipe, $\frac{mbv\,dv}{dx}$, can be produced. And so also the accelerative force in the vessel equals $\frac{mm}{hh}v\,dv \Big/ \frac{m}{h}dx = \frac{mv\,dv}{h\,dx}$, which, applied to the mass ha, gives the motive force $\frac{mav\,dv}{dx}$ for propelling the water in the vessel; and thus the sum of those three motive forces—through the throat, through the pipe, and through the vessel—must equal the total motive force p. This gives us the equation

$$\frac{hh - mm}{2h}vv + \frac{hbv\,dv}{dx} + \frac{mav\,dv}{dx} = p.$$

Therefore, as before, let $p = gah$ be the weight itself of the aqueous column, and let $gz = \frac{1}{2}vv$ be established as in Coroll. 3, §IX; with these being substituted, this equation will result:

$$\frac{hh - mm}{h}z + \frac{hb\,dz}{dx} + \frac{ma\,dz}{dx} = ha,$$

or

$$(hh - mm)z\,dx + (hhb + hma)\,dz = hha\,dx,$$

MOTION OF WATER THROUGH VESSELS

from which $dx = \dfrac{hhb + hma}{hha - hhz + mmz} dz$, which properly treated and integrated logarithmically, will give

$$x = \frac{hhb + hma}{hh - mm} \ln\left(\frac{hha}{hha - hhz + mmz}\right),$$

from which, by progressing to numbers (assuming $1 = \ln e$), one has

$$z = \left(\frac{hha}{hh - mm}\right)\left(1 - 1/e^{(hh-mm)x/(hhb+hma)}\right).$$

But if the water in the vessel (which, for the sake of brevity, is considered to have only an orifice of area m, without the annexed pipe) be animated by a gravity g' different from the natural gravity g, it will be found that

$$z = \frac{g'hha}{g(hh - mm)}\left(1 - 1/e^{(hh-mm)x/(hhb+hma)}\right).$$

COROLLARY. If $x = \infty$, which gives the case of the maximum velocity to which the flow converges, it will be true that

$$1/e^{(hh-mm)x/(hhb+hma)} = 0,$$

and therefore $z = \dfrac{hha}{hh - mm}$ for natural gravity g, which conforms wholly to Coroll. 3, §IX; and if, in addition, m is infinitely small with respect to h itself, then $z = a$, just as is developed in Coroll. 1, §X, all of which confirms the method splendidly.

Section XII

Let us consider now the case where the vessel AF (Fig. 2) does not remain full of water, but, as a measure of the water flowing out, it is emptied gradually, and its surface AE descends continually.

Consider the water in the horizontal pipe to have passed through the length x, hence that a quantity of water mx, that is, equal to an aqueous cylinder of which the base is m and the length is x, has flowed out from it (for I assume the vessel and the pipe to be full at the beginning). But if, accordingly, in EF the portion EI is assumed equal to $\dfrac{m}{h} x$, it is clear that the horizontal HI is the location of the uppermost surface to which the water descends in the vessel after the portion of water mx has flowed out through the pipe. Therefore, there will remain in the vessel the aqueous column $GI = ha - mx$, of which the weight $g(ha - mx)$ is directly that very thing which we called p. Thus if, therefore, the accelerative force of the water remaining in the vessel (which in §XI is generally found equal to $\dfrac{mv\,dv}{h\,dx}$) is applied to the aqueous mass, which now is $ha - mx$, we will have the motive force $\dfrac{mv\,dv}{h\,dx}(ha - mx)$, which is compatible with the water flowing down through the vessel; from which now by collecting the three forces—through the throat, through the pipe, and through the vessel—and by equating the sum to p itself, that is, to $g(ha - mx)$, we will acquire this equation:

$$\frac{hh - mm}{2h}\,vv + \frac{hbv\,dv}{dx} + \frac{mv\,dv}{h\,dx}(ha - mx) = g(ha - mx).$$

By substituting $g\,dz$ for $v\,dv$, and gz for $\tfrac{1}{2}vv$, as we did in Coroll. 3, §IX, we will change our equation to this other:

$$\frac{hh - mm}{h}\,z + \frac{hb\,dz}{dx} + \frac{m\,dz}{h\,dx}(ha - mx) = ha - mx;$$

and, by multiplying by $h\,dx$, into this:

$$(hh - mm)z\,dx + hhb\,dz + m\,dz(ha - mx) = (hha - hmx)\,dx.$$

This is the very equation from which, if the value of z itself is found, the height will be obtained through which a heavy weight, having fallen freely, will acquire the sought velocity, certainly equal to that which the water in the pipe will have after the quantity mx has flowed out.

The derived equation in which the unknowns are found to be interrelated can also be integrated according to our rules with the support of the Lemmata following shortly, and thus the value of z itself may be known in finite terms. However, at this point one should not tarry any longer on this matter: it suffices to me to have reduced the Problem to a differential equation by using purely mechanical principles, which might have been presented by someone else before me, although I do not recall ever having seen it. Indeed, it should be known that this very equation can be derived through the method of live forces, so that their use and their validity hence are substantiated against Adversaries.

COROLLARY 1. In order to determine the maximum velocity of the liquid flowing out and, furthermore, that of the liquid descending in the vessel, one need merely set $dz = 0$; after this has been done, our equation will furnish $(hh - mm)z = hha - hmx$, from which $z = \dfrac{hha - hmx}{hh - mm}$, which, since until now it contains the unknown x itself, indeed determines nothing unless the value of z itself is also found at the same time from the general equation.

COROLLARY 2. If m is very small with respect to h itself, the general equation assumes this form: $z\,dx + b\,dz = a\,dx$; from which $dx = \dfrac{b\,dz}{a - z}$, which gives $z = a - a/e^{x/b}$. Thus, in this case, in order that z be a *maximum*, it is necessary that x be infinite, and then it will occur that $z = a$, which certainly can be gathered at once from $dx = \dfrac{b\,dz}{a - z}$, or from $dx(a - z) = b\,dz$; for by making $dz = 0$, on account of z itself being a maximum, one will have $a - z = 0$, and from that, $z = a$. From there, in turn, it is clear that in a very wide vessel the water flowing out through a very narrow pipe immediately acquires a maximum velocity, always constant thereafter and equal to that which a weight falling freely from the height of the vessel would acquire, as we saw above in Coroll. 1, §X. Obviously, in this case the vessel can be considered as always full because, on account of the comparatively infinite area of the vessel with respect to the constricted [one] of the pipe, an almost infinite time as well would certainly be required before the water descended noticeably in the former.

SECTION XIII

Now take another case. Let the pipe (which is assumed full of water right up to C at the beginning before the flow) be extended indefinitely; thus it is certain that while the water is descending in the vessel, nothing can flow out of the pipe, but always some of the liquid descends from the vessel into the pipe, and this together with that which is already assumed to be present there is forced to flow, propelled jointly within the pipe. The law of acceleration and the velocity itself are sought after an arbitrary space has been flowed through within the cavity of the pipe. The accelerative force in the pipe, as was shown in §XI, here also will be $v\,dv/dx$, but the mass of water to be propelled now is $mb + mx$, which, multiplied by the accelerative force $v\,dv/dx$, gives the motive force in the pipe $(mbv\,dv + mxv\,dv)/dx$; this, translated to the area of the vessel, gives the equivalent motive force in the vessel equal to $(hbv\,dv + hxv\,dv)/dx$. And so the three motive forces—through the throat, the pipe, and the vessel—having been joined, and these having been equated to the total motive force p, for the vessel always filled by new water flowing in, this equation (see §XI) will result:

$$\frac{hh - mm}{2h} vv + \frac{hbv\, dv + hxv\, dv}{dx} + \frac{mav\, dv}{dx} = p = gha;$$

but for the vessel accepting no new liquid, this other will appear (see §XII):

$$\frac{hh - mm}{2h} vv + \frac{hbv\, dv + hxv\, dv}{dx}$$
$$+ \frac{mv\, dv}{h\, dx}(ha - mx) = p = g(ha - mx).$$

With gz substituted for $\tfrac{1}{2}vv$, the prior equation gives this:

$$(hh - mm)z\, dx + (hhb + hma + hhx)\, dz = hha\, dx;$$

but the latter gives this:

$$(hh - mm)z\, dx + (hhb + hma + hhx - mmx)\, dz = (hha - hmx)\, dx.$$

However, either equation can be integrated through the Lemma promised above, which I now show.

Section XIV

LEMMA. Let the equation to be integrated (and indeed without the necessity of separating the unknowns) be

$$\alpha z\, dx + (\beta + \gamma x)\, dz = (\epsilon + \theta x) \cdot dx.$$

I write y for $\beta + \gamma x$, from which $dx = dy/\gamma$, and the equation is changed into this: $\dfrac{\alpha}{\gamma} z\, dy + y\, dz = (\epsilon + \theta x)\, dx;$ after this has been multiplied by $y^{\alpha/\gamma - 1}$, there will be obtained

$$\frac{\alpha}{\gamma} z y^{\alpha/\gamma - 1}\, dy + y^{\alpha/\gamma}\, dz = (\epsilon + \theta x)\, dx \cdot y^{\alpha/\gamma - 1}$$
$$= (\epsilon + \theta x) \cdot \frac{1}{\gamma} \cdot (\beta + \gamma x)^{\alpha/\gamma - 1} \gamma\, dx.$$

After integration there will result

$$y^{\alpha/\gamma} z = \int (\epsilon + \theta x) \frac{1}{\gamma} (\beta + \gamma x)^{\alpha/\gamma - 1} \gamma\, dx$$
$$= \frac{1}{\alpha} (\beta + \gamma x)^{\alpha/\gamma} \cdot (\epsilon + \theta x) - \int \frac{\theta}{\gamma \alpha} (\beta + \gamma x)^{\alpha/\gamma} \gamma\, dx$$
$$= \frac{1}{\alpha} (\beta + \gamma x)^{\alpha/\gamma} (\epsilon + \theta x) - \frac{\theta}{\alpha\alpha + \gamma\alpha} (\beta + \gamma x)^{\alpha/\gamma + 1}$$
$$- \frac{\epsilon}{\alpha} \beta^{\alpha/\gamma} + \frac{\theta}{aa + \gamma\alpha} \beta^{\alpha/\gamma + 1}.$$

MOTION OF WATER THROUGH VESSELS

Let it be noted here that the last two terms given are added for the purpose of rectifying the equation, as is customary; for, with x vanishing, z certainly vanishes also. Now let the equation be divided by $y^{\alpha/\gamma}$, that is, by $(\beta + \gamma x)^{\alpha/\gamma}$, and the correct value of z itself results, namely

$$z = \frac{1}{\alpha}(\epsilon + \theta x) - \frac{\theta}{\alpha\alpha + \gamma\alpha}(\beta + \gamma x)$$
$$+ \left(\frac{\theta}{\alpha\alpha + \gamma\alpha}\beta^{\alpha/\gamma + 1} - \frac{\epsilon}{\alpha}\beta^{\alpha/\gamma}\right) \cdot (\beta + \gamma x)^{-\alpha/\gamma}.$$

Section XV

Therefore, in order that the application of this might be made to the prior equation, $(hh - mm)z\,dx + (hhb + hma + hhx)\,dz = hha\,dx$, here there will be $\alpha = hh - mm$, $\beta = hhb + hma$, $\gamma = hh$, $\epsilon = hha$, and $\theta = 0$, by substitution of which there will be obtained

$$z = \frac{hha}{hh - mm} - \frac{hha}{hh - mm}(hhb + hma)^{(hh-mm)/hh}$$
$$\times (hhb + hma + hhx)^{(-hh+mm)/hh},$$

or, which is the same thing,

$$z = \frac{hha}{hh - mm}\left(1 - \left(\frac{hb + ma}{hb + ma + hx}\right)^{(hh-mm)/hh}\right).$$

But if on the contrary it is applied to the latter, where $\alpha = hh - mm$, $\beta = hhb + hma$, $\gamma = hh - mm$, $\epsilon = hha$, and $\theta = -hm$, there results

$$z = \frac{hha - hmx}{hh - mm} + \frac{hm}{2(hh - mm)^2}(hhb + hma + hhx - mmx)$$
$$+ \left[\frac{-hm}{2(hh - mm)^2}(hhb + hma)^2 - \frac{hha}{hh - mm}(hhb + hma)\right]$$
$$\cdot (hhb + hma + hhx - mmx)^{-1};$$

after these have been separated in order, by proceeding as usual, one has at last

$$z = \left(\frac{hhax - \tfrac{1}{2}hmxx}{hhb + hma + hhx - mmx}\right).$$

COROLLARY I. If m is very small with respect to h, there will be, for the case of the vessel always full, $z = \dfrac{ax}{b + x}$; similarly, for the other case, there results $z = \dfrac{ax}{b + x}$, which certainly must happen thus generally, because, indeed, on account of m being infinitely small, the water must flow out for an infinite time before its uppermost surface descends noticeably in the very large vessel. In any case, it is evident that it is the same as if the vessel remained always full, and accordingly these two cases certainly reduce to the same.

COROLLARY 2. If $b = 0$, that is, if no water is contained in the indefinitely long horizontal pipe FB at the beginning of flow, for the case of the vessel always full,

$$z = \frac{hha}{hh - mm}\left(1 - \left(\frac{ma}{ma + hx}\right)^{(hh-mm)/hh}\right);$$

but for the other case in which no new liquid is received,

$$z = \frac{hhax - \tfrac{1}{2}hmxx}{hma + hhx - mmx}.$$

In this last case the following is also noteworthy: at that moment at which the surface of the liquid will have descended all the way to the bottom of the vessel, which is done by assuming $x = \dfrac{h}{m}a$, one will have $z = \tfrac{1}{2}a$, that is, the velocity of the water in the pipe after total depletion of the vessel will be that which a weight would acquire by falling from half the height of the vessel.

CONCERNING A CONDUIT OF THREE OR MORE PIPES

SECTION XVI

Let there now be (Fig. 3) the conduit AL consisting of three pipes, AD, GC, and BL, all full of water. And let there be a motive force p

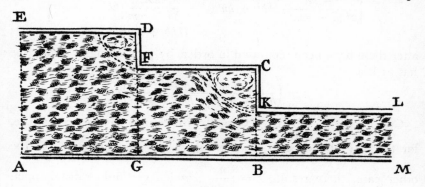

FIGURE 3

MOTION OF WATER THROUGH VESSELS

which, expanded uniformly over the surface AE, drives or presses on the same. The acceleration and the actual velocity with which the water flows out of the pipe BL are sought.

First of all it should be noted here that two very short throats are formed, one in the transition at GF, the other in the transition at BK, which individually require their own motive forces which are to be translated to the area AE, to which then are to be added the motive forces of the aqueous columns contained in the individual pipes, after translation of these forces to the area AE; after this has been done, the sum of all these translated forces is to be equated to the total motive force p, from which the desired equation will result.

Section XVII

Consequently, let the lengths of the pipes be $AG = a$, $GB = b$, $BM = c$, and their areas be $AE = h$, $GF = m$, and $BK = n$. Here also let the velocity in the last pipe BL be designated as v, the velocity in the second pipe GC as $u = \frac{n}{m} v$. Thus there will be, through the reasoning furnished in §IX, a motive force at the surface AE, required for forming the throat through GF, equal to

$$\frac{hh - mm}{2h} uu = \frac{hhnn - mmnn}{2hmm} vv$$

(by substitution of the value of uu itself, which is $\frac{nn}{mm} vv$); likewise the motive force in the pipe GC required for the throat through BK is $\frac{mm - nn}{2m} vv$, which indeed, after having been translated to the area AE, by making $\frac{mm - nn}{2m} vv$ into $\frac{hmm - hnn}{2mm} vv$, in proportion as m is to h, gives the motive force in the first pipe AD for producing the throat through BK; and so both forces added together give $\frac{hhnn - mmnn}{2hmm} vv + \frac{hmm - hnn}{2mm} vv$, that is, $\frac{hhmm - mmnn}{2hmm} vv$ or $\frac{hh - nn}{2h} vv$, equal to p. And thus is determined the velocity of flow through the three pipes after the former has reached constancy.

Corollary. Hence it is evident that the water is moved through the three pipes in the same manner as if, the second having been removed, the third were immediately attached to the first, after it has been stipulated, of course, that the flow has reached the greatest and constant velocity; finally, it is now clear that henceforth, no matter how many pipes are considered, the motive forces at the individual throats, translated to the first pipe and added together, are equivalent to that unique motive force in the first pipe to be applied to the unique throat which would be made by attaching the last pipe immediately to the first pipe. And thus the same constant velocity to which the flow converges is obtained in each case, whether the water goes through the entire conduit composed of all the pipes, or, the intermediate ones having been omitted, through the first and the last connected to each other directly. Everything, therefore, which we have shown above concerning constant velocity through two pipes is to be applied to a conduit consisting of as many pipes as one might wish.

Section XVIII

Now there comes up to be considered the acceleration in a conduit of many pipes when indeed the flow of water begins from rest, with the first pipe, however, remaining always full by means of the influx of new water following the descending [water] at the same velocity. In this matter, nothing else is to be done than to translate the motive force, considered to be in proportion to the aqueous mass to be driven through the individual pipes, to the area of the first pipe. If the sum of these translated motive forces is added to the motive force through the throats, that is, through that single one which would be formed if the last pipe were attached directly to the first, the force of all will result, which is to be made equal to p itself.

Section XIX

Thus let us apply this rule to a conduit of three pipes the lengths of which are a, b, and c, and areas h, m, and n. Let x be the length of the space through which the water, beginning from rest, travels in the last or third pipe and v be the velocity acquired in this pipe. For the purpose of imitating the process in §XI, $\frac{n}{m}x$ will be the distance which the water travels in the same time in the second pipe, and $\frac{n}{m}v$ its acquired velocity. Likewise $\frac{n}{h}x$ will be the distance traveled in the first pipe, and $\frac{n}{h}v$ the acquired velocity. Hence the accelerative force in the third pipe is $\frac{v\,dv}{dx}$, and this multiplied by the aqueous mass nc in this pipe will give the motive force $\dfrac{ncv\,dv}{dx}$, which, translated to the first pipe, will give the equivalent $\dfrac{hcv\,dv}{dx}$. Thus also the accelerative force in the second pipe is $\dfrac{nn}{mm}v\,dv\Big/\dfrac{n}{m}dx = \dfrac{nv\,dv}{m\,dx}$, which, applied to the mass of water mb of the second pipe, gives the motive force $\dfrac{nbv\,dv}{dx}$, which, translated to the first pipe, yields $\dfrac{hnbv\,dv}{m\,dx}$. Thus the accelerative force in the first pipe, $\dfrac{nn}{hh}v\,dv\Big/\dfrac{n}{h}dx = \dfrac{nv\,dv}{h\,dx}$, applied to the mass ha of the first pipe, gives the motive force of the water in the first pipe $\dfrac{nav\,dv}{dx}$, which, since it is already in the first pipe, is not to be translated farther. Those three forces are therefore $\dfrac{hcv\,dv}{dx}$, $\dfrac{hnbv\,dv}{m\,dx}$, and $\dfrac{nav\,dv}{dx}$, the sum of which, added to the force caused by the throats, will be found to be $\dfrac{hh-nn}{2h}vv + \left(hc + \dfrac{hnb}{m} + na\right)\dfrac{v\,dv}{dx}$, equal to the total force p.

Section XX

Now let there be four pipes, of which the lengths are a, b, c, and e and the areas are h, m, n, and q; let x be the distance traveled in the last pipe and v the velocity acquired in the last pipe. In order to observe the uniformity and the law of progression from one pipe to another, I will begin at the first, in which the accelerative force is $\dfrac{qq}{hh}v\,dv\Big/\dfrac{q}{h}dx$, the velocity is $\dfrac{q}{h}v$, and the element of velocity is $\dfrac{q}{h}dv$. And as the element of distance to be traveled is $\dfrac{q}{h}dx$, so there results from the law of acceleration the accelerative force

$$\frac{qq}{hh}v\,dv\Big/\frac{q}{h}dx = \frac{qv\,dv}{h\,dx};$$

and by multiplying this by the mass of water to be moved, [one causes] this motive force $hqav\,dv/h\,dx$ to appear, which, because it is already in the first pipe, does not require further translation. But in the second pipe, the accelerative force $\dfrac{qq}{mm}v\,dv\Big/\dfrac{q}{m}dx = \dfrac{qv\,dv}{m\,dx}$, applied to the aqueous mass mb, gives the motive force in the second pipe $mqbv\,dv/m\,dx$, which, translated to the first pipe, gives the equivalent $hqbv\,dv/m\,dx$. In the same manner, the motive force translated from the third pipe to the first will be $hqcv\,dv/n\,dx$, and the motive force translated from the fourth to the first will be $hqev\,dv/q\,dx$. Therefore, all added together equal

$$\frac{hqav\,dv}{b\,dx} + \frac{hqbv\,dv}{m\,dx} + \frac{hqcv\,dv}{n\,dx} + \frac{hqev\,dv}{q\,dx} = \left(\frac{a}{h} + \frac{b}{m} + \frac{c}{n} + \frac{e}{q}\right)\frac{hqv\,dv}{dx}.$$

Generally, therefore, for any number of pipes whatever, the lengths of which are $a, b, c, \ldots, \pi$ and areas $h, m, n, \ldots, \omega$, the sum of all the motive forces translated to the first pipe will be equal to

$$\left(\frac{a}{h} + \frac{b}{m} + \frac{c}{n} + \frac{e}{q} \cdots + \frac{\pi}{\omega}\right)\frac{h\omega v\,dv}{dx};$$

to which if the motive force $\dfrac{hh - \omega\omega}{2h}\,vv$ for all the throats is added, there emerges the total motive force to be set equal to p itself. This equation results therefrom:

$$\frac{hh - \omega\omega}{2h}\,vv + \left(\frac{a}{h} + \frac{b}{m} + \frac{c}{n} \cdots + \frac{\pi}{\omega}\right)\frac{h\omega v\,dv}{dx} = p;$$

or, by writing gz for $\frac{1}{2}vv$, this other:

$$\frac{hh - \omega\omega}{h}\,z + \left(\frac{a}{h} + \frac{b}{m} + \frac{c}{n} \cdots + \frac{\pi}{\omega}\right)\frac{h\omega\,dz}{dx} = \frac{1}{g}p,$$

or,

$$(hh - \omega\omega)\,z\,dx + \left(\frac{a}{h} + \frac{b}{m} + \frac{c}{n} \cdots + \frac{\pi}{\omega}\right)hh\omega\,dz = \frac{h}{g}p\,dx.$$

COROLLARY 1. If the lengths a and π of the first and last pipes and the lengths of the intermediate ones as well remain invariable, the first certainly through continual influx, the last through efflux, the sum of the series $\dfrac{a}{h} + \dfrac{b}{m} + \dfrac{c}{n} \cdots + \dfrac{\pi}{\omega}$ will be constant, which may be called M, and $p = gha$, from which this equation appears:

$$\frac{hh - \omega\omega}{h}\,z + \frac{Mh\omega\,dz}{dx} = ha,$$

or,

$$(hh - \omega\omega)z\,dx + Mhh\omega\,dz = hha\,dx,$$

of which x is computed by logarithms given in z, z itself by numerals given in x.

COROLLARY 2. But if now, with no new water flowing in, the first pipe be depleted by the flowing out through the last, of given length—just as would be done if the first pipe in the form of a vertically erected vessel were to contain liquid pressed by its own weight during the time in which it would be expelled through the horizontal conduit which the remaining pipes form—and if the distance traveled through the last pipe from rest be called x, the height of the liquid remaining in the cylindrical vessel will be $a - \frac{\omega x}{h}$; and therefore from the series $\frac{a}{h} + \frac{b}{m} + \frac{c}{n} \cdots + \frac{\pi}{\omega}$ there is now to be removed $\frac{\omega x}{hh}$, and for $\frac{1}{g}p$ there must be written $ha - \omega x$, which gives this equation:

$$(hh - \omega\omega)z\,dx + Mhh\omega\,dz - \omega\omega x\,dz = (hha - h\omega x)\,dx,$$

which through the Lemma of §XIV, can be integrated.

COROLLARY 3. Furthermore, if the last pipe be prolonged indefinitely, such that with the uppermost surface of the water descending in the vessel, the water indeed does not flow out of the last pipe but is continually thrust forward in it more and more, there is to be written in the series not only $a - \frac{\omega x}{h}$ for a, but also $\pi + x$ for π, and so for this case we will acquire this other equation:

$$(hh - \omega\omega)z\,dx + Mhh\omega\,dz - \omega\omega x\,dz + hhx\,dz = (hha - h\omega x)\,dx.$$

This is integrable through the same Lemma.

COROLLARY 4. If by consideration of the vessel itself as the first pipe there results $\frac{a}{h} = \frac{b}{m} = \frac{c}{n} \cdots = \frac{\pi}{\omega}$, that is, if the lengths of the pipes of which the number be N be everywhere proportional to their respective sizes, our general equation is changed to this:

$$(hh - \omega\omega)z\,dx + Nha\omega\,dz = \frac{1}{g}hp\,dx.$$

COROLLARY 5. But if, with exception of the vessel or the first pipe, there results $\frac{b}{m} = \frac{c}{n} \cdots = \frac{\pi}{\omega}$, and if the number of remaining pipes be N, surely there will be

$$(hh - \omega\omega)z\,dx + ha\omega\,dz + \frac{Nhh\omega\,dz}{m} = \frac{1}{g}hp\,dx.$$

COROLLARY 6. There shall now be an infinite number of pipes, but each one of them, except the first, of infinitely short length, such that all of them combined represent a truncated conoidic conduit of which the anterior area equals m, and posterior equals ω, such as $RSTV$ (Fig. 4). If this is conceived to be cut by two closely spaced planes,

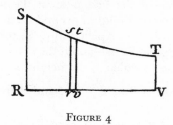

FIGURE 4

sr and tv, parallel to SR and TV themselves, $srvt$ will be one of those short pipes, having for a length an element of the length RV of the whole conduit, and for an area the plane sr. From this, in order to obtain the sum of the series $\dfrac{b}{m} + \dfrac{c}{n} \cdots + \dfrac{\pi}{\omega}, \dfrac{vr}{sr}$ must be integrated, which in many examples can be done algebraically: for instance, if ST be a straight line, that is, if $SRVT$ be an ordinary truncated cone; likewise if ST be the arc of a hyperbola of any kind whatever asymptotic to RV.

Section XXI

Let us illustrate the very matter in the prior example. Let $SRVT$ be a truncated cone of which the anterior area $SR = m$, the posterior area $TV = \omega$. Furthermore let their semidiameters be proportional to $\sqrt{m}$ and $\sqrt{\omega}$. Henceforth let its abscissa Vv be equal to t, its element $vr = dt$, the semidiameter of the area $tv = y$, and the total length of the pipe $RV = L$; y will be found proportional to $(t\sqrt{m} - t\sqrt{\omega} + L\sqrt{\omega})/L$; but the area sr itself, which is proportional to yy, is $(t\sqrt{m} - t\sqrt{\omega} + L\sqrt{\omega})^2/L^2$, because of which

$$\frac{vr}{sr} = \frac{L^2\, dt}{(t\sqrt{m} - t\sqrt{\omega} + L\sqrt{\omega})^2};$$

the integral of which, rectified in an appropriate manner, is $\dfrac{Lt}{t\sqrt{m\omega} - t\omega + L\omega}$; and therefore, through the whole conduit $RSTV$, taking Vv, or t, $= VR = L$, the required integral is found as $\dfrac{L}{\sqrt{m\omega}} = \dfrac{b}{m} + \dfrac{c}{n} \cdots + \dfrac{\pi}{\omega}$. And thus our general equation of §XX,

$$(hh - \omega\omega)z\, dx + \left(\frac{a}{h} + \frac{b}{m} + \frac{c}{n} \cdots + \frac{\pi}{\omega}\right)hh\omega\, dz = \frac{h}{g}p\, dx$$

MOTION OF WATER THROUGH VESSELS

will give the following equation for the conduit or conic pipe the length of which is L and the two extreme areas of which are m and ω, for the height of the vessel being a and the area being h:

$$(hh - \omega\omega)z\,dx + ha\omega\,dz + \frac{hhL\omega\,dz}{\sqrt{m\omega}} = \frac{1}{g}hp\,dx.$$

Section XXII

For the case of Coroll. 1, §XX, there will result as well

$$(hh - \omega\omega)z\,dx + Mhh\omega\,dz = hha\,dx,$$

where $M = \dfrac{a}{h} + \dfrac{L}{\sqrt{m\omega}}$. For the case of Coroll. 2 of the same section, the same having been assumed for M, there will result

$$(hh - \omega\omega)z\,dx + Mhh\omega\,dz - \omega\omega x\,dz = (hha - h\omega x)\,dx.$$

For the case of Coroll. 3, it must be understood that the conic pipe has a cylindrical pipe of indeterminate length and of area ω attached to it at its extremity so that the propelled water may always be contained in it and travel through the distance x from the beginning of motion; there will then result

$$(hh - \omega\omega)z\,dx + Mhh\omega\,dz - \omega\omega x\,dz + hhx\,dz = (hha - h\omega x)\,dx.$$

Section XXIII: General Scholium

Many other corollaries could be derived from these, useful no less than curious and elegant. Certainly those which pertain to this matter have their entire basis in the ones already transmitted and explained; it is clear that I did not indicate this in expressed words. For example, we have supposed that some water or any other liquid gravitates in the first pipe only, just as in a vessel, and from there is forced through a conduit having a horizontal position, through which the water, while it is being moved, is deserted, so to speak, by its own gravity. Meanwhile if in this conduit, or in the pipes which compose it, it also retains its own gravity, whether the total or only a part, as would happen if the pipes were not horizontal but either vertical or inclined to the horizontal in some diverse manner, this causes no difficulty at all. For instance, the very weight of the water from any pipe whatever, by means of §II, can be translated to the vessel or the first pipe so that the water in the remaining pipes may be considered as being without gravity. But the translated gravities, added together with the weight of the water in the vessel or the first pipe, can be considered in place of that which we called p or the fundamental motive force by which the flow of the total aqueous mass is generated. Thus if, arbitrarily, the conduit $EGBL$ (Fig. 5) were to

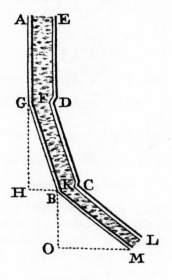

FIGURE 5

consist of three pipes, AD, GC, and BL, of different areas, the first of which, AD, would have an area AE or GD, the second GC an area GF or BC, the third BL an area BK or ML; and if the first were vertical, the second were to make the angle GBH with the horizontal, the third the angle BMO; and if the areas be $AE = h$, $GF = m$, and $BK = n$; then the force of gravity or the natural accelerative force is g, and the motive force in the pipe AD, full of water, is $gh \cdot AG$. Likewise $GB/GH = g \Big/ \dfrac{g \cdot GH}{GB}$ is the accelerative force of the liquid in pipe GC. Similarly $BM/BO = g \Big/ \dfrac{g \cdot BO}{BM}$ is the accelerative force in the pipe BL. Thus $\dfrac{g \cdot GH}{GB} m \cdot GB$ or $gm \cdot GH$ will give the motive force of the water in the second pipe; similarly $gn \cdot BO$ gives the motive force of the water in the third pipe. But the motive forces are now to be translated from the oblique pipes GC and BL to the vertical one by setting $m/h = gm \cdot GH/gh \cdot GH$, and $n/h = gn \cdot BO/gh \cdot BO$. And so in this way all the water in the pipes can be considered as being without gravity, but in its place the first column AD can be considered pressed by the

MOTION OF WATER THROUGH VESSELS 377

motive force expanded uniformly over the surface AE, which force would be equal to $gh(AG + GH + BO) = gh \cdot A = p$ (because $AG + GH + BO$ equals the total vertical height of the conduit, which let be A). And so we have reduced this case and other similar ones to our general method.

Note: If one or more of the oblique pipes is directed upward, there will be, on account of it or them, a negative motive force translated to the first pipe, and there will have to be assumed for A an excess by which the sum of the positives is greater than the sum of the negatives, or vice versa. In a word, A will be the excess or defect by which the surface of the water in the first pipe is higher or lower than the horizontal which is the surface of the water in the last pipe. This serves in the determination of the law according to which liquids oscillate in pipes curved in any shape whatever. At this point also refer to the following Problem, proposed to me by my Son six or seven years ago but expressed slightly more generally.

HYDRAULICS PROBLEM

Section XXIV

ABCD [Fig. 6] *is a vessel filled with water to EF. GI is a cylindrical pipe the section KI of which is also full of water. With the thumb spread over the orifice GO, the pipe is immersed in the water contained in the vessel, but just to the extent that the section MI of the pipe, greater than KI, penetrates the external water up to MN. The thumb now having been removed, the surface KL will scend (on account of the prevailing pressure of the external water), and on account of the impulse received, it will reach above the surface EF up to PQ. The height MP or NQ is sought to which without question the water in the pipe certainly can rise.*

SOLUTION. Let the immersed part HM of the pipe be set equal to a, its portion HK originally full of water equal to b (less than a), the area EF of the vessel h, the area GO or HI of the pipe m. But the water now surrounding the pipe and pressing down by its own weight tries to enter through the opened orifice HI and to ascend by propelling the portion of water HL lying above. I understand that action and effect in this way. Let another pipe, facing downward, having an area h equal to the area EF of the vessel and having a height $HM = a$, be attached to the orifice HI. Let this pipe be full of water, but of water such that it would be raised, that is, would be pushed upward, and certainly by a force precisely as large as that by which the water gravitates downward in the vessel of height MH (in place of which that in the pipe is substituted through a fiction of the mind). Accordingly, the motive force of the water in this fictitious pipe extending upward will be *gha*, and thus, with respect to this, the negative motive force of the water *HL* in the pipe *HO* will be *gmb*, which translated to the fictitious pipe gives *ghb*, which, of course, being opposite to that *gha*, is to be subtracted from the same, and there remains $gha - ghb$ or $gh(a - b)$ for the motive force which we called *p*, to which, therefore, are to be equated the motive forces which are generated by the flow through the throat to be formed at the entrance *HI* in flowing through the pipe *HO* and rising in the fictitious pipe. Thus if, beginning at *KL*, the distance traveled by the water within the pipe *HO* is called *x*, and as well the distance through which the surface of the water in the fictitious pipe travels in rising is $\frac{m}{h}x$, for the sake of an imitation of the reasoning of §XIII we shall have the accelerative force in the pipe *HO* equal to $\frac{v\,dv}{dx}$, which, multiplied by the mass of the water to be pressed upward, $mb + mx$, gives the motive force in this pipe as $(mb + mx)\frac{v\,dv}{dx}$, [which force is] to be translated to the fictitious pipe in order that from there we might obtain the equivalent motive force $(hb + hx)\frac{v\,dv}{dx}$. And since in addition in the fictitious pipe (in which the water ascends at the velocity $\frac{m}{h}v$ through the distance $\frac{m}{h}x$) its own motive force, which is not to be translated any further, is $(ha - mx)\frac{mv\,dv}{h\,dx}$, therefore, after that which is required for forming the throat has been added to these two forces, we will obtain the total motive force

$$\frac{hh - mm}{2h}vv + (hb + hx)\frac{v\,dv}{dx} + (ha - mx)\frac{mv\,dv}{h\,dx}.$$

However, since here

$$p = gh\left(a - \frac{m}{h}x - b - x\right) \quad \text{or} \quad g(ha - hb - mx - hx),$$

there will result the equation for determining the velocity *v*, namely this:

$$\frac{hh - mm}{2h}vv + \left(hb + hx + ma - \frac{mm}{h}x\right)\frac{v\,dv}{dx} = g(ha - hb - mx - hx);$$

which having been reduced, and gz having been written for $\tfrac{1}{2}vv$, will yield

$$(hh - mm)z\,dx + (hhb + hhx + hma - mmx)\,dz$$
$$= (hha - hhb - hmx - hhx)\,dx,$$

which through the Lemma of §XIV is integrable. If the vessel AC or the fictitious pipe is of exceedingly great size (which is the tacit understanding of the Problem), there emerges, obviously, this much simpler equation (after the terms in which m appears have been neglected, and the remaining ones divided by hh)

$$z\,dx + (b + x)\,dz = (a - b)\,dx - x\,dx;$$

this, integrated, gives $(b + x)z = (a - b)x - \tfrac{1}{2}xx$, from which, if $z = 0$, that is, if the surface KL ceases to rise, which happens when it reaches the maximum height PQ to which it can ascend, it is necessary then that also $(a - b)x - \tfrac{1}{2}xx$ becomes $= 0$, wherefore $a - b = \tfrac{1}{2}x$ or $x = 2a - 2b$; therefore $KP = 2KM$.

Section XXV

The same Problem can be solved more easily if it is considered as the case in §XIII. By understanding, of course, that the vessel AF in Fig. 2, full of water at the beginning of flow, has a height $a = MH$ in Fig. 6, and the pipe FC, which is horizontal in Fig. 2, is now verti-

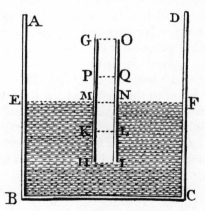

Figure 6

cally erect and continued indefinitely, in the latter let the lowest part of length $b = HK$ be full of water at the beginning in Fig. 6. Now, therefore, if because of the prevailing pressure of the aqueous column in the vessel the water in the pipe ascends above b through the distance x, and likewise that in the vessel descends through the distance

$\frac{m}{h}x$, we will have the motive force about to arise in the vessel due to the weight of the water lying above equal to $g(ha - mx)$, and the motive force opposite to the prior [force] in the vertical pipe coming from the weight of all the water existing in the pipe equal to $g(mb + mx)$, which, translated to the vessel, gives $g(hb + hx)$, to be subtracted from the former $g(ha - mx)$. And so there remains $p = g(ha - hb - hx - mx)$, to which must be equated the sum of the three motive forces generated by the motion through the throat, the pipe, and the vessel, as we found in §XIII; after that has been done, the following equation is obtained:

$$\frac{hh - mm}{2h}vv + \frac{(hbv\,dv + hxv\,dv)}{dx} + \frac{mv\,dv}{h\,dx}(ha - mx)$$
$$= p = g(ha - hb - hx - mx);$$

which, after corresponding terms have been joined together, will have this form:

$$\frac{hh - mm}{2h}vv + \left(hb + hx + ma - \frac{mmx}{h}\right)\frac{v\,dv}{dx} = g(ha - hb - hx - mx),$$

exactly that which we found just above.

Section XXVI

From our Theory set forth so far, the physical reason can be given (which, certainly, neither Newton nor anybody else gave correctly from purely dynamical principles) as to why obviously a solid cylindrical body which is moved uniformly in a continuous infinite fluid of the same density as the body, with its own base directed forward, suffers a resistance equal to the weight of the cylindrical body, with the assumption, of course, that the velocity of the body is equal to that which a heavy weight can acquire by falling freely from a height equal to the side of the cylinder. From a number of proofs, which are mine, it is pleasing to give the following support to our Hydraulics Theory in this writing.

Let the cylinder $RMNS$ (Fig. 7), which may be moved in the direction of side MN, be in a standing fluid, equally dense, continuous, and infinite. Let the velocity of the cylinder be v, the side MN be a, and the base, or the area NS, be h. Let us imagine that in place of the solid cylinder there is the pipe MS full of the same fluid matter, and through this stationary pipe (where beyond the boundary configuration I consider nothing additional) there flows at a continuous and constant velocity v an integral fluid cylinder, such that the pipe remains always full, and that as much as flows out through *NS* is brought in through *MR* as a new supply at the same speed. To an observer it becomes manifest at once that the fluid cylinder in efflux through

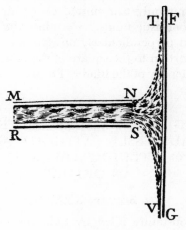

FIGURE 7

NS meets, of course, the same resisting force by the approach to the standing fluid external and opposed to the motion as a solid cylinder itself would meet, because the fluid cylinder, while it is moved through the pipe, can be considered as solid, and all the other circumstances are equal. Therefore, it is only to be seen how great the resistance is which the fluid going out is experiencing at the moment of egress itself. However, it is evident that this resistance develops from the throat *TNSV* which is formed behind the orifice *NS* of the pipe; the shape of this throat must be such that at an arbitrary small distance it has the asymptote *FG* perpendicular to the direction of the axis of the pipe for the reason that, on account of the very rapidly decreasing and wholly vanishing motion of the fluid which has come out, the areas of the throat must increase in turn, and in a very short time they must be spread out practically to infinity. I assume, to be sure, that the fluid coming out of the pipe is not miscible with the other at rest outside. Thus through those matters which were explained in §IX, and because the ultimate velocity in the throat is v, the force through the throat will be $\frac{1}{2}hvv$, and as well, on account of the constant velocity in the pipe, there will result, through Coroll. 2, §IX, $\frac{1}{2}hvv = p = gha$, that is, $\frac{1}{2}vv = ga$. Thus by writing gz for $\frac{1}{2}vv$, there will result $z = a$. It is proper, therefore, in order that the resistance become equal to the weight of the cylinder, that the required velocity of the fluid in the pipe should be that which a heavy object falling freely from the height *a* would acquire. Q.E.D.

COROLLARY. From the fundamental property shown (previously not sufficiently accurately established), there follow all things beyond this which are commonly transmitted concerning the resistances of continuous and nonelastic fluids. Certainly the resistances in fluids of this sort exerted perpendicularly on opposite planes of bodies are composed in proportion to the square of the relative velocity and the first power of the density of the fluid. From this at last the remaining are deduced.

CONCERNING THE PRESSURE ON THE BASE OF A CYLINDRICAL VESSEL (WITHOUT AN ATTACHED PIPE) DUE TO A FLUID FLOWING OUT THROUGH AN ORIFICE

SECTION XXVII

Let there be constantly filled by fluid the cylindrical vessel AF (Fig. 8), of which the area $AE = h$, the length AG or $EF = a$, and the

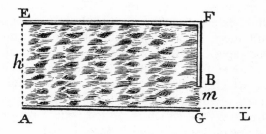

FIGURE 8

area GB of the orifice $= m$. Let the discharging fluid, after [fluid] has already been flowing out for some time, have a velocity v, so that in the vessel itself it has the velocity $\frac{m}{h} v$. Let x be the length GL of a cylinder of which the base is m, which cylinder may define the quantity of fluid already having flowed out. Now, furthermore, let the velocity which will exist afterwards be u, and the length of the previously mentioned cylinder of fluid which will flow out further be y, and also the total length of that which has flowed out and that which will flow out be $x + y$. Let us also consider the fluid to be free from all gravity, and likewise to have no other force for pressing the base than that which stems from the motion. This force meets an equal

MOTION OF WATER THROUGH VESSELS

resistance from the opposition of the base on account of the equality between action and reaction. But the resistance is found if the retardative force, which diminishes the velocity of the column of fluid, is found in the usual manner, and that [force], multiplied by the mass of the column, that is, by ha, will give the resistance of the pressure on the base. I accomplish the matter as follows: the equation shown in §XI,

$$\frac{hh - mm}{2h} vv + \frac{hbv\, dv}{dx} + \frac{mav\, dv}{dx} = p,$$

is changed in the present case (where the length b of the pipe, since it is absent, is 0, and the weight p of the column of fluid in the vessel is 0) into this particular equation,

$$\frac{hh - mm}{2h} vv + \frac{mav\, dv}{dx} = 0,$$

and by replacing v by u, into this similar one,

$$\frac{hh - mm}{2h} uu + \frac{mau\, du}{dx} = 0.$$

Section XXVIII

Through reduction, and after dy has been written for dx (for x is now constant, while $x + y$ is the indeterminate and variable length of the cylinder of fluid flowing out), the equation appears in this form:

$$\frac{hh - mm}{2h} dy + \frac{ma\, du}{u} = 0;$$

and after integration,

$$\frac{hh - mm}{2h}(x + y) + ma \ln u = ma \ln v + \frac{hh - mm}{2h} x.$$

I write this so by adding the last two constant terms for the sake of rectification, to the end that, for y vanishing and u beginning at v, the equation itself becomes an identity. And so there will result $ma \ln \left(\frac{u}{v}\right) = -\left(\frac{hh - mm}{2h}\right) y$, from which, by going to numbers and by putting $1 = \ln e$, one has $uu = vv e^{-(hh - mm)y/hma}$.

Section XXIX

After proper differentiation of this derived equation (after certainly having assumed v as constant), there will result

$$u\, du = -vv\, dy \left(\frac{hh - mm}{2hma}\right) \bigg/ e^{(hh - mm)y/hma}.$$

However, the accelerative force in the vessel is negative, that is, it is transformed into the retardative force $-\dfrac{mu\,du}{h\,dy}$, which accordingly will be $vv\left(\dfrac{hh-mm}{2hha}\right)\Big/e^{(hh-mm)y/hma}$.

Section XXX

This force, which is useful to us at the very first moment after the abolition or supposed cessation of the gravity which previously the column of fluid in the vertically erect vessel had, will be $vv = 2gz$, as certainly $y = 0$; and so that force which was found will be $gz\left(\dfrac{hh-mm}{hha}\right)$ (see §XI, where

$$z = \left(\dfrac{hha}{hh-mm}\right)\cdot(1 - 1/e^{(hh-mm)x/hma})),$$

and hence by multiplying by the mass ha of the fluid, the resistance or the pressure on the base arising from the motion of the fluid alone is found to be $gha(1 - 1/e^{(hh-mm)x/hma})$, to which if in addition is added the weight gha of the column of fluid, which in a vertical position acts constantly on the base whether the fluid is moving or at rest, there will result a total pressure $gha + gha(1 - 1/e^{(hh-mm)x/hma})$.

Corollary. If $x = \infty$, the total pressure will be $gha + gha/e^0 = 2gha$, as $e^0 = 1$. But if $x = 0$, the pressure at the base will be $gha + gha(1-1) = gha$, which is also obviously true from the fact that initially only the weight of the fluid cylinder acts on the base, and afterwards, with x increasing, the pressure also increases, and in such a way that it never attains $2gha$, much less exceeds it, although it may approach this quantity rather closely.

Scholium. On the other hand, let no one believe that the heavy liquid in the vessel perhaps presses on the base differently when it is being moved from when it is at rest; notwithstanding that the contrary is readily evident to an observer of the nature of immaterial forces; just as for example the cause of gravity was considered to be apart from a body, so these forces act instantly throughout the total mass to be set in motion, and thus they act in the same manner and they exert the same pressure on an obstacle as if the heavy liquid lying over it were at rest. Nevertheless, I will prove the truth of the matter in our case through calculus. Certainly it is clear at once that the heavy liquid is accelerated by descending in the vessel. Moreover, its motive force, however great it may be, must have two parts, of which one is given over to the retardative force at the base opposed to that to be considered, but the other part remaining is employed in the acceleration of the actual descent. But this latter is exactly that which results from the equation of §XI which is to be solved, that is,

$$z = \frac{hha}{hh - mm}\left(1 - 1/e^{(hh-mm)x/hma}\right),$$

by differentiating which and multiplying it by g we will have

$$g\,dz = v\,dv = \frac{h}{m}g\,dx/e^{(hh-mm)x/hma}.$$

Therefore, the remaining accelerative force in the vessel, or $\dfrac{mv\,dv}{h\,dx}$, is equal to $g/e^{(hh-mm)x/hma}$, to which if there is added the [accelerative] force which the above-found retardative force diminishes, $g(1 - 1/e^{(hh-mm)x/hma})$, together they constitute an accelerative force g, and therefore they constitute a pressure gha developing from the weight, that is, equal to the weight itself. Q.E.D.

And indeed one will have to proceed thus in the remaining cases where one or more pipes are attached to the vessel, so that, naturally, first of all one may find the retardative force on the base of any one of the given pipes by supposing that the fluid suddenly parts with its gravity, and then one may add to the discovered retardative force the pressure provided by the weight only of the fluid (by considering it as being at rest and stagnant) and propagated either directly to the first base or indirectly through the preceding pipes to whatever base we wish.

APPENDIX:

OUTLINE OF THE CALCULATION TO BE EMPLOYED FOR DETERMINING IN A SINGULAR WAY THE VELOCITIES OF WATER FLOWING THROUGH MANY PIPES FROM ONE TO ANOTHER, AND ESPECIALLY IF IT SHOULD FLOW OUT THROUGH SEPARATE INDIVIDUAL PIPES, AND HENCE TO BE EMPLOYED FOR FINDING THE PRESSURES EXERTED UPON THE BASES OF THE INDIVIDUAL PIPES.

In advance it is necessary to point out that we assume this conduit to be composed of several pipes attached to each other one by one, having any position whatever, vertical, horizontal, or inclined. We next assume that the conduit is constantly full of water and that the flow has come to steadiness, during which time as much liquid flows out of any pipe as is necessary for supplying the next smaller pipe, such that therefore every one is constantly full, and that they may be thus considered individually as if they were isolated and set apart.

Let the length of the first and largest pipe be a, that of the second and next smaller be b, that of the third be c, etc.; the area of the first be h, of the second be m, of the third be n, of the fourth be q, etc.; the orifice of the first pipe, equal to the area of the second pipe, be m, the orifice of the second be n, the orifice of the third be q, etc. The natural gravity be g, the natural gravity in different oblique directions be γ', γ'', γ''', etc. But for the vessel or first pipe let the gravity arising from the mutual action in the attached pipes be g', for the second g'', for the third g''', etc. Let the length of the aqueous cylinder having flowed out through the first orifice be x', that through the second x'', that through the third x''', etc. Let the height from which a weight having fallen naturally acquires the velocity of the water flowing out through the first orifice be z', that which pertains to the second z'', to the third z''', etc.

And so with these things having been set forth, and the remaining things as they are in hydraulics Literature, certainly there results, for pipes attached to each other in succession, through the translation of forces, and indeed for a conduit of two pipes $gha + \gamma'hb = g'ha + g''hb$ or $ga + \gamma'b = g'a + g''b$; for a conduit of three pipes $ga + \gamma'b + \gamma''c = g'a + g''b + g'''c$. I call these equations fundamental.

Next it is clear that there results $z' = \dfrac{nn}{mm} z'' = \dfrac{qq}{mm} z'''$, etc. Likewise $x' = \dfrac{n}{m} x'' = \dfrac{q}{m} x'''$, etc. On the other hand, through §XI, the following equations result for the individual pipes being vertically erect:

For the first pipe,
$$g(hh - mm)z'\,dx' + ghma\,dz' = g'hha\,dx'$$

For the second pipe,
$$g(mm - nn)z''\,dx'' + gmnb\,dz'' = g''mmb\,dx''$$

For the third pipe,
$$g(nn - qq)z'''\,dx''' + gnqc\,dz''' = g'''nnc\,dx''',$$

and so on in succession.

Note that, if any one of the pipes were horizontal, the γ pertaining to it would vanish in the fundamental equation. Thus for example if there were three pipes, of which the first were vertical but the remaining two were horizontal, the fundamental equation would be this: $ga = g'a + g''b + g'''c$; but if in fact all three were vertical, this fundamental equation would result: $g(a + b + c) = g'a + g''b + g'''c$. Now since that is the prize achievement in this investigation, it is necessary to define the forces of the gravities g', g'', g''', etc., resulting from the mutual action of the natural gravity, from which later the velocities as well as pressures on the bases of the pipes become known. Moreover, I describe this so. By imitation of the operations applied in §XI, that which follows is found for the individual pipes:

For the first pipe,
$$z' = \frac{g'}{g}\left(\frac{hha}{hh - mm}\right)(1 - 1/e^{(hh-mm)x'/hma})$$

For the second pipe,
$$z'' = \frac{g''}{g}\left(\frac{mmb}{mm - nn}\right)(1 - 1/e^{(mm-nn)x''/mnb})$$

For the third pipe,
$$z''' = \frac{g'''}{g}\left(\frac{nnc}{nn - qq}\right)(1 - 1/e^{(nn-qq)x'''/nqc})$$

And so on.

And accordingly, since z', z'', z''', etc., as well as x', x'', x''', are given, respectively, in terms of each other—namely $z' = \frac{nn}{mm} z'' = \frac{qq}{mm} z'''$, and also $x' = \frac{n}{m} x'' = \frac{q}{m} x'''$—let the values of the individual z and x be expressed by one of them, and there will result as many equations, less by one, as there are pipes, or as there are hypothetical gravities g', g'', g''', etc. Certainly, for example, for three pipes, z' having been retained to which the remaining z'', z''' are to be reduced, and also x'', x''' to the retained x', there will result these two equations:

$$z' \text{ or } \frac{g'}{g}\left(\frac{hha}{hh - mm}\right)(1 - 1/e^{(hh-mm)x'/hma}) = \frac{nn}{mm} z''$$

or
$$\frac{nng''}{mmg}\left(\frac{mmb}{mm - nn}\right)(1 - 1/e^{(n/m)(mm-nn)x'/mnb}),$$

and that first one is also the same as
$$\frac{qqg'''}{mmg}\left(\frac{nnc}{nn - qq}\right)(1 - 1/e^{(q/m)(nn-qq)x'/nqc}).$$

But since there are three hypothetical gravities g', g'', g''', to be sought, another equation is required at this point for the solution of the Problem. Moreover, this must be sought from the fundamental equation $ga + \gamma'b + \gamma''c = g'a + g''b + g'''c$, or (if indeed the two pipes are assumed horizontal) from this there is merely $ga = g'a + g''b + g'''c$, since γ' and γ'' vanish.

Let us make an application, for the sake of brevity, to the very simple case of two pipes constantly full of water, of which the first may be vertical, the other horizontal, and let us stipulate that the flow has reached uniformity, that is, x', x'', etc., $= \infty$. There will result one equation drawn from z',

$$g'\left(\frac{hha}{hh-mm}\right) = \frac{nng''}{mm}\left(\frac{mmb}{mm-nn}\right),$$

another from the fundamental $ga = g'a + g''b$, from which, by proceeding customarily, one has

$$g' = \frac{gnn(hh-mm)}{mm(hh-nn)} \quad \text{and} \quad g'' = \frac{ghha(mm-nn)}{mmb(hh-nn)}.$$

Hence all the remaining are derived, as certainly $z' = \dfrac{hhnna}{mm(hh-nn)}$ and $z'' = \dfrac{hha}{hh-nn}$, agreeing absolutely with those which we have proven and given above. Likewise the pressures on the base of any pipe are most easily determined. Since surely the individual pipes can be considered as if they were solitary, the formula must be employed which we found above for the first and only pipe with the writing of only the letters which are suitable for any other pipe considered as if alone or separate. Accordingly, since for that one alone the total pressure was found as $gha + gha[1 - 1/e^{(hh-mm)x/hma}]$, here the total pressure will have to be written:

For the first pipe, $g'ha + gha[1 - 1/e^{(hh-mm)x'/hma}]$,
For the second, $g''mb + gmb[1 - 1/e^{(mm-nn)x''/mnb}]$,
For the third, $g'''nc + gnc[1 - 1/e^{(nn-qq)x'''/nqc}]$;

and so, after the values of g', g'', and g''' themselves have been substituted—or, because we make the application to only two pipes, and indeed where $x = \infty$, merely the values of g' and g'' must be substituted, which are $g' = gnn(hh - mm)/mm(hh - nn)$, and $g'' = ghha(mm - nn)/mmb(hh - nn)$—the first pressure will result as $2gha\dfrac{nn(hh-mm)}{mm(hh-nn)}$, and the second pressure as $\dfrac{2ghha}{m}\left(\dfrac{mm-nn}{hh-nn}\right)$. If in addition $h = \infty$, but m and n are finite, the first pressure will be

$2gha\dfrac{nn}{mm} = \infty$, as is proper, but the second pressure will be $\dfrac{2ga}{m}(mm - nn)$, which is finite.

This method having been observed properly, the gravities for 5 pipes, g', g'', g''', g^{iv}, g^{v}, are found as follows:

$$g' = ghhssa(hh - mm)/hhmma(hh - ss)$$
$$g'' = ghhssa(mm - nn)/mmnnb(hh - ss)$$
$$g''' = ghhssa(nn - qq)/nnqqc(hh - ss)$$
$$g^{iv} = ghhssa(qq - rr)/qqrrd(hh - ss)$$
$$g^{v} = ghhssa(rr - ss)/rrsse(hh - ss).$$

From this foundation, the law of progression for any number of pipes develops more satisfactorily. And so in the truncated conoidic conduit FB (Fig. 9) attached to the cylindrical vessel AF, which conduit

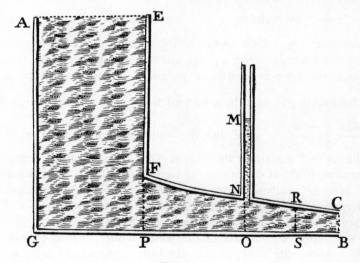

FIGURE 9

is considered as being made up of innumerable pipes of infinitely small length, there will be found, for any area NO, the hypothetical gravity by which the layer of water of infinitely small thickness is set into motion when the flow will have reached constancy. By designating the area NO by y and the thickness of the aqueous layer by dx, and by employing the rest of the nomenclature which we have used so far, [one finds that] the gravity setting this layer into motion will be $\dfrac{ghh\omega\omega a \cdot 2y\,dy}{y^4\,dx(hh - \omega\omega)} = \dfrac{ghh\omega\omega a \cdot 2\,dy}{y^3\,dx(hh - \omega\omega)}$. Further, the weight itself of this layer, or the pressure by which it is agitated, will result if one multiplies by the quantity of matter $y\,dx$; therefore this pressure will be $\dfrac{ghh\omega\omega a\cdot 2dy}{yy(hh-\omega\omega)}$. To this then there are to be added the pressures of all the following layers from O right through to the limit B, but gathered by translation to the position O, as our method explained at the beginning requires. In the meantime, to this end let y be assumed to be constant and the other area, $RS = t$, to be variable; the pressure of the latter layer $t\,dx$ will be $\dfrac{ghh\omega\omega a\cdot 2dt}{tt(hh-\omega\omega)}$, which may be translated to the invariable position NO by making $\dfrac{ghh\omega\omega a\cdot 2dt}{tt(hh-\omega\omega)}$ into $\dfrac{ghh\omega\omega ay\cdot 2dt}{t^3(hh-\omega\omega)}$ as t is to y, the properly correct integral of which gives

$$\dfrac{ghh\omega\omega ay}{\omega\omega(hh-\omega\omega)} - \dfrac{ghh\omega\omega ay}{tt(hh-\omega\omega)},$$

where now, by putting $t = y$, one has $\dfrac{ghha(yy-\omega\omega)}{y(hh-\omega\omega)}$ as the total pressure by which certainly the water at NO is compressed. Therefore, in order that z, the altitude of the aqueous cylinder of which the base is y and the weight is equal to this pressure, may be found, gyz is to be set equal to $\dfrac{ghha(yy-\omega\omega)}{y(hh-\omega\omega)}$, from which $z = \dfrac{hha(yy-\omega\omega)}{yy(hh-\omega\omega)}$. And thus the water in the pipe inserted at the position N will remain at this height NM. But it is to be noted that the conduit $FPBC$ is considered as having the diameters of the maximum area FP and minimum area CB small enough with respect to the length PB so that certainly the tangent to the curve FNC at any point N makes a very small angle with the horizontal PB. For otherwise, from the impact of the water while moving toward the overly curved side FNC of the conduit, there would arise a new force of pressure (which is here being neglected as accidental) which, augmenting the prior [force], would increase the height NM, just as actually occurs if the conduit FB ends in a plate, perforated by an orifice of area ω, on which the water, impinging perpendicularly, can increase the compression in the regions near the orifice. In the more remote regions that increase becomes less noticeable, and it varies as the curvature of the throat postulates, which moreover I consider to depend upon the peculiar nature of the water or other fluid flowing through, and so to be generally indeterminable.

SECOND PART

Containing the Direct and Universal Method for Solving all Hydraulics Problems whatsoever which can be Formed and Proposed Concerning Water Flowing through Conduits of any Shape

Section I

Consider any conduit whatever; let it be straight or curved, or let it be continuous or composed of many cylindrical pipes, or, finally, let it be vertical, or horizontal in part, or inclined differently in its different parts. Let this conduit be full of water, or some other heavy liquid, homogeneous and very fluid. Moreover, let it begin and continue to flow by accelerating (as much and as long as it can), and indeed in such a way that the conduit remains constantly full, with new water, of course, entering from another source; this replaces that escaping from the last orifice at any moment by flowing in through the first orifice with that velocity with which the uppermost surface would subside if the inflow were suddenly stopped. This condition is added for the sake of easier calculation; the method is surely valuable if no new liquid were to enter, right up to the complete depletion of a vessel or conduit. First, the velocity of the liquid flowing out is sought for any given quantity of liquid already having flowed out. Then is sought how much the sides of a conduit are pressed at individual points by the liquid flowing through, or, since it comes back to the same thing, to what vertical height a liquid of the same kind as that flowing through must remain elevated in a pipe inserted at any point and erected vertically.

Section II

And so let the arbitrary conduit *ECce* be given (Fig. 10), the vertical line *AB* considered as the axis of the abscissas, attached to which in order are *AEe*, *PFf*, *TNn*, and *BCc*, the parts *Ee*, *Ff*, *Nn*, and *Cc*, of which may define the areas or horizontal sections of the conduit. Let the liquid contained in it be considered as divided into horizontal

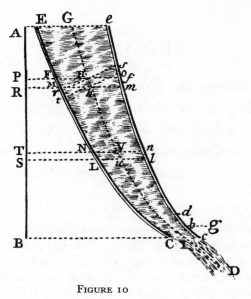

FIGURE 10

layers of infinitely small thickness, *FMmf*, *NLln*, etc., the intermediate points of which, or the centers of gravity *G*, *H*, *V*, *I*, etc., may form the line *GHVI*, either straight or curved, which I shall call the *center line*, or simply the *centric*; the latter certainly will be given, on account of the given curves *EFC*, and *efc*, which are determined from the given Shape of the conduit.

Let the first area *Ee* be h, the last area *Cc* be ω, some intermediate area *Ff* be y, another intermediate one *Nn* be r, and the thickness of the individual layers *PR* or *TS* be dt. The layers themselves will be $Fm = y\,dt$, $Nl = r\,dt$. Let, furthermore, some indeterminate straight line *ID*, which is of length x, touching the centric at *I*, be equal to the length of the oblique cylinder of liquid flowing out in the direction of *ID* itself, the base of which cylinder is *Cc*, and let this contain the quantity of liquid already emitted. The velocity of the liquid flowing out at that very moment is v. Let the gravity by which bodies are naturally set in motion be g. Now by designating the elements *Hh*, etc., of the centric line as ds, the gravities by which the layers are set in motion in the directions *Hh*, *Vu*, etc., will be $\dfrac{g\,dt}{ds}$, and therefore directions will be $\dfrac{gy\,dt^2}{ds}$, $\dfrac{gr\,dt^2}{ds}$, etc.; but the absolute motive forces themselves [acting] in the vertical direction will be $gy\,dt$, $gr\,dt$, and so on.

Section III

By translating these absolute forces (through the Hydrostatic Principle as it was shown in the first Part) to the first area h, the former will be $gh\,dt$ for the individual areas. Therefore, by integrating through all dt, that is, through the total height AB (let this be equal to a), gha will be equal to the total pressure to be applied vertically to Ee, equivalent to the sum of the absolute motive forces in all the sections. And this total pressure gha to be applied to the first area is that which is customarily called p by me.

Now let the tangent to the center line GHI at I be to its vertical subtangent [i.e., vertical projection] as α is to 1, and the tangent G to its subtangent as β is to 1, but the tangent at any intermediate point H to its subtangent as ds is to dt. Certainly through the resolution of the motion, v, or the actual efflux velocity of the liquid at I, will also be to its vertical subvelocity [i.e., vertical velocity component] as α is to 1, and therefore that subvelocity is $\dfrac{v}{\alpha}$. Similarly, by calling u the actual velocity at H in the direction of Hh, its subvelocity will be $\dfrac{u\,dt}{ds}$. In order that the actual velocity u may be found, however, it is to be noted that the subvelocities of the layers are in reciprocal ratio to their areas, to the end that they transmit equal quantities of liquid in the same elementary time interval; and so, by setting $y/\omega = \dfrac{v}{\alpha}\bigg/\dfrac{v\omega}{\alpha y}$, $\dfrac{v\omega}{\alpha y}$ will be equal to the subvelocity of the layer Fm. Furthermore, by setting $dt/ds = \dfrac{v\omega}{\alpha y}\bigg/\dfrac{v\omega\,ds}{\alpha y\,dt}$, $\dfrac{v\omega\,ds}{\alpha y\,dt}$ will be equal to u, or the actual velocity of the layer Fm in the direction Hh. Hence the actual velocity of the first layer adjacent to the area Ee (where y is inserted for h, and $ds : dt = \beta : 1$) will be $\dfrac{\beta\omega v}{\alpha h}$.

By no means is one to reason differently in finding the actual displacement of the layer Fm in the direction Hh. For by calling dx the instantaneous displacement of the last layer Cc in the direction the weights or the motive forces of the layers themselves in those

ID, its subdisplacement in the vertical direction will be $\dfrac{dx}{\alpha}$; moreover, here the subdisplacement of the layers are also in reciprocal ratio to the areas, and so, by setting $y/\omega = \dfrac{dx}{\alpha} \Big/ \dfrac{\omega\, dx}{\alpha y}$, [then] $\dfrac{\omega\, dx}{\alpha y}$ will be equal to the subdisplacement of the layer *Fm*. Therefore by setting $dt/ds = \dfrac{\omega\, dx}{\alpha y} \Big/ \dfrac{\omega\, dx\, ds}{\alpha y\, dt}$, $\dfrac{\omega\, dx\, ds}{\alpha y\, dt}$ will be equal to the actual displacement of the layer *Fm* in the direction of its motion, *Hh*.

Section IV

With the water already in motion, its layers act mutually on each other in different ways by pushing and resisting, and certainly with different forces, on account of the diversity of the surroundings with respect to position as well as speed. And so, in the meantime, let the indeterminate accelerative force which arises from the mutual action be called γ, and let the acquired velocity which some layer *Fm* has in the direction *Hh* be called u. Therefore $\gamma\, ds = u\, du$, from which $\gamma = \dfrac{u\, du}{ds}$. Let this be multiplied by the mass of the layer, $y\, dt$, and its motive force $\gamma y\, dt = \dfrac{yu\, du\, dt}{ds}$ will develop in the direction *Hh*. However, in order that this may be obtained in the vertical direction, from which the former can be produced, one has to set

$$dt/ds = \frac{yu\, du\, dt}{ds} \Big/ yu\, du.$$

This will be the motive force required in the vertical direction, which, therefore, translated to the first area h, gives the equivalent $hu\, du$. Let this be integrated so that $\tfrac{1}{2}huu$ is obtained, which, through the necessary correction, is to be applied to all layers contained and added together in the entire conduit *ECce*. Therefore (since the velocity of the last layer is v, and that of the first is $\dfrac{\beta\omega v}{\alpha h}$), the correct integral is $\dfrac{h}{2}\left(vv - \dfrac{\beta\beta\omega\omega}{\alpha\alpha hh}vv\right)$ or $\dfrac{vv(\alpha\alpha hh - \beta\beta\omega\omega)}{2\alpha\alpha h}$; this is equal to the equivalent vertical force to be applied at *Ee* by which, of course, the individual layers obtain their own particular forces for pushing each other mutually, to the effect that they merely conserve their own effort at the moment at which the water is flowing out at the velocity v; thus, since it occurs not continuously but in an indivisible instant and

METHOD FOR SOLVING HYDRAULICS PROBLEMS 395

[since it] depends upon the shape of the conduit only, this force, having arisen from the translation, can be called the *static force* or *static potential*, or, if it be more pleasing, the *hydrostatic potential*, since it consists in the effort alone of making the transition from one layer to the next lower, no attention having been given to the actual accelerative force.

Section V

Further, the other force is to be sought which arises from the actual acceleration of the liquid flowing through. To this end I set the actual accelerative force of any arbitrary advancing layer $Fm = \gamma'$; (on account of the actual displacement of the last layer Cc through the small space dx) the displacement of the layer Fm will be $\dfrac{\omega\, dx\, ds}{\alpha y\, dt}$, and therefore (§III) $\dfrac{\gamma' \omega\, dx\, ds}{\alpha y\, dt} = u'\, du' = \dfrac{\omega \omega v\, dv (ds)^2}{\alpha \alpha yy (dt)^2}$, from which $\gamma' = \dfrac{\omega v\, dv\, ds}{\alpha y\, dx\, dt}$, and the actual motive force in the direction of the layer Hh, or $\gamma' y\, dt$, is $\dfrac{\omega v\, dv\, ds}{\alpha\, dx}$, and so the vertical motive force from which the latter can be produced is $\dfrac{\omega v\, dv (ds)^2}{\alpha\, dx\, dt}$, which, translated to the first area h, gives the equivalent $\dfrac{h \omega v\, dv (ds)^2}{\alpha y\, dx\, dt}$; in order that this may be integrated through the total length AB of the axis corresponding to the whole conduit, for any layer and for any acquired velocity v, herein not only h and ω but also $\dfrac{v\, dv}{\alpha\, dx}$ must be considered as constant; and so, by integrating, we will have $\dfrac{h \omega v\, dv}{\alpha\, dx} \int \dfrac{(ds)^2}{y\, dt}$ equal to the other force that is to develop from the actual acceleration of the liquid flowing out, which it may please one to call the *hydraulic force*, as distinguished from the *hydrostatic force*, which consists of effort alone, or pressure employed instantaneously, and this at every instant, however the liquid may be moved.

Section VI

These two forces, the *hydrostatic* and the *hydraulic*, compose the total force which certainly is generated by the action of the primitive p, which was found in §III to be gha. Accordingly, by equating this to the sum of those two found in §§IV and V, we will obtain the most general equation for the determination of the velocity with which the liquid flows at any moment, which equation is this:

$$\frac{vv(\alpha\alpha hh - \beta\beta\omega\omega)}{2\alpha\alpha h} + \frac{h\omega v \, dv}{\alpha \, dx} \int \frac{(ds)^2}{y \, dt} = gha.$$

Here it is to be understood that $\int \frac{(ds)^2}{y \, dt}$ denotes the sum of all the $\frac{ds^2}{y \, dt}$ which are contained not only between Cc and Ff but everywhere between the extremes, by including all from one [extreme] to the other.

Section VII

If one now wishes to express the equation in z, or the height from which a given body, by falling under natural gravity g, would acquire the desired velocity v, then one should write, according to dynamic principles, $2gz$ for vv and $g \, dz$ for $v \, dv$, which will give this equation:

$$\frac{gz(\alpha\alpha hh - \beta\beta\omega\omega)}{\alpha\alpha h} + \frac{gh\omega \, dz}{\alpha \, dx} \int \frac{(ds)^2}{y \, dt} = gha;$$

or, after a reduction has been made, this:

$$(\alpha\alpha hh - \beta\beta\omega\omega)z \, dx + \alpha hh\omega \, dz \int \frac{(ds)^2}{y \, dt} = \alpha\alpha hha \, dx;$$

or, because $\int \frac{(ds)^2}{y \, dt}$, to be taken through the total length of the axis, can be considered as constant, and thus as given, at least through quadratures, let that be called M; and the equation will be reduced to the following form:

$$(\alpha\alpha hh - \beta\beta\omega\omega)z \, dx + \alpha Mhh\omega \, dz = \alpha\alpha hha \, dx.$$

From the resolution of this equation, z will be found in quantities given in x and the constants M, a, h, ω, α, β.

COROLLARY 1. For an existing uniform velocity of efflux, to which it reasonably arrives very quickly, and almost in one wink of the eye, as will be shown in the proper place in this writing, dz vanishes; therefore, after this has been neglected, the following algebraic equation develops: $(\alpha\alpha hh - \beta\beta\omega\omega)z = \alpha\alpha hha$, from which the desired

$$z = \frac{\alpha\alpha hha}{\alpha\alpha hh - \beta\beta\omega\omega}.$$

COROLLARY 2. Hence let two vessels or two conduits have any shapes whatever, even though very different from each other, provided only that they have the same vertical height a, and that they have the uppermost and the lowest areas, or the first and the last, h and ω, in the same ratio, and furthermore that they have α and β in either case proportional to each other. The water will flow out from either conduit or vessel at the same velocity after the flows in both cases have become uniform.

COROLLARY 3. If the centric line GHI is a straight line, whether it be vertical or oblique, there will be $\beta = \alpha$, and $ds : dt = \alpha : 1$. Hence, $ds = \alpha\, dt$, and $\int \dfrac{(ds)^2}{y\, dt}$, or $M, = \alpha\alpha \int \dfrac{dt}{y}$, which changes the general equation $(\alpha\alpha hh - \beta\beta\omega\omega)z\, dx + \alpha Mhh\omega\, dz = \alpha\alpha hha\, dx$ to this:

$$(hh - \omega\omega)z\, dx + \alpha hh\omega\, dz \int \frac{dt}{y} = hha\, dx.$$

But, whatever might be the position of the straight center line, whether vertical or oblique, in the case of uniform efflux z will always be $\dfrac{hha}{hh - \omega\omega}$.

COROLLARY 4. Now, with the shape of the conduit or vessel being maintained, and both the uppermost and lowermost areas Ee and Cc as well, if a little change is made in the direction of the inflowing and outflowing liquid, that change, even if it be hardly noticeable, can produce a marked change in velocity. In Fig. 11, for example, if borders or lips $Emne$ and $Cpqc$, having very little height Em and cp, are added to the vessel or conduit $ECce$ such that the areas mn and pq are kept the same as the previous values Ee and Cc, and such that the total height of the vessel is not increased noticeably, no one will easily believe how much the change in velocity is going to be due to this operation. Whereas now certainly the water flows in and out no longer obliquely but vertically, on account of the vertical direction of the lips, which even gives a vertical orientation to the extreme tangents of the center line, and so yields $\alpha = \beta = 1$, it is clear that the general equation

$$(\alpha\alpha hh - \beta\beta\omega\omega)z\, dx + \alpha Mhh\omega\, dz = \alpha\alpha hha\, dx$$

now at once assumes this form:

$$(hh - \omega\omega)z\, dx + Mhh\omega\, dz = hha\, dx;$$

and, for a uniform velocity, there will be $z = \dfrac{hha}{hh - \omega\omega}$; I consider it worthwhile to show this, lest otherwise, if, in experiments yet to be performed, one does not attend accurately enough to trifling circumstances which seem to be of no importance, and then, what is produced

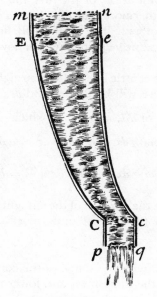

FIGURE 11

seems, falsely, to agree less accurately with ours; lest, I say, our theory be immediately suspected of being in error. And so it happened once to the illustrious Poleni, otherwise a well-regarded and industrious Man in experimental matters, who, about to see what different quantities of water in a given time openings of different size would emit, [the openings] having been affixed to the same vessel full of water, had arranged for this purpose that there should be several thin plates not altogether compact but that every one be pierced by an orifice of some particular size, so that first one and then another [plate] would cover the aperture at the bottom of the vessel. Moreover, it occurred, unless I am perchance mistaken, that he repeated the experiment twice with some of those plates, and diligently many times thereafter, where, to his astonishment, he observed that that one and the same plate, through its same orifice, had emitted a sometimes greater or sometimes lesser amount of water in the same time, accordingly as one or another face of that plate was looking outward. Finally the form of the orifice was examined more carefully, and then it was observed that the shape of the orifice, obviously having been cut in a thin plate, was not exactly cylindrical, but rather like a small truncated cone, having one base a little larger than the other, which was already sufficient for revealing the reason why, with the larger base of the orifice being open outwards, the water would flow out more fully than in the contrary sense; and this on account of a double reason, for both the aqueous stream springing forth was wider, and its velocity was greater, just as is evident from our formula, $z = \dfrac{hha}{hh - \omega\omega}$, where it is clear that the value of this fraction would be greater if ω were greater, h and a remaining fixed, and contrarily, it would be less if ω were less.

COROLLARY 5. In the case in which $\alpha h = \beta\omega$, or where $\alpha : \beta = \omega : h$, it will be true that $M \cdot \omega \cdot dz = \alpha a\, dx$, from which $z = \dfrac{\alpha\, ax}{M \cdot \omega}$, whence it is clear that, with the efflux x increasing to infinity, z also increases to infinity, and thus that the velocity never converges to uniformity. This certainly appears also from the very formula of Corollary 1, for $z = \dfrac{\alpha\alpha hha}{\alpha\alpha hh - \beta\beta\omega\omega} =$ (in this case)

$$\dfrac{\alpha\alpha hha}{\alpha\alpha hh - \alpha\alpha hh} = \dfrac{a}{0} = \infty.$$

Section VIII

SCHOLIUM 1. It is to be noted that in conduits and pipes [that are] not very wide and not sufficiently long, it is commonly observed, as I already hinted in the Preface,* that the sections Fm (Fig. 10), being in a state of flow, easily adjust themselves from a horizontal position to a position perpendicular to the sides, or rather to the centric line GHI, which is clearly evident, in any case, from the motion of the uppermost surface Ee (if no other liquid follows), as for example in barometric tubes and other syphons of that sort having diameters of not more than one or two lines; whether this is accomplished on account of the adhesion of the fluid to the sides, which must be uniform around the boundary in a circuit of the layers, in order that the fluid may be moved as aptly as possible and without noticeable friction, or it occurs on account of some other physical reason, is not to be determined at this point. It is sufficient to insert here that this circumstance does not oppose our Theory. For, because, through the general rule, the center of gravity of bodies urged into motion by

* p. 355.

any cause whatever is moved in the same manner and at the same velocity in its initial direction as if their entire material were concentrated at the center of gravity itself, certainly the material of any layer *Fm* can be considered as concentrated at the center of gravity *H* or *h*. As, therefore, in not very wide oblong conduits any arbitrary small portion of them can be taken as quasi-cylindrical or prismatic, it is evident that every layer *Fm*, for the slightest reason, can change its horizontal position *Ff* to *rs*, perpendicular to *Hh*, during which *Hh* retains the same length, and the quantity of the new layer *rtos* is equal to [that of] the layer *FMmf*. Therefore, let us consider how individual remaining layers *Nl* (without any other change either in velocity or in direction with respect to *Vu*) may adjust themselves into a position perpendicular to the sides of the conduit or, preferably, to the centric line. If now we consider further what would happen if the exit *Cc* were closed and in its place the orifice *cd*, of the same area as *Cc*, were opened in the wall of the conduit, with no difficulty we understand that the water must go out through the aperture *cd* under the same inclination to *cd* under which it was going out through *Cc*, and thus that its direction *bg* will be horizontal. Since in addition the aperture *cd* is set equal to the area *Cc*, and the tendency of flowing through *Cc* already is diverted towards *dc* (through the common hydrostatic law), it is necessary, certainly, that the velocity of the water flowing out through *cd* will be the same as we determined for *Cc*. Whence it is also realized that if to the orifice *cd* were attached a new horizontal conduit, in which certainly the centric line would be horizontal, the motion and the velocity of the water flowing through it and flowing out will obtain in the same manner as they would obtain if that same new conduit (*cd* being closed) were attached to *Cc* in the direction *ID*, but in which the water flowing would have to be considered as deprived of its own gravity. Thus, in order that here as well the weights of the layers translated to the area *Ee* yield the same sum *gha*, just as if the new conduit were not present, nothing in the expressed (§VII) general equation

$$(\alpha\alpha hh - \beta\beta\omega\omega)z\,dx + \alpha Mhh\omega\,dz = \alpha\alpha hha\,dx$$

should be changed other than that M, or $\int \frac{(ds)^2}{y\,dt}$, now expresses the sum of all the $\frac{(ds)^2}{y\,dt}$ which are contained in both conduits. But the uniform velocity on either hand will be the same, whether in the simple or in the combined conduit (because the term in which M is found vanishes in the case of uniformity), as it is always that which is found through $z = \dfrac{\alpha\alpha hha}{\alpha\alpha hh - \beta\beta\omega\omega}$.

CONCERNING THE PRESSURES WHICH THE SIDES OF A VESSEL SUSTAIN BECAUSE OF LIQUID FLOWING THROUGH

Section IX

In order that we might comprehend clearly and correctly in what that force consists which is exerted on the sides of a conduit while liquid flows in it, it is to be understood that that force is nothing more than that which takes its origin from a compression force by which, certainly, consecutive portions of the fluid, for example *EFfe* and *CFfc*, are driven one against the other; whence at *Ff* by this very contact there arises through action and reaction an intermediate force which I customarily call *immaterial*, because it is apart, so to speak, from the portions pressing each other and yet intermediate between the two, and it does not pertain to one more than to the other. It is characteristic of this force to drive the preceding portion of liquid *forward*, or in the direction in which it is going, but the following portion *backward*, or in the direction from which it comes, and to make the following portion of liquid, which is propelled by translated forces, and the preceding portion of liquid, against which some of the acceleration must press, acquire at this very contact an equality of accelerative forces; just as we showed a short while ago,* the same effect occurs in solid bodies, where, after they have been animated individually by different accelerative forces, there arises in their contact, when they begin to act on each other, an intermediate *immaterial* force, appearing to be truly common to each body, which thus would regulate the particular accelerative force of each, the one by diminishing, the other by increasing, in order that thence in the total mass combined from these two bodies one common accelerative force may result.

Section X

This, however, is the distinction in the manner of acting: that in solid bodies acting directly upon each other, that immaterial force acts forward and backward like some elastic straight line which,

* Nos. CLXXVII, p. 262, and CLXXIX, pp. 333, 340. [*Opera Omnia*, Book IV.]

placed between the bodies, tries to expand itself; but in portions of fluid acting mutually on each other, the immaterial force lying between must be considered just as elastic air, which extends itself not only in opposite directions, but into all surrounding regions; from which now it is easily understood that from this immaterial force itself the pressure, which is the subject here, develops. This certainly is exerted on the walls of a conduit, by which in turn it must be confined while it acts freely forward and backward on the portions of the liquid wherein it exists.

Section XI

It therefore remains that, according to this given idea concerning the immaterial force, we determine its quantity or measure. Let that [which is] to be sought, which we may say is π, be anywhere in Ff [Fig. 10]. Now I proceed thus: for the time consider a part of the conduit $EFfe$ (during flow) to be removed suddenly, the remaining $CFfc$ staying in its place with all its environs, and [consider] at that same moment that there is placed at the area Ff a new motive force equal to π itself. One understands at any rate that in this way the efflux of liquid flowing out of the truncated conduit is to be accelerated (at least in the first instant of time) just as if the conduit had remained whole. Therefore, I will already consider the residual conduit $CFfc$ as an integral conduit, the uppermost or first area of which is y, or Ff, any variable intermediate area Nn is r, and the adjacent section Nl is $r\, dt$. Thus if (§IV) for h I substitute y, I will have $\dfrac{vv(\alpha\alpha yy - \omega\omega(ds)^2/(dt)^2)}{2\alpha\alpha y}$ as the hydrostatic force; indeed that which at the first point G was called β is ds/dt at point H, the ratio, of course, of the tangent to the subtangent, and (§V) $\dfrac{y\omega v\, dv}{\alpha\, dx}\int \dfrac{(ds)^2}{r\, dt}$ is the hydraulic force, where, in the integration, r is considered to be continuous from ω right on to y.

Section XII

The sum of these two forces, the hydrostatic and the hydraulic, should be equated to the original force p, which here should be (§§ III and VI) gyt, if indeed this alone were acting on the liquid contained in the truncated conduit; but, because π acts together with gyt, it is necessary by all means to establish this equation:

$$\frac{vv(\alpha\alpha yy - \omega\omega(ds)^2/(dt)^2)}{2\alpha\alpha y} + \frac{y\omega v\, dv}{\alpha\, dx}\int \frac{(ds)^2}{r\, dt} = gyt + \pi.$$

METHOD FOR SOLVING HYDRAULICS PROBLEMS

From this emerges at once the sought value of π itself: namely, for gyt having been transposed, there results

$$\frac{vv(\alpha\alpha yy - \omega\omega(ds)^2/(dt)^2)}{2\alpha\alpha y} + \frac{y\omega v\, dv}{\alpha\, dx}\int\frac{(ds)^2}{r\, dt} - gyt = \pi,$$

where also it is to be warned that, in the integration $\int\frac{(ds)^2}{r\, dt}$, r must be taken as variable from B to P, whence, for any assumed y, $\int\frac{(ds)^2}{r\, dt}$ will be given. Therefore, let this be set equal to N, and there will result

$$\frac{vv(\alpha\alpha yy - \omega\omega(ds)^2/(dt)^2)}{2\alpha\alpha y} + \frac{Ny\omega v\, dv}{\alpha\, dx} - gyt = \pi.$$

Since, therefore, from the resolution of the general equation (§VII) there evolves the value of vv itself, or $2gz$, this substituted in the latter will give the value of π itself in terms of g and purely linear quantities. Thus, now,

$$\frac{gz(\alpha\alpha yy - \omega\omega(ds)^2/(dt)^2)}{\alpha\alpha y} + \frac{gNy\omega\, dz}{\alpha\, dx} - gyt = \pi.$$

Section XIII

If now, furthermore, it is desired to know, if some tube open at both ends is introduced at some arbitrary place f in the conduit and is erected to a vertical position, how far the liquid must ascend in it on account of this pressure π which makes it ascend, it is agreed to consider that π is equal to the weight of some cylinder formed from the liquid, animated by the natural gravity g, which has for a base the area Ff or y, and for a height that very [height] of the liquid standing in the tube; whence $\frac{\pi}{gy}$ will be this height at which the suspended liquid will stand in the tube, invariable indeed, after the velocity of the liquid flowing out will have reached reasonable uniformity; but before this occurs (although it happens in an instant, more or less), the liquid proceeds to ascend in the tube until it will have acquired the appropriate stable location, when indeed the liquid flowing out is no longer noticeably accelerated.

Section XIV

It happens in certain cases that the value of π itself might be negative, either when, naturally, the negative quantities $\frac{-vv\omega\omega(ds)^2}{2\alpha\alpha y(dt)^2} - gyt$

prevail over the positive $\frac{1}{2}vvy + \frac{y\omega v}{\alpha}\frac{dv}{dx}\int\frac{(ds)^2}{r\,dt}$, or, with the velocity already being uniform, so that $dv = 0$, whenever $\frac{vv\omega\omega(ds)^2}{2\alpha\alpha y(dt)^2} + gyt$ is greater than $\frac{1}{2}vvy$; this can happen not only in those cases where αy is less than $\frac{\omega\,ds}{dt}$, but also in those where αy is greater than $\frac{\omega\,ds}{dt}$, if only at the same time gyt be great enough that its excess over

$$\frac{y\omega v}{\alpha}\frac{dv}{dx}\int\frac{(ds)^2}{r\,dt}$$

be greater than the defect of the prior. However, in whatever manner it may occur, it is plain that in cases of this sort the pressure is changed into relaxation, which causes the walls of the conduit around *Ff* not only not to be pressed outward, but to be contracted inward everywhere (if the rigidity of the walls is not interfered with). Hence it follows that the liquid can be elevated upward as if by suction to the height $\frac{\pi}{gy}$ through a tube implanted in the conduit as before but directed vertically downward to a very low point where it opens into a little vessel full of liquid.

Section XV

Scholium 2. Thus far we have not attended to certain particular and accessory causes (not always having importance) which can change either the pressures or the suctions π determined by our method. Among such causes, one occurs principally which acts such that the water, being in motion and striking an immobile surface in its path, impresses a force upon it during [its] approach which is called the *force of fluid resistance*, unquestionably proportional to both the square of the velocity and the square of the sine of the obliquity of the incidence, as is known. And so, for this very reason, that force becomes unnoticeable in rather narrow oblong conduits; for in these, on account of *FM* being almost parallel to *hH* itself, which is the direction of the motion of the fluid when it reaches *Ff*, just as at any other place *Nn* where the direction *Vu* is almost parallel to *NL* and *nl*, and the sine of incidence can be regarded as negligible. In a conduit made up of cylindrical pipes that sine is absolutely zero, because the direction of the fluid is everywhere parallel to the sides of the cylinders throughout the total length of the conduit. Besides, another accessory cause which can confound the effect arising from the pressure π is found in an exceedingly curved conduit, in which of course swiftly flowing liquid acquires a centrifugal force (which we discussed elsewhere); this centrifugal force would render a greater pressure π than is correct in the convex part of the conduit, but a lesser one in the concave part of the same. And therefore, if one would like to conduct an experiment with the aid of a tube to be implanted in this conduit, the insertion should be made neither in the convexity nor in the concavity of the curve, but at the side, such that the tube extends out from the conduit perpendicularly to the plane of convexity and concavity, and therefore, if that plane be not horizontal, the tube as well as the structure is to be turned until the tube attains a vertical position.

GENERAL COROLLARIES ABOUT VELOCITIES AND PRESSURES

Section XVI

In the uniform and constant efflux of liquid, the general equation (§VI) is changed, on account of $dv = 0$, to this:

$$\frac{vv(\alpha\alpha hh - \beta\beta\omega\omega)}{2\alpha\alpha h} = gha, \quad \text{or,} \quad vv = \frac{2\alpha\alpha ghha}{\alpha\alpha hh - \beta\beta\omega\omega}.$$

Hence this elegant Theorem is deduced: If there are two vessels or conduits having equal vertical heights and equal uppermost and lowest areas, no matter what shapes the remaining portions of the conduits may have and however different [they may be] from each other, if only their centric lines be so comparative that the ratio between α and β in one be the same as between α and β in the other vessel or conduit, I say that from either (understandably both being full) the liquid, after the efflux will have reached equilibrium, will flow out on either hand at the same velocity. This surely is evident from the very value of vv itself, which is $\dfrac{2\alpha\alpha ghha}{\alpha\alpha hh - \beta\beta\omega\omega}$, inasmuch as herein the intermediate areas y are not obtained at all. Let there be, for example, two vessels of which the shapes are $ABCD$ and $EFGH$ (Fig. 12), of which the centric lines are straight, and indeed it matters not whether they be vertical or oblique, or one more or less oblique than the other, since in all of these cases on either hand it will always be that $\alpha = \beta$, provided only that those two vessels have equal vertical heights a, likewise equal extreme areas $AD = EH = h$ and $BC = FG = \omega$, or, which suffices, only that $AD : EH = BC : FG$, and

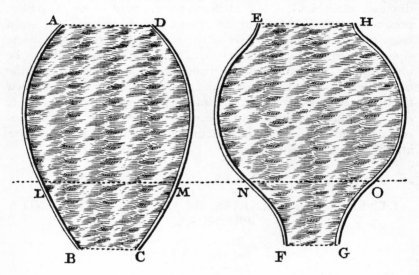

FIGURE 12

that those vessels be constantly full of water or some other homogenous liquid in which on either hand there is, of course, the same natural gravity g animating the sections; the maximum and uniform velocity of the water flowing out through BC will be equal to the maximum and uniform velocity of the water flowing out through FG. Indeed in such a case, on either hand (§VI), on account of $dv = 0$, and $\alpha = \beta$, there results, I say, $(hh - \omega\omega)z = hha$, and thus $z = \dfrac{hha}{hh - \omega\omega}$, conforming to Corollary 2, §VII, and such others as we found in this way in the first Part for a cylindrical vessel. Meanwhile, it is manifest that the value $\dfrac{hha}{hh - \omega\omega}$ for a given height a of each vessel is the same if the ratio between h and ω is the same for each, that is, if $AD : EH = BC : FG$. But it is to be noted that here we avoid the contraction of the aqueous stream which is customarily observed somewhere beyond the orifice, particularly in those vessels which suddenly terminate in an orifice opened in a rather wide base, contrary to what occurs in those having the shape of $EFGH$, converging, so to speak, into a cylindrical pipe in which no noticeable contraction of the stream is apparent. Meanwhile, if the understanding of this is also to be obtained, the vessel should be considered as if continued to the maximum contraction of the stream where it stops drawing together, and then the area of the *vena contracta* should be taken as the smaller orifice ω itself, and its distance from the greatest area as the true vertical height.

Section XVII

With the same conditions existing, as in the preceding, of two vessels of equal height, $ABCD$ and $EFGH$ (Fig. 12), [being] equally large at the extremities and having straight centric lines, if now in addition to this they [each] have a third area LM and NO mutually equal to each other and equally distant from the orifices BC and FG, then not only will the maximum velocity of each (by the preceding) be equal, but also the pressures at LM and NO will be equal, and therefore in tubes inserted in these places and bent to a vertical position, the water in each will stand at the same height. The truth of this is evident from the equation (§XII) which, on account of $\alpha = \beta = \dfrac{ds}{dt}$, in the case of uniform velocity results in this simpler one,

$$\frac{vv(yy - \omega\omega)}{2y} - gyt = \pi,$$ or (by writing $2gz$ for vv) in this,

$$\frac{gz(yy - \omega\omega)}{y} - gyt = \pi,$$

in which, because for either vessel z, y, ω, g, and t are the same, the same value of π itself must result by all means, and hence also of

$$\frac{\pi}{gy} = \frac{z(yy - \omega\omega)}{yy} - t = \frac{hha(yy - \omega\omega)}{yy(hh - \omega\omega)} - t.$$

Section XVIII

Thus if all the LM and NO at equal vertical distances from BC and FG were equal—which would occur if those two vessels $ABCD$ and $EFGH$ were, for example, truncated conoids of the same type, one right, the other scalene—in this case not only would the uniform velocities at which the water would flow out from each vessel be equal but also the pressures at the individual equal heights would be identical, and thus even the suspensions of water standing in the tubes would have the same height in each vessel.

Section XIX

The equation found in §VII for determining velocity in general, whether it be already steady or not yet steady, gives

$$dx = \frac{\alpha M h h \omega \, dz}{\alpha\alpha hha - \alpha\alpha hhz + \beta\beta\omega\omega z}.$$

If this value is substituted in the equation for the pressures π (§XII) and $2gz$ for vv, we will have:

$$\frac{gz(\alpha\alpha yy - \omega\omega(ds)^2/(dt)^2)}{\alpha\alpha y} + \frac{(\alpha\alpha hha - \alpha\alpha hhz + \beta\beta\omega\omega z)gNy}{\alpha\alpha Mhh} - gyt = \pi.$$

Thus the height of the liquid in the tube, or $\dfrac{\pi}{gy}$, is

$$\frac{z(\alpha\alpha yy - \omega\omega(ds)^2/(dt)^2)}{\alpha\alpha yy} + \frac{N(\alpha\alpha hha - \alpha\alpha hhz + \beta\beta\omega\omega z)}{\alpha\alpha Mhh} - t,$$

which, therefore, for any determined z whatever expresses generally the height of the liquid in the tube. Next, which is interesting, that initial height is found at once, that is, that which would be observed in the tube at the instant that the orifice below is opened and the liquid is about to flow. Since indeed at the first instant of time $z = 0$, it will surely occur that (z having been deleted) $\dfrac{\pi}{gy} = \dfrac{Na}{M} - t$.

Section XX

Previous to this proof someone might have doubted whether or not perhaps at that instant that the orifice BC is opened and before the liquid pours forth into actual motion, whether or not, I say, the pressures at any location LM were still the same, at least for an instant, as they had been previously when the orifice BC had still been closed or blocked off. Truly, the often-used and common hydrostatic law accepted everywhere shows, for the case of the vessel closed at BC, that the liquid in a tube inserted somewhere in the circumference of some section LM, and vertically erect, will stand suspended at some height $a - t$, that is, at the same horizontal as the uppermost surface AD of the liquid contained in the vessel. But now we see that the situation develops in one way in the case of the orifice BC closed and in another in the case of the same opened, even if the liquid is not yet actually flowing out. Since N, as a part, is certainly less than M, the whole, $\frac{Na}{M}$ will be less than a, and therefore as well $\frac{Na}{M} - t$ less than $a - t$. Hence it is evident that in the first instant in which the orifice BC is opened some of the liquid is already lost, so to speak, or, rather, emits because of its own gravity: this does not contribute to pressing the sides but to propelling the liquid to such an extent that it no longer presses the sides of the vessel as greatly as it had done before the aperture was formed. Meanwhile, surely, it should be indicated here that I avoid the accessory causes which can alter the determined height in the tube, $\frac{Na}{M} t$. For example, something should be mentioned concerning the shape of the vessel; for instance, if it were very wide, and suddenly it converged into a narrow orifice, then undoubtedly our Theory could be out of tune with what experience would show. The reason is that the Theory supposes that the sections Fm and Nl (Fig. 10) aspire to this arrangement with respect to the motion (although they are not yet moved by the impulse): that [on the one hand] the areas Ff, Nn, etc., through the whole traverse retain a horizontal position, extended throughout to the sides, and [on the other] that the centric line GHI everywhere passes through the midpoints G, H, V, and I, which certainly must thus prevail accurately enough, as the Theory shows it, in those vessels and conduits the sides of which gradually, not suddenly, converge or diverge toward the orifice below. But in others, being very large and ending in a large base which contains a narrow orifice for an exit, in these, as one would expect, the layers do not spread throughout entire cross sections of the vessel, but rather to various [radial] extents, in proportion as the quality of the liquid, not perfectly fluid but more or less viscous, requires this, that it may experience the minimum possible resistance from friction, with the remaining part of the liquid near the sides of the vessel certainly being at rest or without sufficient disposition to motion. From this it can happen that in such a vessel a continuous throat or some sort of cataract is formed, such as Newton fully conceived, although not according to that law which he indicated as necessary. It is understood from the foregoing remarks that in vessels of this sort it is not the external and artificial shape of them that is to be considered, but that internal [shape] of the continuous throat formed by nature, not such as Newton conceived, but that which is best suited to the quality or the constitution of the liquid. Therefore, if in the enclosing surface of this throat a tube could be implanted, the height of the liquid suspended in it would always be observed entirely as our rule postulates, whether the liquid is already in motion or is beginning to be moved in the vessel.

Section XXI

But since it cannot be easily noticed by observation just when a throat or cataract is terminated in a very wide vessel or just where its enclosing surface is located, it will be safer if the tube penetrates perpendicularly within the vessel as far as the middle, that is, right to the centric line, and then the other part of the tube, bent upwards outside the vessel, is vertical. In this way, then, the height of the liquid in the tube shows uniquely how great is the compression of the liquid at the place on the centric line with which the orifice of the tube is in contact, where those accessory causes altering the effect of the compression, concerning which we have treated above (§XV), have no effect. Nevertheless, care is to be taken that at least the portion of the tube to be projected into the liquid in the vessel be slender enough, lest otherwise its very great thickness present some impediment to the free motion of the liquid. These precautions having been properly taken, I do not doubt that the following will be most accurately observed: 1, that the general height of the liquid in a tube, with the acceleration of the efflux continuing to this time, will be (§§XII & XIII),

$$\frac{\pi}{gy} = \frac{z(\alpha\alpha yy - \omega\omega(ds)^2/(dt)^2)}{\alpha\alpha yy} + \frac{N\omega \, dz}{\alpha \, dx} - t;$$

2, but with the noticeable acceleration ceasing, at which time certainly the velocity will have reached reasonable uniformity, there will be $\frac{\pi}{gy} = \frac{z(\alpha\alpha yy - \omega\omega(ds)^2/(dt)^2)}{\alpha\alpha yy} - t$; 3, that the initial height will be (§ XIX) $\frac{\pi}{gy} = \frac{Na}{M} - t.$

APPLICATION OF OUR THEORY TO EXAMPLES OF VESSELS AND CONDUITS ALWAYS FULL

Section XXII

Let the center line of some vessel be in a vertical position. For this rather simple case, where $\alpha = \beta = 1$, it is true that (§VII)

$z = \dfrac{hha}{hh - \omega\omega} (1 - 1/e^{(hh-\omega\omega)x/Mhh\omega})$, which is found from the reduction of the equation $(hh - \omega\omega)z\, dx + Mhh\omega\, dz = hha\, dx$ developed there and applied to this case, after, of course, $\ln e = 1$ has been agreed upon; and thus for $x = \infty$, that is, in the case of steady efflux or uniform velocity, there will result $z = \dfrac{hha}{hh - \omega\omega}$, from which follows the Theorem already demonstrated in §XVI. Therefore, let the vessel $ABCD$ (Fig. 13) be of any shape whatever, its centric or

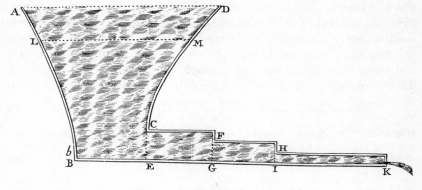

FIGURE 13

vertical height be a, and let [the vessel] have attached to it the conduit CK composed of many pipes, for example, of three cylindrical pipes CG, FI, and HK, placed in a horizontal position. Let the uppermost area AD (to which the vessel with the pipes is assumed [to be] always full) be h, the areas $CE = m$, $FG = n$, and $HI = q$, and the orifice of the last pipe be ω. I say that z will always be $\dfrac{hha}{hh - \omega\omega}$, and thus that neither the shape of the vessel nor the number of pipes nor their sizes come into consideration, provided only that the first h and the last ω be given, and the height a of the vessel be given. And also it is not of concern to know whether the throat or cataract extends throughout the whole internal region of the vessel or only occupies some portion of it around the centric line. The matter is clear by §XV, because the conduit CK is assumed horizontal, and thus the uniform velocity is always the same as if the orifice ω were attached immediately to CE by applying some perforated section to the aperture CE.

COROLLARY. If ω is very small with respect to h, there will result $z = a$, and accordingly the velocity of the water flowing out uniformly, which is the greatest velocity that it can acquire, will be equal to that which a heavy body acquires by falling from a height a.

Section XXIII

For finding the height of the liquid in a tube to be implanted somewhere in the horizontal conduit CK, let it be noted that in this case $t = 0$, since t signifies the excess of vertical height of the place where the tube is inserted above the height of the orifice through which the liquid is flowing, or, which is the same, t signifies the height of the place of insertion of the tube above the place of efflux. Here, moreover, on account of the horizontal position of the conduit, the centric itself is also considered as horizontal, especially if its pipes, of which the conduit is composed, are not at all large enough that (§XXII) the sections of liquid flowing through them may become vertical. Therefore, the height of the liquid in the tube (§XIX) will be $\frac{\pi}{gy} = \frac{z(yy - \omega\omega)}{yy} + \frac{N(hha - hhz + \omega\omega z)}{Mhh}$, since $\alpha = \beta = \frac{ds}{dt}$. In the case of uniform velocity, where (preceding paragraph) one has $z = \frac{hha}{hh - \omega\omega}$, after this value has been substituted for z, the latter term $\frac{N(hha - hhz + \omega\omega z)}{Mhh}$ vanishes, and the former $\frac{z(yy - \omega\omega)}{yy}$ becomes $\frac{hha(yy - \omega\omega)}{yy(hh - \omega\omega)}$. Therefore, the height in the tube will be $\frac{\pi}{gy} = \frac{hha(yy - \omega\omega)}{yy(hh - \omega\omega)}$. Let the results which we have already found in the Appendix* at the end of the first Part be compared, although in another way. Indeed, for the first pipe CG, where $y = m$, that height will be $\frac{hha(mm - \omega\omega)}{mm(hh - \omega\omega)}$; for the second pipe FI, where y is n, the height will be $\frac{hha(nn - \omega\omega)}{nn(hh - \omega\omega)}$; for the third pipe HK, where $y = q$, the height in the tube will be $\frac{hha(qq - \omega\omega)}{qq(hh - \omega\omega)}$, and so on for however many pipes there may be composing the conduit. All of this corresponds most accurately to the experiments performed concerning this matter.

Corollary. The initial height in the tube is $\frac{Na}{M}$, as we found above for vessels themselves without conduits.

* Above, p. 390.

Section XXIV

And thus this is related to the height of the liquid in the tube to be implanted somewhere in the vessel itself (and if it is desired, to be extended right to the center line, which for the future we always assume as a vertical straight line, but with the conduit attached to the vessel, we assume the other [center line] as horizontal); let the place of insertion be at some point in the indeterminate section LM, the distance of which from the lowest horizontal is t, and the area of the throat (if it is not LM itself), whatever it may be in the experiment, should be taken as r. The height of the liquid in the tube (by §XI, applied to this) will be obtained as

$$\frac{z(rr - \omega\omega)}{rr} + \frac{N(hha - hhz + \omega\omega z)}{Mhh} - t,$$

where $N = \int \frac{dt}{r}$ contained between BE and LM, and $M = \int \frac{dt}{r}$, but contained between BE and AD (§VII). For the velocity of the liquid flowing out uniformly, where $z = \frac{hha}{hh - \omega\omega}$, let this value be substituted for z, and the height in the tube will appear as $\frac{hha(rr - \omega\omega)}{rr(hh - \omega\omega)} - t$ (since the term in which N and M appear vanishes), and thus, after the height t of the place of insertion has been added, the total height, above the lowest horizontal BE, of the whole of the liquid standing in the tube will be $\frac{hha(rr - \omega\omega)}{rr(hh - \omega\omega)}$. Thus if ω be infinitely small with respect to h and r, that total height will be a, that is, the whole of the liquid in the tube is at the same level as the uppermost area AD; that this must so occur we certainly should be able to understand also from the fact that the liquid in the vessel is, so to speak, at rest. Finally, the initial total height in the tube, with $z = 0$ certainly, here is also $\frac{Na}{M}$. All these results are in excellent accord with each other.

Section XXV

Now let us assume that the horizontal conduit CK converges in a truncated cone, or any conoidic whatever, and has its greater base directed toward the vessel. For uniform velocity of the outflowing water, the height in the tube implanted at any place whatever between F and H (by designating the area $FG = y$) will be, I say, that height (as expressed in §XXIII) $\dfrac{hha(yy - \omega\omega)}{yy(hh - \omega\omega)}$. Hence if the smaller base were attached to the vessel, and the oblong conduit were not suddenly to diverge too greatly, lest the water diffuse in it, but the sections succeeding in order follow the preceding ones as the customarily established Theory supposes, then, on account of y being less than ω, the pressure on the sides will be negative, and therefore it is changed to suction, by which it occurs that, with the water descending vertically in the tube and discharging into the water contained in the vessel below, [the water in the tube] is raised through suction to the height $\dfrac{hha(\omega\omega - yy)}{yy(hh - \omega\omega)}$. But if also ω be greater than h, then the numerator and denominator of the fraction become negative, and thus its value is again positive, which indicates that pressure is present. Whereby, the vessel $ABCD$ being always full, so that the uppermost area AD would be less than the orifice of the divergent conoidic, through the larger base of which the water emits, it is to be observed again that the water will continually and without end rise in the upwardly erected tube. In fact, in such a case the acceleration of the flowing water never ceases, hence it never reaches a constancy of velocity, which is evident from the general equation (from Art. VII applied here) $(hh - \omega\omega)z\, dx - Mhh\omega\, dz = hha\, dx$; or, more clearly, from the equation shown in Art. XXII in finite terms,

$$z = \frac{hha}{hh - \omega\omega}\left(1 - 1/e^{(hh - \omega\omega)x/Mhh\omega}\right),$$

which is equivalent to $z = \dfrac{hha}{\omega\omega - hh}\left(e^{(\omega\omega - hh)x/Mhh\omega} - 1\right)$, from which it is clear at once that in the case in which ω is greater than h, it turns out that $z = \infty$, and therefore that the velocity is infinite when x is infinite, contrary to what occurs if h is greater than ω.

CONCERNING THE SHORTNESS OF TIME FROM THE BEGINNING OF EFFLUX RIGHT UP TO THE ESSENTIALLY CONSTANT OR UNIFORM VELOCITY

Section XXVI

Although, accurately speaking, an infinite time is required before the flow of water springing forth from vessels through orifices would arrive gradually to perfect and geometric uniformity, experience nevertheless shows daily that water, especially from rather wide vessels, even though hardly three or four feet high, converges with such rapidity from the first moment of flow to its maximum and constant velocity of flow, while it flows through an admittedly somewhat narrow orifice, that the gradual increments of velocity through which it goes from rest to the uniform and maximum possible velocity which it can essentially attain cannot be perceived by observation. In order that we might interpret the reason for this phenomenon from our Theory, let us consider a cylindrical or prismatic vessel of sufficiently large area h and of suitable height a, from which water rushes forth in a horizontal direction through a narrow orifice ω which may be formed either immediately beyond the vessel itself or [beyond] an intermediate conduit having the orifice ω at its extremity. However, let us first consider, for the sake of brevity, that, of course, the orifice ω is formed near the bottom in the very wall of the vessel.

Section XXVII

The general equation (§VI) for the determination of the velocity increasing to this point was this:

$$\frac{vv(hh - \omega\omega)}{2h} + \frac{hh\omega v\, dv}{dx} \int \frac{dt}{y} = gha;$$

which, in our case, where $\int \frac{dt}{y} = \frac{a}{h}$, and $\omega\omega$ as compared to hh can be neglected, is changed to this:

$$2a\omega v\, dv = 2gha\, dx - hvv\, dx, \quad \text{or} \quad dx = \frac{2a\omega v\, dv}{2gha - hvv}.$$

And thus the element of time $d\theta$, or $\dfrac{dx}{v}$, will be equal to

$$\frac{2a\omega\, dv}{2gha - hvv} = \frac{2a\omega\, dv/h}{2ga - vv} = \frac{a\omega}{h\sqrt{2ga}}\left(\frac{dv}{v + \sqrt{2ga}} + \frac{dv}{-v + \sqrt{2ga}}\right).$$

By integrating one has

$$\theta = \frac{a\omega}{h\sqrt{2ga}} \ln\left(\frac{v + \sqrt{2ga}}{-v + \sqrt{2ga}}\right)$$

$$= (\text{since } v = \sqrt{2gz})\; \frac{a\omega}{h\sqrt{2ga}} \ln\left(\frac{\sqrt{z} + \sqrt{a}}{-\sqrt{z} + \sqrt{a}}\right)$$

$$= (\S\text{XXII})\; \frac{a\omega}{h\sqrt{2ga}}$$
$$\times \ln\,(1 + \sqrt{1 - 1/e^{hx/a\omega}})/(1 - \sqrt{1 - 1/e^{hx/a\omega}}).$$

Indeed in this case $z = a(1 - 1/e^{hx/a\omega})$. Hence

$$\frac{h\sqrt{2ga}}{a\omega} \theta = \ln\left(1 + \sqrt{1 - 1/e^{hx/a\omega}}\right)/\left(1 - \sqrt{1 - 1/e^{hx/a\omega}}\right).$$

By transferring from logarithms to numbers and proceeding in the customary way, one finds that $e^{hx/2a\omega}$ is equal to the fraction

$$(e^{h\theta\sqrt{2ga}/a\omega} + 1)/2e^{h\theta\sqrt{2ga}/2a\omega}.$$

Section XXVIII

Certainly from the dynamic principle for the free fall of heavy bodies, by letting C equal the height through which a freely falling weight travels in the given time θ, one finds that $\theta = \sqrt{\frac{2C}{g}}$. Let this value be substituted in the fraction found above, and it will result that $e^{hx/2a\omega}$ equals this other fraction $(e^{2h\sqrt{aC}/a\omega} + 1)/2e^{h\sqrt{aC}/a\omega}$.

Now, because the area h of the vessel is assumed to be much greater than the area ω of the orifice, and the height C of free fall to be traveled through in one second is 15 feet, and in addition, from the nature of the logarithmic curve, e is greater than two, it is manifest that for any ordinary height a of the vessel, the number $e^{2h\sqrt{aC}/a\omega}$ is immensely greater than unity, such that the latter can be disregarded in the numerator of our fraction. Therefore there will be, essentially,

$$e^{hx/2a\omega} = e^{2h\sqrt{aC}/a\omega}/2e^{h\sqrt{aC}/a\omega} = \tfrac{1}{2}e^{h\sqrt{aC}/a\omega},$$

or

$$2e^{hx/2a\omega} = e^{h\sqrt{aC}/a\omega},$$

or, by taking the logarithms, $\frac{hx}{2a\omega} + \ln 2 = \frac{h\sqrt{aC}}{a\omega}$, from which $x = 2\sqrt{aC} - \frac{2a\omega \ln 2}{h} = 2\sqrt{aC}$ (on account of ω being incomparably less than h), which is equal to $2\sqrt{60}$ feet (by putting $a = 4$ feet and $C = 15$ feet), or roughly 16 feet. Thus if, therefore, in the equation $z = a(1 - 1/e^{hx/a\omega})$ which determines the velocity for any efflux of water of length x, we substitute 4 for a, 16 for x, 2 for e (although, which would prove the case better, e is greater than 2), and if we consider the area h of the vessel to be to the area ω of the orifice as 100 is to 1, we will have $z = 4(1 - 1/2^{400})$, which, on account of the extremely small value of the fraction $1/2^{400}$, is reckoned to be not different from four feet, which defines the height of the vessel, and likewise that height from which a weight having fallen acquires a velocity equal to that which the efflux has when it will have come to uniformity; from this it is evident that after one second of time has elapsed, the water flowing out already has essentially that uniform velocity.

But in order that it be more evident with what promptness the velocity of efflux may converge to uniformity, let us see how insignificantly the velocity of the outflowing water acquired after one tenth of a second has passed should be out of accord with the maximum velocity which it could acquire if the efflux were to endure for an infinite length of time. Let us reduce feet to inches, and we will have $a = 48$ inches, and C is found to be about 2 inches, from which x or $2\sqrt{aC}$ is about 20 inches, and $e^{hx/a\omega} = e^{2000/48}$, for which I write merely 2^{40}. Thus there will be $z = a(1 - 1/2^{40})$, which still, on account of the imperceptible smallness of the fraction $1/2^{40}$, is to be considered as differing not at all from that very a defining the uniform velocity.

COROLLARY. The efflux of water from rather wide vessels through narrow orifices can safely be considered as constant the instant after the beginning of motion.

GENERAL HYDRAULICS THEOREM DEDUCED DIRECTLY FROM THE PRINCIPLES OF HYDRODYNAMICS, PROVED THROUGH THE INDIRECT METHOD OF LIVE FORCES

SECTION XXIX

For a more substantial confirmation of the validity of our direct and universal method, it is pleasing now to propose an indirect solution, to be derived from the Theory of the conservation of live forces, of the principal Proposition concerning the velocity of water flowing out of a vessel or conduit which is always full, just as we established through the equation shown in §VII.

Let us consider that the water flowing out through Cc (Fig. 10) is directed immediately to a horizontal position, so that it can be regarded as without ascent and without descent in its displacement. Let x actually be the length along the oblique direction ID of the aqueous cylinder having Cc for a base, which cylinder contains as great a quantity of water as has already flowed out. That quantity will be $\frac{\omega x}{\alpha}$, the differential $\frac{\omega\, dx}{\alpha}$ of which defines the elementary particle of water about to flow out from Cc immediately after the quantity $\frac{\omega x}{\alpha}$ has been emitted. Let z be the vertical height from which some heavy body having fallen freely may acquire the sought velocity which of course that elementary particle of water $\frac{\omega\, dx}{\alpha}$ must have. Through the principle of its live forces, the velocity will be $\sqrt{z}$ and its subvelocity will be $\frac{1}{\alpha}\sqrt{z}$, from which the subvelocity at G is $\frac{\omega\sqrt{z}}{\alpha h}$; but the actual velocity itself in the tangential direction at G is $\frac{\beta\omega\sqrt{z}}{ah}$. Similarly the subvelocity at any point H is $\frac{\omega\sqrt{z}}{\alpha y}$, and therefore the actual velocity itself at H is $\frac{\beta\, dz\sqrt{z}}{ay\, dt}$.

Section XXX

But since, through the individual cross sections in the entire conduit at the same instant, the same quantity $\frac{\omega\, dx}{\alpha}$ of water must flow, such an elementary quantity $\frac{\omega\, dx}{\alpha}$ is to be considered as stationed above the uppermost surface Ee at a vertical height equal to $\frac{\beta\beta\omega\omega z}{\alpha\alpha hh}$ in order that, beginning to fall by its own gravity at a suitable time, it might arrive at the area Ee, and there replace, at the same moment and at the same velocity $\frac{\beta\omega\sqrt{z}}{\alpha h}$, the uppermost particle $\frac{\omega\, dx}{\alpha}$ descending in the conduit, and in this way the conduit will be kept continually full, as the statement of the Problem requires.

Section XXXI

Thus, if therefore each individual $\frac{\omega\, dx}{\alpha}$ has been set at the proper height $\frac{\beta\beta\omega\omega z}{\alpha\alpha hh}$ above Ee, and if all have fallen successively and are about to enter through Ee, maintaining the conduit always full, it is evident that an equal quantity of water $\int \frac{\omega\, dx}{\alpha}$ or $\frac{\omega x}{\alpha}$, which we understand to be moving in the extended plane Cc, must have flowed out through the orifice Cc, and certainly in such a way that each of its individual particles $\frac{\omega\,dx}{\alpha}$ has the proper acquired velocity $\sqrt{z}$. For that reason any one of those particles $\frac{\omega\,dx}{\alpha}$ is to be considered as having fallen from the original place of rest all the way to the lowermost level, BCc extended, from the height $\frac{\beta\beta\omega\omega z}{\alpha\alpha hh} + AB = \frac{\beta\beta\omega\omega z}{\alpha\alpha hh} + a.$ Accordingly, it is necessary to multiply the descents by the descending particles in order to obtain $\frac{\beta\beta\omega^3 z\,dx}{\alpha^3 hh} + \frac{a\omega\,dx}{\alpha}$, which integrated gives $\frac{\beta\beta\omega^3}{\alpha^3 hh}\int z\,dx + \frac{a\omega x}{\alpha}$ as the live force acquired from the universal descent of the heavy particles. This, moreover, must be equal to its own effect, which consists of the sum of the products which are formed by multiplying the individual particles by the squares of their respective velocities.

Section XXXII

For this reason I multiply the particle $\frac{\omega\,dx}{\alpha}$, which has already flowed out of the conduit, by the square of its velocity, which is z, and I will have $\frac{\omega z\,dx}{\alpha}$, the integral $\frac{\omega}{\alpha}\int z\,dx$ of which expresses the live force arising from the velocities of the entire aqueous matter flowing out of the conduit; at this point one must add to this [the live force] that all the matter flowing within the conduit has, and which is determined by multiplying the individual layers $y\,dt$ by the squares of their respective ultimate velocities $\frac{\omega\omega z(ds)^2}{\alpha\alpha yy(dt)^2}$, so that for any layer $y\,dt$, the live force arising from its own motion, $\frac{\omega\omega z(ds)^2}{\alpha\alpha yy(dt)^2}y\,dt = \frac{\omega\omega z(ds)^2}{\alpha\alpha y\,dt}$, is found. Therefore, the live force of all the layers to be considered through the entire conduit is $\frac{\omega\omega z}{\alpha\alpha}\int \frac{(ds)^2}{y\,dt} =$ (on account of $\int \frac{ds^2}{y\,dt}$, which may be called M, having been given throughout the entire conduit) $\frac{M\omega\omega z}{\alpha\alpha}$. Therefore, by considering the sum of both these live forces arising from the motion, we will have the live force of the entire aqueous systems equal to $\frac{\omega}{\alpha}\int z\,dx + \frac{M\omega\omega z}{\alpha\alpha}$.

Section XXXIII

Thus, by equating this force, determined from the motion, with that which we determined immediately above from the descent of the particles, the following equation will be yielded to us:

$$\frac{\beta\beta\omega^3}{\alpha^3 hh} \cdot \int z\, dx + \frac{a\omega x}{\alpha} = \frac{\omega}{\alpha} \cdot \int z\, dx + \frac{M\omega\omega z}{\alpha\alpha},$$

which, differentiated and freed of fractions, furnishes this:

$$\beta\beta\omega\omega z\, dx - \alpha\alpha hh z\, dx = M\alpha hh\omega\, dz - \alpha\alpha hh a\, dx,$$

or finally (after the reduction has been completed), this:

$$(\alpha\alpha hh - \beta\beta\omega\omega) z\, dx + \alpha M hh\omega\, dz = \alpha\alpha hh a\, dx,$$

just as we found through the direct method (§VII).

COROLLARY. If h or Ee be of very great size with respect to ω or Cc, the equation found reduces to this: $\alpha z\, dx + M\omega\, dz = \alpha a\, dx$; and for uniform efflux to this: $z = a$. If, however, Ee indeed be not of very great size with respect to Cc, let us assume, nevertheless, that new water follows by continuously descending within the conduit, not with some acquired velocity, but that it begins to follow from rest,

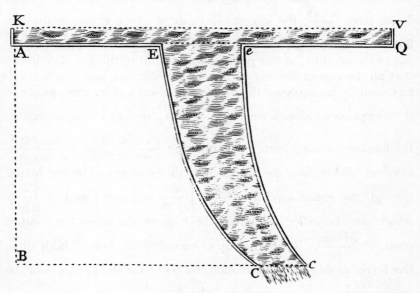

Figure 14

such that also in this manner the conduit is kept always full. This should be applicable to the result if to Ee (Fig. 14) is attached a very wide vessel, but of very small height, which is full of water. Certainly the water flows out therefrom taking its motion from rest, and yet, flowing into the conduit with the required velocity, it will continually maintain the fullness of it. The very matter is evident from the figure, where the conduit $ECce$ has the very wide cylindrical vessel $AQVK$ attached, the height AK or QV of which is assumed very small, so that essentially it does not increase the vertical height AB of the conduit Ec, so that KB can be taken for AB, and still the volume AV of this cylindrical vessel may enclose a very great amount of water. For this reason, in order that the velocity of the water already emitted through Cc may be determined, no longer Ee but KV is to be taken for the first area h, with the vertical height a of the conduit nevertheless being maintained, because according to the hypothesis AB does not differ noticeably from KB. This having been agreed upon, we will always have $z = a$ for uniform velocity, that is, that velocity which a weight would acquire by falling freely from the height $a = AB$ or KB.

A SINGULAR EXAMPLE OF DETERMINING THE MOTION OF WATER DESCENDING VERTICALLY IN A CONOIDIC CONDUIT WHERE NOTHING FLOWS OUT AND NO NEW WATER FOLLOWS THE DESCENDING [WATER]

Section XXXIV

Let the Hyperbola BEG (Fig. 15) exist between orthogonal asymptotes, the one AM vertical, the other AH horizontal, the ordinates DI, EK, FL, GM, etc. of which let define the very areas of the conoidic conduit, continued to infinity, which is known to be generated if another hyperbola described between the same asymptotes (the ordinates of which are in proportion to the roots of the first ordinary hyperbola) is revolved about the vertical asymptote just as about an axis. Let it be understood that at some place in the conduit designated by the hyperbolic area DK a portion of water is furnished, beginning to descend from rest, and that by descending it has arrived at some other place FM, so that, as a consequence, $FM = DK$. The velocities at GM, FL, etc., and the velocity at any other intermediate section $POop$ are sought.

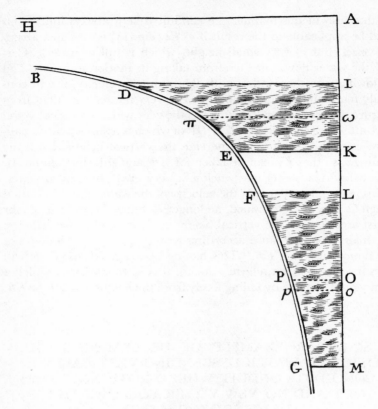

FIGURE 15

Section XXXV

Let every individual rectangle of coordinates, that is, the product of each area PO by the altitude AO of the conoid, be aa; likewise let the abscissas be given as $AI = b$, $AK = c$, and any descent AL of the upper area be assumed as x. There will result, from the nature of the hyperbola, the area $DI = \dfrac{aa}{b}$, $EK = \dfrac{aa}{c}$, $FL = \dfrac{aa}{x}$, and in addition (on account of $DK = FM$) $AM = \dfrac{cx}{b}$, from which $GM = \dfrac{aab}{cx}$, and the very area, or rather the solid DK or FM, is $aa(\ln c - \ln b)$. Further (with the calculus proving it), the distance of the center of gravity of the area DK from the horizontal AH will be $\dfrac{c-b}{\ln c - \ln b}$ and the distance of the center of gravity of the area FM (note that I always understand the solid to be represented by the area) from AH will be $\dfrac{x(c-b)}{b(\ln c - \ln b)}$. Thence let the descent of the center of gravity from the position DK to the position FM, $\dfrac{(x-b)(c-b)}{b(\ln c - \ln b)}$, be multiplied by the quantity of descending water, defined by $aa(\ln c - \ln b)$, and the product $\dfrac{aa(x-b)(c-b)}{b}$ will be equal to the live force produced from the descent.

Section XXXVI

Now if the velocity at GM be set equal to $\sqrt{z}$, the velocity at FL will be $\dfrac{b}{c}\sqrt{z}$. Also, if any AO equal y, [then] PO will be $\dfrac{aa}{y}$, the section Po will be $\dfrac{aa\,dy}{y}$, and its velocity will be precisely $\dfrac{by}{cx}\sqrt{z}$. Therefore, the square of this applied to the layer Po gives $\dfrac{aabbzy\,dy}{ccxx}$, equal to the live force of the layer Po, the properly corrected integral of which (by taking a, b, c, z, and x as constants) is $\dfrac{aabbz}{2ccxx}yy - \dfrac{aabbz}{2cc} =$ (in case y, or AO, is $\dfrac{cx}{b}$, or AM) $\tfrac{1}{2}aaz - \dfrac{aabbz}{2cc} = \dfrac{aaz}{2cc}(cc - bb)$, equal to the live force arising from the motion of the entire aqueous mass. By comparing this with the preceding, we will have

$$\frac{aa(x-b)(c-b)}{b} = \frac{aaz}{2cc}(cc - bb),$$

from which, by reduction, there is found $z = \dfrac{(x-b)2cc}{(c+b)b}$, equal to the square of the velocity at GM; correspondingly, the square of the velocity at FL is $\dfrac{(x-b)2b}{c+b}$, and the square of the velocity at any intermediate section Po whatever (pertaining to the abscissa AO or y) is $\dfrac{(x-b)2byy}{(c+b)xx}$.

Corollary 1. The lowermost layer EK, descending from rest to the position GM, acquires a greater velocity than a heavy body falling freely from the height KM. For $\dfrac{(x-b)2cc}{(c+b)b}$ is greater than KM or than $\dfrac{cx}{b} - b$.

COROLLARY 2. But the uppermost layer DI, descending to the position FL, acquires a lesser velocity than a weight falling freely from the height IL. For $\dfrac{(x-b)2b}{c+b}$ is less than IL or than $x - b$.

COROLLARY 3. Therefore, the lower parts of the aqueous mass are accelerated more vigorously and the upper parts more sluggishly than if they were to descend freely, animated only by natural gravity. This could also have been foreseen before the calculation from the fact that the portions of water in the narrower places are pressed upon by those lying above and so are incited to greater acceleration. But on the other hand, they resist those portions which occupy the wider places, and thus the upper [portions] are retarded in relation to their own natural acceleration.

COROLLARY 4. Hence, somewhere an intermediate layer Po is given which is neither incited nor retarded but which is accelerated in the same manner as if it were to descend freely. In order that this may be determined, I make AL to AO, or x to y, in proportion as AI or b is to $A\omega$, which will be $\dfrac{by}{x}$; and $\pi\omega$ will be the original position of the layer Po. Thus ωO, or $y - \dfrac{by}{x}$, is the height through which the layer Po descends. Therefore, in order that the acceleration of this layer be equal to the natural [acceleration], it is necessary only to make $y - \dfrac{by}{x} = \dfrac{(x-b)2byy}{(c+b)xx}$; now this will yield $y = \dfrac{(c+b)x}{2b}$, which shows that the distance AO is the arithmetic mean between AL and AM, just as $A\omega$ is the arithmetic mean between AI and AK. And similarly, $LO = OM$ and $I\omega = \omega K$. And so in these places the intermediate layer Po is pressed downward by the water FO lying above just as much as it is pressed upward by the water pM lying below, such that it descends by no means differently than if it were to descend freely, animated by natural gravity alone. Furthermore, this is also to be observed, that in these same places the compression of the water becomes a maximum, from which we conclude that if at the position of the section Po a tube were inserted vertically, the water in it would ascend to a greater height from the place of insertion than if it were inserted in a section at any other place between FL and GM. For truly the height in the tube depends upon the pressure of the water alone, as is evident from that explained above.

COMPARISON OF THIS SOLUTION THROUGH LIVE FORCES WITH THAT WHICH IS SHOWN THROUGH OUR DIRECT METHOD OBTAINED MERELY FROM DYNAMIC PRINCIPLES

Section XXXVII

In the second Part of this hydraulics Paper, §VII,* we gave through dynamic principles the most general equation for the determination of velocity, namely this:

$$(\alpha\alpha hh - \beta\beta\omega\omega)z\,dx + \alpha Mhh\omega\,dz = \alpha\alpha hha\,dx.$$

But there it is supposed that the vessel or conduit is always full, indeed with new liquid following continuously at the same velocity [and] adjoining to that which is already descending at the uppermost area, and the conduit itself is of a given and determined vertical height. But since in the present example this conduit is of indefinite height, and of course, only the portion FM is filled by liquid which always changes its position, and accordingly has its height LM variable at any instant, it may not immediately be seen that the case of this example is included by that formula found in §VII, especially since no new liquid here succeeds that descending, but its one and the same quantity DK or FM is always retained in the conduit, occupying one place after another.

Section XXXVIII

Meanwhile, we will show by what means, through some fiction of the mind, the present case can be resolved according to the hypothetical laws established in §VII. Certainly the original space $DEKI$ [Fig. 15] is to be considered as a conduit of given height IK, the upper area of which is DI, the lower EK, both given and determinate. Now while liquid flows from EK, about to occupy the lower positions in the prolonged conduit, I consider some fluid to be flowing in through the uppermost area DI, free from gravity as well as from all inertia or resistance, which, although this may not exist in the nature of things, nevertheless can be assumed such that it does nothing other than fill the space which would be left empty by the descending liquid. Thus, this having been presupposed, the actual liquid will have descended from DK to the position FM. I translate the forces of these individual layers PO, hydrostatic as well as hydraulic, to the

---

* Page 396 above.

uppermost area *DI*, where the fictitious fluid which occupies the space *DFLI* presses and [where it] must maintain the same effect that the force arising from the real liquid *FM*, similarly translated to the uppermost area *DI*, maintains by gravity, and this is that force which I called p or gha, [determined] by applying all these things to the reasoning of our Theory explained in this hydraulics Paper. And thus no forces come into account other than those which result from gravity and the motion of the real fluid, the fictitious fluid contributing absolutely nothing and serving no other purpose than to transmit the translated forces [required] for expelling the real liquid *FM*.

Section XXXIX

Therefore nothing remains other than that the required application be made of the method explained in §§IV and V, to the end that it may be adapted to the proposed example, where it is immediately evident that $\alpha = \beta = 1$. But here the remaining letters define those things which follow: surely $h = DI$; $\omega = EK$; a, or the height of the real liquid, LM; M equals the sums of all $\frac{Oo}{IO}$ contained in the height LM; further, z is the height from which a weight having fallen freely acquires the velocity v with which, immediately after the beginning of the descent, the real liquid or, afterwards, the fictitious fluid flows out through EK; and, finally, dq is the instantaneous displacement from the area EK.

Section XL

These things having been properly attended to, I now proceed as follows: after the velocity through EK has been set equal to v, the velocity through FL will be $\frac{xv}{c}$; the velocity through GM is $\frac{xv}{b}$; the instantaneous displacement through GM is $d(AM) = \frac{c\,dx}{b}$; the displacement through PO is $d(AO) = \frac{y\,dx}{x}$; the displacement through EK, or dq, is $\frac{c\,dx}{x}$; all of these displacements, since they must be simultaneous, are therefore in reciprocal proportion to the areas, just as are the velocities themselves, and thus in direct proportion to the distances from the horizontal AH. Now there is to be sought, by repetition of §IV (since $\alpha = \beta = 1$), the hydrostatic force which

METHOD FOR SOLVING HYDRAULICS PROBLEMS 427

is expressed through half of the uppermost area DI multiplied by the difference of the squares of the maximum velocity through GM and the minimum velocity through FL. Moreover, $DI = \frac{aa}{b}$, the velocity through GM is $\frac{xv}{b}$, and the velocity through FL is $\frac{xv}{c}$, from which the total hydrostatic force is

$$\frac{aa}{2b}\left(\frac{xxvv}{bb} - \frac{xxvv}{cc}\right) = \frac{aaxxvv}{2b^3cc}(cc - bb).$$

Section XLI

By repetition of §V, I determine the hydraulic force in the following manner. I multiply the accelerative force of the indefinite layer Po, which I call γ', by its displacement $\frac{y\,dx}{x}$, and I will have, through the dynamical principle, $\frac{\gamma' y\,dx}{x} = u'\,du' = \frac{yyv\,dv}{cc}$, from which the progressive accelerative force $\gamma' = \frac{xyv\,dv}{cc\,dx}$, and the motive force itself of the layer Po, that is, $\gamma' Po$, is $\frac{aaxv\,dv\,dy}{cc\,dx}$, which, translated to the area DI, or to $\frac{aa}{b}$, forms $\frac{aaxv\,dv\,y\,dy}{bcc\,dx}$, which integrated gives $\frac{aaxv\,dv\,yy}{2bcc\,dx} =$ (by rectifying it, or by applying it to all yy that are contained in the interval LM) $\frac{aax^3\,v\,dv\,(cc-bb)}{2b^3cc\,dx}$ equal to the hydraulic force. The sum of the hydrostatic and hydraulic forces will be

$$\frac{aaxxvv(cc-bb)}{2b^3cc} + \frac{aax^3v\,dv(cc-bb)}{2b^3cc\,dx},$$

or

$$\frac{aaxx}{2b^3cc}\left(vv + \frac{xv\,dv}{dx}\right)(cc-bb),$$

which sum of forces must be equal to the original translated force about to arise from the gravity of the layers. Moreover, the original translated force of any layer Po whatever is found by changing $g(Po)$ or $\frac{gaa\,dy}{y}$ into $\frac{gaa\,dy}{b}$ just as PO is to DI, or as AI is to AO, that is, just as b is to y, which properly integrated through the interval LM gives $\dfrac{gaax(c-b)}{bb}$ for the total basic translated force.

Section XLII

Thus we obtain the equation between the sum of the hydrostatic and hydraulic forces and the original translated force, which equation thus appears as $\dfrac{aaxx}{2bcc}\left(vv + \dfrac{xv\,dv}{dx}\right)(cc-bb) = \dfrac{gaax(c-b)}{bb}$, from which, by dividing by $\dfrac{aax}{bb}(cc-bb)$, there results:

$$\frac{x}{2bcc}\left(vv + \frac{xv\,dv}{dx}\right) = \frac{g}{c+b};$$

or, by reducing:

$$xvv\,dx + xxv\,dv = \frac{2gbcc\,dx}{c+b};$$

and so, by integrating and organizing in the proper manner (in order that v itself equals 0 when AL, or x, equals AI or b) there will result $\tfrac{1}{2}xxvv = \dfrac{2gbccx - 2gbbcc}{c+b}$, or by writing, according to the dynamic law, $2gz$ for vv, and by dividing by g, there will be $xxz = \dfrac{2bccx - 2bbcc}{c+b}$, and thus $z = \dfrac{(x-b)2bcc}{xx(c+b)}$, which determines the velocity through EK, from which now the velocity through any other area whatever is determined. And in fact, by changing $\dfrac{(x-b)2bcc}{xx(c+b)}$ into $\dfrac{(x-b)2cc}{b(c+b)}$ just as $(GM)^2$ is to $(AM)^2$, that is as cc is to $\dfrac{cc\,xx}{bb}$, or just as bb is to xx, this will be equal to the height from which a weight falling freely acquires the velocity which the liquid has at the lowermost point M. Further, by changing $\dfrac{(x-b)2bcc}{xx(c+b)}$ into $\dfrac{(x-b)2b}{c+b}$ just as cc is to xx, [this result will be] equal to the height from which a weight having fallen freely will have that velocity which pertains to the liquid at the uppermost point L. And finally, by changing $\dfrac{(x-b)2bcc}{xx(c+b)}$ into $\dfrac{(x-b)2byy}{xx(c+b)}$ just as cc is to yy, this will indicate the height from which a weight must fall freely in order to acquire the appropriate velocity which the liquid will have at any given intermediate point O. All this agrees marvelously with those [results] which we found through the Theory of live forces.

Section XLIII

On the other hand we can now show that which cannot be shown as easily through live forces, that is, how to find how much the liquid at any section may be pressed during descending. We certainly saw above in Coroll. IV that the liquid, after it descends from its initial or original position DK to the position FM, undergoes different pressures at its individual sections PO, and that the maximum of these occurs where PO cuts LM in half; truly, to determine its magnitude [there] and to compare it with some given weight—much less in other places inasmuch as PO certainly divides LM in some other proportion—would be a matter of more profound involvement if someone would wish to show this from the nature of live forces. Through our direct method, explained in the chapter on pressures, it is by no means difficult to obtain what is desired, even for this special example.

Section XLIV

And so let it remain to be investigated by how large a force the mass FM of the liquid is pressed at any place PO whatever, which force to be sought here I will now call π. I showed in §§XI and XII that, if only the portion PM, the remainder FO having been removed, were to proceed to descend so that it was driven not only by its own gravity, but in addition by π as well, it will (except for the first instant) be accelerated in the same manner as the total mass FM must be accelerated by its own gravity alone. Let IK be cut at ω in a similar ratio as LM has been cut at O, so that $AI : A\omega = AL : AO$; $\pi\omega$ will be the initial position of P itself. And $\pi K = PM$. Let AL be to AO, or AI to $A\omega$, as 1 is to n, from which $AO = nx$ and $A\omega = nb$. And let us imagine that the liquid contained in πK descends by the force of its own gravity and, in addition, by the force π which pertains to it at any position of descent, in order that the acceleration occur just as if the total mass DK were to descend by the force alone of its own gravity. Therefore, that which was AL, or x, now is AO, or nx, and that which was AI, or b, now is nb. And so the hydrostatic and hydraulic forces are found, if one writes nx for x, $n\,dx$ for dx, and

430   HYDRAULICS, PART II

likewise nb for b; and thus there will result $\dfrac{aannxxvv}{2n^3b^3cc}(cc-nnbb)$ or $\dfrac{aaxxvv}{2nb^3cc}(cc-nnbb)$, equal to the hydrostatic force, and also

$$\dfrac{aan^3x^3v\,dv}{2n^3b^3ccn\,dx}(cc-nnbb) \quad\text{or}\quad \dfrac{aax^3v\,dv}{2nb^3cc\,dx}(cc-nnbb),$$

equal to the hydraulic force. Moreover, the force of gravity arising through the translation, which was $\dfrac{gaax(c-b)}{bb}$, now indeed is $\dfrac{gaanx}{nnbb}(c-nb)$ or $\dfrac{gaax}{nbb}(c-nb)$.

Section XLV

Now, by taking the sum of the hydrostatic and hydraulic forces, and by equating that to the original translated force arising from gravity, $\dfrac{gaax}{nbb}(c-nb)$, to which must be added the force of pressure translated from PO to $\pi\omega$, which is obtained if we change π into $\dfrac{x\pi}{b}$ just as PO is to $\pi\omega$, or as AI is to AL, that is, just as b is to x, we get this equation:

$$\left(\dfrac{aaxxvv}{2nb^3cc}+\dfrac{aax^3v\,dv}{2nb^3cc\,dx}\right)(cc-nnbb)=\dfrac{gaax}{nbb}(c-nb)+\dfrac{x\pi}{b},$$

which, adjusted, assumes this form:

$$\dfrac{aa}{2nbbcc}[d(\tfrac{1}{2}xxvv)(cc-nnbb)]=\dfrac{gaa\,dx}{nb}(c-nb)+\pi\,dx.$$

Since, moreover, the velocity of the diminutive mass PM, which, however, is pressed by the force π, must be the same as the velocity of the whole mass FM, for which velocity we found just above that $\tfrac{1}{2}xxvv=\dfrac{2gbccx-2gbbcc}{c+b}$, let us write the differential of this, which is $\dfrac{2gbcc\,dx}{c+b}$, for $d(\tfrac{1}{2}xxvv)$, and this will yield

$$\dfrac{gaa\,dx}{nb(c+b)}(cc-nnbb)=\dfrac{gaa\,dx}{nb}(c-nb)+\pi\,dx.$$

By dividing by dx and transposing, we will obtain

$$\pi=\dfrac{gaa[cc-nnbb-(c-nb)(c+b)]}{nb(c+b)}=\dfrac{gaa(cn-c-nnb+nb)}{n(c+b)}.$$

Therefore, if at *PO* a tube, which should be erected vertically, is inserted in order that it may be known to what height the liquid can stand suspended in it (at least for a rather short time), π is to be divided by $g(PO)$, that is, by $\frac{gaa}{nx}$, in order to obtain

$$\frac{\pi n x}{gaa} = \frac{x(cn - c - nnb + nb)}{c + b},$$

the height of the liquid in the tube, from which height the absolute compression of the liquid at *PO* is to be evaluated. Q.E.I.

Section XLVI

Thus, if it be desired further to determine the point O in any given *LM* where the intensity of the pressure is greatest, that is, where the liquid in the tube will obtain a maximum height, the derived quantity $\frac{x(cn - c - nnb + nb)}{c + b}$ is to be differentiated, with n having been assumed as variable and the rest as constants; this having been done, there will appear $c + b - 2bn = 0$, from which results $n = \frac{c + b}{2b}$, and thus

$$nx \text{ or } AO = \frac{(c + b)x}{2b} = \frac{c + b}{2b} AL = \frac{c}{2b}AL + \frac{1}{2}AL = \frac{1}{2}AM + \frac{1}{2}AL.$$

From this it is evident that the point O of maximum pressure is half way between *M* and *L*, clearly just as we foretold by conjecture in Coroll. IV above.

Section XLVII

In addition, if the derived value of n itself, which is $\frac{c + b}{2b}$, is substituted in the expression $\frac{x(cn - c - nnb + nb)}{c + b}$, the maximum height itself of the liquid in the tube, $\frac{x(c - b)^2}{4b(c + b)}$, will result. Thus, because $\frac{1}{2}LM$ or $LO = \frac{x(c - b)}{2b}$, *LO* or the height of the liquid in the conduit above the point *O* where the tube is inserted will be to the height of the liquid in the tube as $\frac{x(c - b)}{2b}$ is to $\frac{x(c - b)^2}{4b(c + b)}$, or as $2(c + b)$ is to

$c - b$, or as double the sum of the original distances $AK + AI$ is to the simple difference of the same, that is, as $4A\omega$ is to IK.

Section XLVIII

Finally, sound reason alone dictates that the mass of the descending liquid FM undergoes no pressure at the extremities FL and GM, and therefore that the height in a tube inserted either at L or at M must be zero. And this very fact is certainly confirmed by the formula, for in the prior case where $n = 1$ the formula $\dfrac{x(cn - c - nnb + nb)}{c + b}$ is changed to this:

$$\frac{x(c - c - b + b)}{c + b} = 0;$$

but in the latter, where $n = \dfrac{c}{b}$, the same is changed into this:

$$\frac{x(cc - cb - cc + cb)}{b(c + b)} = 0 \text{ also.}$$

Section XLIX

Scholium 3. This example of a liquid descending by its own gravity in an indefinitely continued hyperbolic pipe, which I treated rather extensively as an example, shows in what manner one is to proceed in other cases of this type where a liquid descends within a sufficiently long conduit, with the identical quantity always, so that certainly nothing flows out from it and, equally, no new liquid flows in. Besides, it reveals the access to the solution of Problems concerning the determination of oscillatory motion of fluids in bent or reflex pipes, whatsoever may be their shapes, and of area varying in any way whatever. For, in fact, in such pipes or siphons, while a certain portion of the fluid descends through one leg, through the other leg, albeit very dissimilar, another portion of the fluid equal to the former ascends, that is, descends negatively, so that, just as in pipes continuously inclined downward, the same mass of fluid is always contained within. Thus, if, with the signs in the calculation having been changed for those quantities which require it, one proceeds by the same method which we showed in the example presented, by no means will it be difficult to perform the computation of the velocity of the fluid flowing through in the individual places for any ascent or descent of it whatever; from this all the remaining [relationships] are derived.

CONCERNING VESSELS WHICH, DURING THE TIME WHEN THEY ARE EMITTING LIQUID THROUGH AN APERTURE MADE IN OR NEAR THE BASE, RECEIVE NO NEW LIQUID FLOWING IN FROM ABOVE

Section L

For those cases of vessels which by emitting liquid, but by accepting no other, are finally depleted and evacuated, that method could be applied which I explained in §XLIV and following, where [the text] is concerned with the determination of the motion of a given aqueous mass falling continuously within a hyperbolic-conoidic conduit, certainly with the aid of a fictitious fluid which does nothing other than fill the space left empty by the descending fluid. But, moreover, after that fiction of the mind has been considered and ignored, in the present situation our equation suffices [which was] given in §VII, $(\alpha\alpha hh - \beta\beta\omega\omega)z\,dx + \alpha Mhh\omega\,dz = \alpha\alpha hha\,dx$, which we say to be valid for vessels always remaining full. Of course, that equation will be accommodated easily to these vessels also which are depleted slowly on account of no new liquid flowing in, and in which, hence, the uppermost surface Ee (Fig. 10) continuously descends.

Section LI

In repetition of that which has already been solved in the first Part, §XII,* for the case of a cylindrical vessel, let us consider here a vessel of any shape whatever into which no new liquid is flowing from above, while that which is already within is escaping by continuously passing through the final orifice. And so let us prescribe that the uppermost surface, by descending from the position Ee, has arrived at the position Ff, at which moment the velocity of the flowing liquid is to be determined. For this purpose the area Ff itself, or y, obviously variable, is to be taken as the uppermost area to which the variable height BP or t corresponds (indeed the nature of the translation of motive forces in fluids permits this, as will be evident

* Pages 363-64 above.

to anyone who considers it); but M itself, or $\int \dfrac{ds^2}{y\,dt}$, which was constant in vessels constantly full, is now variable; certainly it is to be applied throughout the variable height BP.

Therefore, in our equation let one write yy for hh, t for a, and $\dfrac{ds}{dt}$ for β, and the following equation, which is the one desired, will be produced:

$$\left[\alpha\alpha yy - \omega\omega \frac{(ds)^2}{(dt)^2}\right] z\, dx + \alpha yy\omega\, dz \int \frac{(ds)^2}{y\, dt} = \alpha\alpha yyt\, dx.$$

But from this, since the element of falling water $\dfrac{\omega\, dx}{\alpha}$ equals the descending layer $-y\, dt$ (I put $-y\, dt$ because t decreases with increasing x), dx will be $\dfrac{-\alpha y\, dt}{\omega}$. Therefore, by substituting this value for dx, the equation will have this appearance:

$$\left[\alpha\alpha yy - \omega\omega \frac{(ds)^2}{(dt)^2}\right] z\, dt - \omega\omega y\, dz \int \frac{ds^2}{y\, dt} = \alpha\alpha yyt\, dt;$$

for vessels having a vertical centric where $\alpha = 1$ and $ds = dt$, this transforms into the rather simple equation

$$(yy - \omega\omega) z\, dt - \omega\omega y\, dz \int \frac{dt}{y} = yyt\, dt.$$

Section LII

From this equation we will now pursue to the end the example shown in the first Part (§XII), where the vessel was assumed cylindrical, and we will add certain worthwhile notes. Here, therefore, h is to be used, contrarily, in place of the constant y, and $\int \dfrac{dt}{y}$ will be $\int \dfrac{dt}{h}$, or $\dfrac{t}{h}$, and the whole term $\omega\omega y\, dz \int \dfrac{dt}{y}$ will be $\omega\omega t\, dz$, whence the equation for the case of the straight cylinder will have this form:

$$(hh - \omega\omega) z\, dt - \omega\omega t\, dz = hht\, dt;$$

then, by the method of integrating used formerly by me, this gives in finite terms the required value of z itself as

$$\frac{hht}{hh - 2\omega\omega}\left[1 - (t/a)^{(hh-2\omega\omega)/\omega\omega}\right].$$

METHOD FOR SOLVING HYDRAULICS PROBLEMS

Here at once it is evident that the velocity, whether initial or final, is zero, that is, in the case in which $t = 0$, as well as [in that] in which $t = a$. Hence it is deduced that somewhere the velocity of the water will be a maximum while flowing out of the vessel or while descending in the vessel itself. In order that this [maximum velocity] be determined, the maximum z is to be found which would result when the surface of the water in the cylinder descends to that distance t from the base which is $a\left(\dfrac{\omega\omega}{hh - \omega\omega}\right)^{\omega\omega/(hh - 2\omega\omega)}$; this may be found in two ways: namely, either by differentiating the ascertained value of a itself in the usual way, or, as is easier, by setting the second term $\omega\omega t\, dz$ in the preceding differential equation, $(hh - \omega\omega)z\, dt - \omega\omega t\, dz = hht\, dt$, equal to zero, from which there results $z = \dfrac{hht}{hh - \omega\omega}$, which, compared with the ascertained general value

$$\frac{hht}{hh - 2\omega\omega}\left[1 - \left(\frac{t}{a}\right)^{(hh - 2\omega\omega)/\omega\omega}\right],$$

will give, as I said,

$$t = a\left(\frac{\omega\omega}{hh - \omega\omega}\right)^{\omega\omega/(hh - 2\omega\omega)}.$$

Section LIII

Meanwhile, in the very special case where $hh = 2\omega\omega$, where the area of the cylindrical vessel is to the area of the orifice as $\sqrt{2}$ is to 1, this inconvenience occurs, that $\dfrac{hht}{hh - 2\omega\omega}\left[1 - \left(\dfrac{t}{a}\right)^{(hh - 2\omega\omega)/\omega\omega}\right]$ becomes $\dfrac{hht}{0}(1 - 1)$, and indeed that the other,

$$a\left(\frac{\omega\omega}{hh - \omega\omega}\right)^{\omega\omega/(hh - 2\omega\omega)}, \text{ becomes } \left(\frac{\omega\omega}{\omega\omega}\right)^{\omega\omega/0}, \text{ or } 1^\infty,$$

from either of which nothing can be concluded. However, this inconvenience is taken up through the rule, applied with some dexterity, which I communicated some time ago to the illustrious L'Hôpital, as is to be seen in the Analysis of infinitesimals, Art. 163.*

First, then, the recently introduced $\dfrac{hht}{hh - 2\omega\omega}\left[1 - \left(\dfrac{t}{a}\right)^{(hh - 2\omega\omega)/\omega\omega}\right]$ is found for the present case as $-2(a - t)\ln(a - t)$, but the other,

* See No. LXXI, p. 401, [*Opera Omnia*] Book I.

$a\left(\dfrac{\omega\omega}{hh-\omega\omega}\right)^{\omega\omega/(hh-2\omega\omega)}$, as $\dfrac{1}{e}$, by assuming, naturally, a as unity and $\ln e = 1$; and so here either quantity, the subtangent of which, $a = 1$, is determined very easily through common Logarithms.

CONCERNING THE MAKING OF CLEPSYDRAS

Section LIV

Until now we treated the shapes of vessels passing water through an orifice below exclusively as if the shapes had been given, so that we might surely bring out the laws according to which the motion of the water would proceed. But now I should like to inquire in reverse order into the shape of the vessel required in order that the uppermost surface of the water might subside according to some proposed law; for example, that it might be lowered at uniform speed, whence from the magnitude of the descent the duration of flow is known immediately; this very frequent use of Clepsydras was instituted long ago among the Ancients for measuring time. Moreover, this can be obtained principally in two ways; one certainly is from the quantity of water having flowed out, the other from the quantity of water still remaining in the vessel; from either, judgment can be passed concerning the interval of time. Let us treat each one separately.

Section LV

Let us consider the rather simple vessels which indeed have their centrics vertical, and of which the equation (§LI) is this:

$$(yy - \omega\omega)z\, dt - \omega\omega y\, dz \int \frac{dt}{y} = yyt\, dt,$$

where the unknown t and y have their origin at the lowest point, or at the orifice ω. Thus, if now we wish to consider that the water flows out at a uniform velocity, z is to be put equal to the constant c, which, this having been done, will give $dz = 0$; and so the second term $\omega\omega y\, dz \int \dfrac{dt}{y}$ vanishes, and the others, divided by dt, will give this algebraic equation:

$$(yy - \omega\omega)c = yyt,$$

from which it is found that $yy = \dfrac{\omega\omega c}{c-t}$ and $y = \sqrt{\dfrac{\omega\omega c}{c-t}}$ or $\sqrt{y} = \sqrt[4]{\dfrac{\omega\omega c}{c-t}}$.

METHOD FOR SOLVING HYDRAULICS PROBLEMS 437

COROLLARY 1. The nature and shape of the vessel are therefore such that it can be generated from the revolution of a plane biquadratic Hyperbola described between two asymptotes, of which one is vertical, the other horizontal [and] distant from the orifice ω by a height c.

COROLLARY 2. Assume that the first height or the initial t equals c, and that the initial area y is infinite; then indeed one has

$$y = \sqrt{\frac{\omega \omega c}{c-t}} = \sqrt{\frac{\omega \omega c}{c-c}} = \infty.$$

COROLLARY 3. The quantity of water remaining in the vessel for any height t, or $\int y\, dt$, is found by integrating and properly organizing $\int dt \sqrt{\frac{\omega \omega c}{c-t}} = 2\omega c - 2\omega \sqrt{cc - ct}$. Thus the total quantity of water from the beginning of flow is $2\omega c$, that is, equal to the cylinder of water of which the base is ω and the height is $2c$.

COROLLARY 4. Therefore, if a cylindrical receptacle of capacity not less than $2\omega c$ is placed under the water about to flow out of the orifice ω, the water that has flowed out and has been collected in that receptacle ascends equally in equal times, and so, after the height of the receptacle has been divided into equal graduations, the Clepsydra will result.

SECTION LVI

SCHOLIUM 4. It should not be concealed that this type of Clepsydra can hardly have any use in practice on account of the immense height which should be given to the hyperbolic vessel in order that the efflux can endure through a period of time, even if it be rather short; this can be understood satisfactorily from the fact that if the vessel were 15 feet high, that is, if $c = 15$ feet, it would contain water $2\omega c = 30\omega$, that is, an aqueous column the height of which is 30 feet above the base ω. Since, moreover, from the orifice ω the water flows out at the uniform velocity required for the height of 15 feet, and since the time of fall through this height is less than one second, it is certainly evident that within one second the aqueous cylinder of area ω and of length twice fifteen, or thirty, feet flows out of the vessel, and therefore in such a short time the whole vessel will be emptied. Let me say nothing yet about the impossibility of the structure of the vessel, since one supposes that its uppermost area is infinite; I would certainly like to correct for this inconvenience by prolonging the vessel near enough to the upper horizontal asymptote that it may acquire an area much greater than that of the orifice ω, which area then could be enclosed by a lip rising up to the asymptote.

Section LVII

The Clepsydra can be made in a more convenient manner by means of constant descent of the uppermost aqueous layer in the vessel itself; a suitable shape of vessel is therefore to be investigated in order that, with water escaping through the orifice ω, the surface of the water remaining in the vessel may descend equally in equal times, so that, in a given interval of time, the water surface passes through a given number of equally spaced divisions into which the vertical axis of the vessel has been divided. But there are two cases in which the intended [effect] can be obtained: namely, either the area ω of the opening is so small that it has no sensible ratio to the area y in the vessel; or the area ω is large enough that it is comparable with any y whatever. We will now treat the first case, since it is easier and more useful in practice; the other will be treated later.

Section LVIII

It is certainly evident that our equation

$$(yy - \omega\omega)z\,dt - \omega\omega y\,dz \int \frac{dt}{y} = yyt\,dt$$

is reduced, in the case in which ω is very small in proportion to y, to precisely these two terms: $yyz\,dt = yyt\,dt$, from which $z = t$; that is, in any vessel whatever having an orifice ω, water certainly emanates from it at that velocity which is acquired by a heavy body falling from the height t which the residual water in the vessel has; of the truth of this matter (found scientifically by us) hydraulics Writers of previous years had knowledge only through experiments. However, once this had been supposed, it was then easy to discover the nature of that conoidic vessel having the orifice ω as if it were its vertex, facing downward, because then it has this effect, that at any moment the uppermost surface y of the residual water descends at a uniform or constant speed. For since the velocities of the fluid flowing in the same quantity and at the same time through two different areas are in reciprocal ratio of the areas, one must set $y/\omega = \sqrt{z} \Big/ \frac{\omega\sqrt{z}}{y}$, and there will result $\frac{\omega}{y}\sqrt{z}$ designating the speed of the descending surface

METHOD FOR SOLVING HYDRAULICS PROBLEMS

y, which speed, since it must be constant, let be set as $\dfrac{\omega}{y}\sqrt{z} = \sqrt{c}$, from which there will be $\dfrac{cyy}{\omega\omega} = z = t$. Moreover, in straight conoids the areas are nothing else but circles, of which the areas y are as the squares of the radii, or of the ordinates of the generating curve which by its revolution about the axis describes the desired conoid. Therefore, by designating the ordinate in the generating curve as s, and the radius of the circular orifice ω as b, and likewise by saying that the area of the circle is to the square of the radius as n is to 1, one will have, certainly, $y = nss$, $yy = nns^4$, $\omega = nbb$, and $\omega\omega = nnb^4$; and so for the equation $t = \dfrac{cyy}{\omega\omega}$ there will result $t = \dfrac{cs^4}{b^4}$ or $\dfrac{b^4 t}{c} = s^4$; this shows the generating curve of the desired conoid to be a biquadratic Parabola, at the lowest point of which, or, at the vertex facing downward, the orifice ω emitting the aqueous stream must be fashioned.

COROLLARY. The parameter of the derived Parabola is $b\sqrt[3]{b/c}$, where c is arbitrary; and therefore c can be assumed so small that $b\sqrt[3]{b/c}$ can be made as great as desired, to the end that the areas of the conoid become incomparably greater than the area of the orifice ω. Accordingly, therefore, for the pleasure of it, the capacity of the vessel can be made so large and the area of the orifice ω can be made so small that the efflux of water may persist for a very long time before the vessel is completely exhausted; attention is paid to this especially in the design of Clepsydras.

Section LIX

Now consider the other case, where the orifice ω is not assumed of so insensible an area that in the universal equation

$$(yy - \omega\omega)z\,dt - \omega\omega y\,dz \int \frac{dt}{y} = yyt\,dt$$

those terms vanish in which ω appears; it is surely necessary that all terms remain, and then that $\dfrac{cyy}{\omega\omega}$ be substituted for z, and $\dfrac{2cy\,dy}{\omega\omega}$ for dz, to the end certainly that the uppermost surface of the water in the vessel descend at uniform speed, which speed is due to the arbitrary height c, whence the resulting equation to be solved is

$$(yy - \omega\omega)c\,dt - 2c\omega\omega\,dy \int \frac{dt}{y} = \omega\omega t\,dt.$$

There are certain indications from which I suspected at once that there exists a certain algebraic curve which agrees with that equation taken in the abstract. Thus after a short investigation, this appeared directly to me:

$$yy = \frac{\omega\omega(t + 3c)}{c};$$

and for the generating curve, this:

$$s^4 = \frac{b^4(t + 3c)}{c},$$

which again is a biquadratic Parabola, but with abscissas t which take their origin not from its very vertex, but indeed from below the same, on the axis, at a distance $3c$. Meanwhile, the derived equation $yy = \frac{\omega\omega(t + 3c)}{c}$, which certainly satisfies in the abstract, cannot for that reason be accepted in this instance because it does not fulfill a tacit condition, which condition consists in this: that for $t = 0$, the entire equation must go to zero, and therefore even $\int \frac{dt}{y}$ must vanish, which does not occur in the case of the equation $yy = \frac{\omega\omega(t + 3c)}{c}$. Of course, it is nothing new that a certain proposition which is true in general is not always correct in particular, especially when it is required to satisfy additional conditions to which it is not necessary to attend when taking the matter in general. Moreover, the true generative curve for the desired shape of the conoidic vessel is ascertained if by this art the equation

$$(yy - \omega\omega)c\, dt - 2c\omega\omega\, dy \int \frac{dt}{y} = \omega\omega t\, dt$$

can be universally solved so that y is determined through t, or vice versa, t through y, either that it be done in finite algebraic terms or in exponentials, or, indeed, through quadratures. But I leave this matter, since it is not important here, to be resolved by others to whom time is available.

Section LX

Scholium 5. An opportune occasion is now given for examining the *Newtonian* cataract, which that Author describes in the second edition of *Principia Mathematica Philosophiae Naturalis*, Prop. 36, Book II, pp. 303 ff. There at once it is to be observed that that form

METHOD FOR SOLVING HYDRAULICS PROBLEMS 441

which Newton gave to his cataract *ABNFEM* (see Fig. 16, which is Newton's from that selected place) certainly is the same as that which

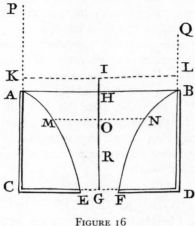

Figure 16

I found above (§LV) for the shape of the Clepsydra which emits water through the orifice ω at a constant speed. Let us note now what Newton himself recognizes, that in such a cataract any layer *MN* whatever descends with that velocity which it would acquire if it would fall freely from a given point *I* through the height *IO*, animated by no force other than its own natural gravity; from this it follows certainly that the layers remain in contact with each other in descending, but nevertheless such that they exert no force on each other, either by pressing or by resisting, just as if individual particles were descending by their own weight. And thus that pressure concerning which I treated above in a particular section will be null through the whole *Newtonian* cataract, and accordingly, concerning the force of pressure which I called π, not even the least will be exerted on the sides *AME*, *BNF*, which also will be evident from my very formula which I gave for π in §XII. If indeed that is applied to the present case, it is obtained, as I said, that the value of π itself is null through the whole height of the cataract. What therefore must be concluded from this? Undoubtedly the following: that if the sides *AME* and *BNF* of the cataract were rigid, resembling those of a funnel, with which Newton compares it, and if at any place whatever, an orifice having been made, a tube erected to a vertical position were inserted, none of the flowing water would enter from the cataract into the tube and ascend, as would happen if the sides were pressed by the water flowing through. Meanwhile, the sides *AME* and *BNF* are pressed inwardly against the axis *HG* by the weight of the water standing in *AMEC* and *BNFD*, through the common hydrostatic law which shows that pressures experienced at the individual places *M* and *N* are proportional to the heights *HO*. Moreover, since the sides of the cataract are not rigid, and those pressures of the standing water have no pressures opposed to them from the water flowing through, it is of course necessary that the water which is considered to stand and which presses continuously obtain its own effect, that is, that it pour into the cataract and mix together with the flowing water itself. Therefore, the shape of the cataract will be destroyed and will be thrown into disorder, and the water descends differently than according to our explanation.

Therefore the *Newtonian* explanation, since it is adverse to the laws of hydrostatics, cannot stand.

EPIMETRUM: CONCERNING THE FORCE BY WHICH A VESSEL IS PUSHED BACKWARD WHEN WATER FLOWS OUT OF IT IN A HORIZONTAL DIRECTION

It is a very well-received truth that *Action is equal to reaction*, that is, that any external force whatever which acts on a certain body or any other obstacle also acts backwards in the directly opposed direction on whatever barrier it may have or find, just as we see it happen when, for example, a stretched elastic placed between two bodies begins to be relaxed at once upon release of the restraint; it propels each body by equal forces, one forward, the other backward. We observe that it does not occur otherwise when an iron shot is fired with great violence from a cannon by the ignition of gunpowder, that similarly the cannon itself is driven back in the opposite direction, however by a much lesser impetus on account of the huge mass of the cannon with respect to that of the shot itself.

In an equal manner, when we see that water is expelled from a vessel in a horizontal direction, whether it occurs immediately through an orifice opened in the side of the vessel or intermediately through some conduit having a horizontal position, we must conclude that some force or power exists which may produce this expulsion, which then acts equally strongly in pressing backward at the wall of the vessel directly opposed to the final orifice of the conduit. Therefore, this effect is to be considered in order that the magnitude or measure of that force driving backwards and forwards may be properly estimated. At once several ideas come to mind which at first glance seem to reduce to the desired result, but, since the various ideas show measures of that quantity differing from each other, it is uncertain whether one or another of them is correct, or, rather, whether or not all of them are out of tune with the truth. And, therefore, the path which we shall follow will be safest and most certain if we deduce the solution immediately from the very principles confirmed in this hydraulics Theory of ours.

To this end let us consider what was stated at the beginning concerning the translation of the motive forces of the individual aqueous sections Fm, Nl, etc. (Fig. 10), to the common uppermost area Ee, where it is shown that those forces, thus collected into one, furnish the same effect in expelling the liquid, assumed free of gravity, through the aperture Cc, and this at the same velocity at all times as that which occurs in the natural manner through the gravitation of the liquid descending without translation. Now indeed we clearly understand that that translation to the uppermost area Ee is arbitrary, inasmuch as it can easily be gathered from the same hydrostatics principle which we assumed that the motive forces of the individual layers can be transferred to any other assumed area whatever, for example to Ff, which may be considered as given or constant; and it is not so that one may say that the forces translated to that [area] act only by pressing the layers below but not those above. For while the continuous layers are thus joined to each other so that the one without the other cannot be moved from a [particular] place, with no difficulty at all do we understand that the propelled aqueous mass below $FCcf$ carries, so to speak, that above, $EFfe$, along with it as well, and that with that [mass] the effect of the pressing potential at Ff must be shared to such a degree that the liquid will flow out through Cc at no other velocity than as if the motive forces of the layers had been translated to the uppermost area Ee.

With these things properly understood, let us now consider that the forces of the layers are translated not to the uppermost area Ee, nor to some intermediate one Ff, but to the lowermost Cc, so that the resultant force from the collection of all the forces of the layers, while it acts immediately only on the lowermost layer, nevertheless must drive the entire aqueous mass $ECce$ into motion precisely by the same law as if the motive forces had been translated to the uppermost surface Ee, or to any other whatever, Ff. Moreover, from the common hydrostatic principle, the sum of the motive forces translated to any area whatever is proportional to the area itself, just by making gha, or the weight of the aqueous cylinder the base of which is Ee and the vertical height is AB, into $g\omega a$, or the weight of the aqueous cylinder which has the same height a or AB and the base ω or Cc, just as Ee is to Cc, or h is to ω; let this weight be designated p, and thus we have the quantity of pressure p by which the water is expelled through the orifice Cc, and which, in pushing forward, at the same time is pushing backward in a common direction.

Let an application be made: Let the vessel ABD (Fig. 13) having any shape whatever be attached to the not very large oblong conduit CEK lying in a horizontal position. For the sake of brevity, let the centric of this vessel be vertical, and it is not important what shape the conduit may have, whether it be a truncated conoid or composed of many pipes, of which [conduit] the extreme orifice at K through which water is expelled equals ω. Now when the vessel has been assumed constantly full, we saw above that for the determination of the velocity v of the water flowing out, this equation results:

$$\frac{vv(hh - \omega\omega)}{2h} + \frac{hM\omega v\, dv}{dx} = gha,$$

if, of course, the motive forces are translated to the uppermost area h, or to AD; whence if we translate the same to the lowermost area ω, there will result $\dfrac{vv\omega(hh - \omega\omega)}{2hh} + \dfrac{M\omega\omega v\, dv}{dx} = g\omega a = p$, [equal] to the weight of the aqueous cylinder the base of which is ω and the height is a, the height of the water above the horizontal BK. Therefore, imagine the small surface Bb in the wall BA of the vessel from the region opposite and equal [in area] to the orifice K; this Bb will endure a similar pressure on itself, equal to p, from the retroaction expelling the fluid.

Certainly this retropressure p exerted at Bb rises to its full intensity immediately as it begins to be moved; that is, from the first instant of motion when the velocity is as yet infinitely small right up until it reaches the level of equilibrium, that force of retropressure is constantly the same. Certainly whatever v is, there always results $g\omega a$ or $p = \dfrac{vv\omega(hh - \omega\omega)}{2hh} + \dfrac{M\omega\omega v\, dv}{dx}$, whence at the beginning of flow, when $v = 0$ or is infinitely small, one will have $p = \dfrac{M\omega\omega v\, dv}{dx}$; but when v will have arrived at uniformity, so that $dv = 0$, then, similarly, $p = \dfrac{vv\omega(hh - \omega\omega)}{2hh}$. From this it follows that

$$\frac{M\omega\omega v\, dv}{dx} = \frac{vv\omega(hh - \omega\omega)}{2hh},$$

assuming, certainly, v for the initial velocity in the first term, but v for the uniform velocity in the other. Further, it is also evident from here that our retropressure is equal to the weight itself of an aqueous cylinder having a base ω and that height which the water has in the vessel above the horizontal BK, because a very well-known hydrostatic principle shows that the area CE through which the water enters into the conduit CK is pressed by the weight of the aqueous column lying above, of which the base is CE itself and the height is that of the uppermost surface AD above EC. And so with the area CE having been set equal to m and the height of the water in the vessel equal to a, the pressure which drives the water in the conduit from CE against K will be gma; with gma having been made into $g\omega a$ in proportion as the area CE is to the area K of the aperture, that is, as m is to ω, the force $g\omega a$ will be, through the same hydrostatics principle, that very [force] which drives the water to eruption through the orifice ω; therefore, the force directly opposed to this, arising from the reaction and action at Bb, will likewise be $g\omega a = p$. Q.E.D.

COROLLARY. It turns out that in vessels constantly full the force of retropressure is invariable from the first instant of efflux right up to the constant velocity, inasmuch as it is always $g\omega a$. Therefore, they err who state that it is variable: namely, less in slower efflux, greater in faster, a maximum in constant efflux.

HYDRAULICS PROBLEM

AB is a cylindrical pipe, uniformly wide and open at both ends A and B [Fig. 17]. Let it be immersed perpendicularly anywhere in an infinite fluid the surface of which is LR, and let the immersed part of the pipe be BC. Let it be understood that the whole pipe AB is full of the same fluid, so that with the thumb held over it at A nothing can flow out. Now there is sought, if freedom is given to the descending fluid by the thumb having been removed, how far it will descend below LR, and afterwards how far it will ascend again above LR; that is, P having been taken for the termination of the descent and O for the termination of the subsequent ascent, the length CP and as well the length CO are sought.

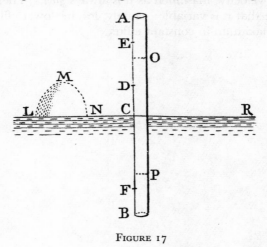

FIGURE 17

SOLUTION

First it is to be considered that the portion of the fluid contained in *CB* has no weight, or rather, that its attempt to descend developing from gravity is eliminated by the opposing pressure of the fluid surrounding the pipe; and thus at once some of the fluid while descending from the projecting part *AC* descends into the submerged part of the pipe; further, the former is to be considered to be separated from its own gravity so that the total accelerative force at any time depends only on the weight of that portion of the fluid which exists as yet in the projecting part of the pipe. Thus at once it is evident that the descent of the fluid must be accelerated as long as some remains in AC and hence that the maximum velocity will occur when it will have descended to the horizontal LR; that then again it must be retarded on account of the prevalent pressure of the ambient fluid; that, when the velocity has been completely eliminated at P, the fluid is compelled to ascend again to O by the external ambient pressure, and that it will have a maximum velocity again at C. Then from O, by another change, it falls, but not to as great a depth as previously, at once about to ascend upward again, and thus it will continue oscillating successively. And so are sought the first descent to P and the first subsequent ascent to O, inasmuch as upon these all the remaining depend.

As far as the determination of the descent is concerned, it is achieved in a two-fold way, either directly from the principles of mechanics or indirectly from the nature of live forces. Accordingly, I accomplish the task by each method.

FIRST METHOD. The principles of mechanics show that the instantaneous increment of velocity is obtained if the element of time is multiplied by the accelerative force; moreover, the element of time is obtained by dividing the element of space to be traversed by the acquired velocity. Thus let the total length of the pipe AB be a, the immersed part CB be 1, any arbitrary part AE passed through by the descending fluid be x, and the natural gravity by which, certainly, bodies are naturally animated or urged to descent be g. Let the velocity acquired by the fluid having fallen from A to E be designated by v. With these [terms] so defined, the weight of the fluid filling up the whole pipe will be ga; the weight of the part CB, $g(1) = g$; and the weight of the part CE, $g(CE) = ga - g - gx$. Now, because equal to the weight of the portion CB is eliminated by the equivalent pressure of the ambient fluid, only the weight of the fluid CE remains, which must accelerate all the remaining fluid EB into descending. Thus the accelerative force will be $\dfrac{g(CE)}{BE} = \dfrac{ga - g - gx}{a - x}$, whence $\dfrac{ga - g - gx}{a - x}\left(\dfrac{dx}{v}\right) = dv$; accordingly

$$v\, dv = \frac{ga\, dx - g\, dx - gx\, dx}{a - x} = g\, dx - \frac{g\, dx}{a - x},$$

and by integration,

$$\tfrac{1}{2}vv = gx + g \ln (a - x) - g \ln a.$$

(*Note:* I add $g \ln a$ here for the sake of correction, so that certainly for x equal to 0, where the velocity is null, the value derived for it vanishes.) And thus, in order that it be known how far the liquid must descend in the pipe so that v again becomes 0, it is necessary to make $x + \ln (a - x) - \ln a = 0$, and, by taking AP as the root of this equation, the point P will be the terminus at which the descent is ended.

Indeed the root x is obtained with the help of a Logarithmic [curve] in this way: let the Logarithmic be HCG [Fig. 18], the subtangent of which equals $1 = BC$, the immersed portion of the pipe.

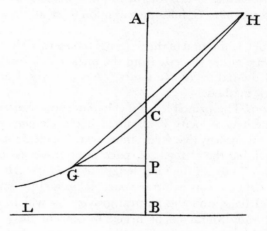

FIGURE 18

In BC extended, let BA be taken equal to the length of the entire pipe, up to which [length] certainly at the beginning it is full of liquid; [the line] AH, parallel to the asymptote BL, is drawn from A, meeting the Logarithmic at H, and from this point H [the line] HG, which cuts the curve at G, is drawn parallel to the tangent at C; whence, further, GP is drawn parallel to the asymptote BL; the point P will be the terminus to which the fluid descends and at which the descent will be completed. The proof is simple. Indeed because $AH = \ln AB = \ln a$, and $PG = -\ln PB = -\ln (AB - AP) = -\ln (a - x)$, and $AH + PG = AP$, there will be $\ln a - \ln (a - x) = x$, from which $x + \ln (a - x) - \ln a = 0$. This is the very same equation [which is] to be developed. Q.E.D.

HYDRAULICS PROBLEM

SECOND METHOD. From the Theory of live forces I find the same in the following way, by indicating here again (Fig. 17) AB by a, CB by 1, and the indeterminate portion AE of the descent by x. Let z be a certain vertical height through which a weight falling freely would acquire a velocity equal to that which the fluid acquires after it has fallen in the pipe from A to E. This velocity will therefore be $\sqrt{z}$. Moreover, by cutting off $BF = AE$, it is clear that through the descent of the fluid from A to E the column AF is translated to EB and that then its individual parts have a velocity $\sqrt{z}$; wherefore all the fluid substance constituting this column EB will have a live force $(a - x)\sqrt{z}\sqrt{z} = az - xz$, to which are to be added the particular collectively taken live forces of the individual particles contained in the space BF, which, with the fluid descending from A to E, have flowed out successively from the orifice B, of which certainly the sum is $\int z \, dx$. Therefore, the magnitude of the live force of the entire substance then in motion is $az - xz + \int z \, dx$.

Because, indeed, this live force must be the effect of the gravity of the portion of the fluid lying above the horizontal LR, therefore it is necessary that that quantity of the live force $az - xz + \int z \, dx$, as the effect, be equal to its own adequate cause, that is, to the sum of the products which are formed by multiplying the individual particles of descending fluid by the appropriate height of each through which any one of them is depressed by its own gravity. Therefore, $CD = AE$ having been assumed, DC will be the descent of the column AD, and therefore $AD(DC)$ expresses the live force arising from the gravity of the substance of the fluid contained in the column AD after it descends to the position EC, where the lower portions that will already have fallen below the horizontal are beginning to be freed of their own gravity. This, moreover, relates to the particles of fluid contained in DC. Individual [particles] make individual descents before they arrive at the horizontal LR according to their different distances from LR; since these distances are expressed through the indeterminate x, and any particle pertaining to it through dx, the product $x \, dx$ will be the live force arising from gravity of any particle whatever, and thus $\int x \, dx$ or $\frac{1}{2}xx$ is the live force of all the particles contained in CD, taken collectively, which, therefore, having been added to $AD(DC)$, that is, to $(a - 1 - x)x$, or to $ax - x - xx$, will become the total live force arising from gravity, $ax - x - \frac{1}{2}xx$, consequently equal to the derived live force gathered from the motion, $az - xz + \int z \, dx$. And thus, by differentiating, there is $a \, dx - dx - x \, dx = a \, dz - x \, dz$, from which $dz = dx - \dfrac{dx}{a - x}$; integrating again,

$z = x + \ln(a-x) - \ln a$. But z is proportional to the square of the velocity sought, vv. Therefore vv, or, if you prefer, $\tfrac{1}{2}vv$, equals $x + \ln(a-x) - \ln a$, wholly as we found above by the direct method, because certainly the square of the velocity of the fluid descending through the height x must be proportional to the same $x + \ln(a-x) - \ln a$.

Now further, in order that one may determine the point O [Fig. 17] to which, after the descent has been completed, it ascends again, it is to be noticed first of all that the liquid will have ascended right up to the uppermost point A if the same particles which had gone out of the orifice B during the descent, their own acquired velocities having been retained, were now again to enter individually in reverse order through the orifice B, and thus any one of them together with the preceding would be thrust on upward by the pressure of the ambient fluid, just as it occurs in every type of oscillation which, when the resistance has been removed, always does the same thing going and returning. But since the particles of fluid flowing out fall within the ambient liquid immediately after discharge and then diffuse from here, it is manifest that those same particles no longer flow into the pipe at their own acquired velocities, but other [particles] in their place, which surround the orifice without motion, are driven upward successively with the preceding column of fluid; these particles therefore, since as yet they have no motion at the beginning of the entering, cause the column of fluid preceding together with those [particles] attaching themselves to ascend more slowly than it would do if the particles were to flow in with some velocity which would aid the ascent. Hence the liquid cannot ascend to A but to some lower point O. In order that the height CO of this point O above the horizontal may be known, let us understand, meanwhile, that the particles which have fallen from the orifice B into the ambient liquid within which they have no gravity are directed upward, and therefore, with no loss in their velocity having occurred, arrive at the horizontal LR, from which afterward the individual [particles], with their gravity recovered, spring upward as much as they can according to the velocity of each, occupying in order a position according to some curve LMN. Indeed now because the amount of live force must be conserved, it is necessary that the sum of the products of the individual particles contained in CO and their respective ascents above the horizontal LR, together with the sum of the products which are formed as well by multiplying the individual particles on the curve LMN by their ascents and distances from LR, be equal to the sum of the similar products of the entire column AC; certainly the liquid possessed this at the beginning of descent. Therefore, if from the sum of the products of the column AC there is subtracted the sum of the products of the particles LMN, there will remain the sum of the products in the column CO, and hence the height CO itself becomes known. For the prior sum is $\frac{1}{2}AC(AC) = \frac{1}{2}(AC)^2 = \frac{1}{2}aa - a + \frac{1}{2}$; the other is $\int z\, dx = ax - x - \frac{1}{2}xx - az + xz =$ (in the case in which x, or the indeterminate AE, becomes AP, where z vanishes) $ax - x - \frac{1}{2}xx$, and therefore $\frac{1}{2}aa - a + \frac{1}{2} - ax + x + \frac{1}{2}xx$ is the sum of the products in CO, or $\frac{1}{2}(CO)^2$; from this,

$$CO = \sqrt{aa - 2a + 1 - 2ax + 2x + xx} = 1 + x - a$$
$$= (\text{in this same case where } x = AP)\ CP.$$

From this it is evident that the fluid in the pipe ascends to a height CO above the horizontal LR [which is] as great as the depth CP to which it had descended below LR immediately before.

COROLLARY. With CO known, one can find, through the construction given above, the second descent and the subsequent second ascent; and then from this the third descent and ascent; and so on. In this way the extents of the individual oscillations can be determined.

INDEX

FOR
Hydrodynamica and *Hydraulica*

Acceleration (in contractions), 370 ff
Acta Eruditorum, 5
Adhesion of water to pipe, 69, 72, 155 ff
Aëreo-aethereal particles, 20
Air,
    homogeneous, height of, 252
    rare, 264
Amontons, Guillaume, 229, 261
Angle of refraction, 247
Aqueducts, 6
Archimedes, 1, 9, 206
Ascent,
    actual, 203
    potential, 13, 35 ff, 104, 124 ff
        loss of, 56, 159 ff, 188
    see also: hindrances, impediments
Attraction, mutual, 21, 96

Barometer, 230 ff
Bernoulli, Daniel, 13, 26, 59, 266, 279, 280, 309, 314, 326, 330, 351
    Jakob, 5, 20
    Johann, 5, 8, 12, 128, 341
    Nicolaus, 341
Bilfinger, Georg Bernhard, 31
Bladders, 6, 23 ff
Borelli, Giovanni Alfonso, 5
Boyle, Robert, 5

Calandrini, Giovanni Ludovico, 224
Capillary effect, 31 ff, 109

Cardano, Girolamo, 213
Castelli, Benedetto, 2
Cataract, 110
    Newtonian, 440 ff
Catherine I, ix
Celerity of sound, 253
Chazelles, Jean Mathieu de, 337
Chimneys, 55
Clairaut, Alexis Claude, 70
Clepsydra, 165, 436 ff
Cohesion of particles, 96
Cochlea, 9
Conduit composed of several pipes, 356 ff
Conservation of live forces, 13, 42
Contraction of a stream, 71 ff, 82, 90, 92, 406
Cramer, Gabriel, 224
Ctesibius, 192

Delisle, Joseph Nicolas, 241
Density of compressed air, 229
Descartes, René du Perron, 275
Descent, actual, 35 ff, 104, 127, 141, 150, 307
Dilation of a stream, 82, 84, 95
Discharge,
    from a cylindrical pipe, 86
    from a dilating pipe, 53, 65, 84, 89, 96, 299
Discontinuity [vaporization] of liquid in a vertical pipe, 54, 303

Displacements, isochronous, 134
Duhamel, Jean Baptiste, 236, 326

Emanuel, Prince of Portugal, 311
Establishment of flow, 76 ff, 96 ff
Euler, Leonhard, ix, 173 ff, 414 ff

Fabretti, Rafaello, 4
Fahrenheit, Gabriel Daniel, 252
Feuillée, Louis, 241
Flow, within an arbitrary conduit, 391 ff, a conoidic conduit, 125, 373, 389, 421, an obliquely oriented conduit, 127, 376
Fluids,
  elastic, 29, 34, 226
  heterogeneous, 166 ff
  within moving vessels, 283 ff
Fontana, Carlo, 55, 217
Force,
  accelerative, 353
  hydraulic, 395
  hydrostatic, 395
  live, 129, 139, 149, 258, 351, 425
    actual, 258
    conservation of, 13, 42
    contained in a compressed elastic body, 257
    loss of, 139 ff, 149
    potential, 258
  motive, 353, 354
  immaterial, 353, 401
  of fluid resistance, 404
  of ignited gunpowder, 9, 264 ff
  of repulsion, 314, 315
  static, 395
Fountain, leaping, 7, 45, 47, 58, 70, 114, 122, 296, 301
Francini, 217
Freezing, 15
Friction,
  in pipes, 42, 50, 63, 66
  mechanical, 203 ff
Frontinus, Sextus Julius, 1, 4, 53

Galilei, Galileo, 5, 12
s'Gravesande, Wilhelm Jacob van, 50, 64
Gravity,
  center of, 12, 20, 130
  component of, 386
  mean specific, 237
Guglielmini, Domenico, 2, 4

Hales, Stephen, 261
Hawkbee, Francis, 247
Height, barometric, 230 ff
Hermann, Jacob, 64, 75, 97
Hindrances to fluid motion, 42, 53, 56, 156, 160, 201, 355
Hire, Philippe de la, 58, 70
Huygens, Christiaan, 12, 275, 276, 281, 324, 351
Hydraulico-statics, 7, 10, 289 ff
Hydraulics, 351
Hydrodynamics, 1, 351, 417
Hydrostatics, 351

Impediments to motion, 42, 63
Impetus, 5, 69, 70, 102, 115, 119, 187, 202, 217, 314 ff
Impulse, 220
[Impulse-momentum law], 328

Jallabert, Louis, 225
Jet, 184, 293 ff, 312 ff
[Jet propulsion,
  of a ship], 331 ff
  of any vessel], 442
Kepler, Johann, 275
Koenig, Emanuel, 241

Laborers, day, 185 ff, 335 ff
Leibniz, Gottfried Wilhelm von, ix
l'Hôpital, Guillaume François de, viii

Machines, hydraulic, 183 ff
Maclaurin, Colin, ix
Mariotte, Edme, 2, 3, 8, 9, 16, 33, 42, 45, 114, 164, 182, 201, 326
Maupertuis, Pierre Louis Moreau de, 70
Moment of a force, 210

Momentum, 317 (*see also:* force of fluid resistance, 404)
Motion, internal, 15, 72

Newton, Isaac, 2, 5, 8, 71, 76, 129, 248, 314, 319, 330, 409, 440

Orifice size in proportion to pipe size, 46
Oscillations,
 center of, 12
 in a curved prismatic tube, 128 ff
 in a straight conical tube, 158,
 in a straight prismatic tube, 179 ff, 377 ff, 446 ff
 isochronous, 8, 132

Papin, Denis, 5
Particles, ultimate heavy, 281
Pascal, Blaise, 5
Pendulum, tautochronous, 133 ff
Percussion, 287
Permanence, state of, 196, 275
Perrault, Claude, 194, 217
Phoronomia, 5
Pipes, obliquely oriented, 49, 61
Plane, inclined, 186, 203
Poleni, Giovanni, 5, 71, 119
Potential,
 absolute, 184
 animated, 217
 moving, 184
 static or hydrostatic, 395
Pressure, 22 ff
 barometric, 237 ff
 effect of moon on, 240
 in underground caverns, 239
 of an aqueous stream, 328
 of flowing water, 29, 382, 401 ff
Pumps, 187 ff

Refraction, angle of, 247
Ricatti, Jacopo Francesco, 320
Rive, Ami de la, 224
Rowers (oarsmen), 337

Saulmon, 278
Scheuchzer, Johann Jacob, 240

Shape of vessel, effect on discharge, 405
Siphon, 119, 303, 339
Smoke, 55
Specific gravity, mean, 237
Spiess, Otto, xiv, xv
Straub, Hans, xiv, xv

Thermometer, air-mercury, 230 ff
Throat (*gurges*), 352
Time of depletion, 77 ff, 363
Time of establishment of flow, 76 ff, 96 ff, 115 ff
Time, periodic,
 of a particle in a vortex, 277
 of a planet, 282
Torricelli, Evangelista, 2, 5
[Trajectory of a jet], 95 ff, 313
Translation, 355
Treadmill, 187, 204
Trireme (galley), 337 ff
Truesdell, Clifford Ambrose, viii, xiii

Vaporization of fluid in a vertical pipe, 303
Varignon, Pierre, 5
Velocity, height of fall corresponding to, 2, 4, 5, 7, 12, 41 ff, 56 ff, 76, 102 ff, 142, 161, 289 ff, 298, 315, 377, 385, 397, 405, 421
 during depletion, 363 ff, 433 ff
 in a clepsydra, 436 ff
 in a vortex, 278
 in composite (manifold) conduits, 50 ff, 64 ff, 165 ff, 368 ff, 385 ff
 for compressed air, 257 ff
Vena contracta, 407 (*see also:* contraction of a stream)
Vessel,
 composite, 356 ff
 manifold, 176 ff
 submerged, 119, 139
Vis viva, (*see:* force, live)
viscosity, 42

Vitruvius Pollio, Marcus, 206 ff
Viviani, Vincenzo, 5
Vlacq, Adriaan, 242
Vortex, 94, 275 ff

Waterscrew (cochlea) of Archimedes, 206 ff

Waterwheel, 204
Weidler, Johann Friedrich, 205, 261
Whorl, 92, 94
Windlass, 186
Windmill, 221 ff
Wool, resilient, 274